AF323923

The Unleashing Of
Evolutionary Thought

The Unleashing Of Evolutionary Thought

By

OSCAR RIDDLE

VANTAGE PRESS
New York

Men have never fully used the powers they possess to advance the good in life, because they have waited upon some power external to themselves to do the work they are responsible for doing.

JOHN DEWEY

As the author's nephew I, George S. Riddle, felt it necessary to go to great lengths to get his book, which I believe is very important to humanity, printed again.

With the above in mind, this book is dedicated to the Maximillian Allen Family, Imogene his wife, Rex his first son, and Danny D. Allen, my best friend since age ten, and now the only living person in the Allen family; and who made the reprinting and publishing of this edition possible.

Acknowledgments

THIS BOOK HAS PROFITED MUCH FROM THE CRITICISM ITS various chapters received while in manuscript. Usually the criticism applied to early, not final, drafts. To those readers and critics the author is most grateful. Chemist Harold C. Urey read Chapters 1 and 2. Paleontologist William K. Gregory read seven chapters. Anthropologist Robert Redfield read Chapters 4 and 5. Physiologist Anton J. Carlson read eight chapters. Zoologist-educator Fernandus Payne read four chapters. Historian David S. Muzzey read Chapter 7. Educator-economist Alvin Johnson read Chapters 9 and 10. Psychologist Harvey A. Carr read the entire manuscript. All chapters were read by at least two of these colleagues in science, education and history. The author alone, however, is responsible for the contents of the book.

The great number of quotations used in this work, from very dissimilar sources, is in itself a statement of debt to a great many authors and publishers. As their individual names appear, let the writer's cordial acknowledgments attach to them.

Contents

Foreword

THIS BOOK IS NOT CONCERNED WITH WHAT SCIENCE HAS done or will do to change the material world in which we live. That kind of change is partly and more directly the child of technology; and, though science fathers technology, the endless issue of new goods and gadgets—from healing drug to atom bomb—is a task shared by both. The unshared task of science is to tell us what we are, and to answer questions man has always asked about the world of things he sees, hears and feels. Science has already provided the outlines of essential answers to these groups of basic questions. The answers have transformed the world of advanced—but effectively isolated—thought. If peoples, or any people, knew and shared this modern insight, an entirely new society—a genuinely modern society—would be brought into existence. It can be shown, however, that the entrenched and growing forces of organized religion, regrettably aided by some scientists and by temporal circumstance, now prevent all the earth's peoples from access to this society-transforming thought. At stake in this struggle is the kind or level of society that accords with what we now know about man and his world; and here all hinges on rejection of the supernatural by free and informed peoples. These pages provide their readers with the unhedged answers that science gives, and thereafter they examine the quite successful aims and tactics of those who want no people to know these unsugared answers. This book deals with prevalent, religion-based subversion of essential culture-building thought—with thought control on a scale otherwise unapproached in the history of man. The practical question that emerges from this discussion thus becomes: Can, or should, an advanced society tolerate the currently cumulative influences of existing organized religions?

This book is not primarily philosophy nor science nor a report on a social inquiry. It is nevertheless a blend of those three things—with sustained accent on the third. In general, it is an account of the cultural impasse in which *available* evolutionary insight, useful or necessary to social progress and racial survival, is made *unavailable* to society by contemporary organized religion. Its subject matter scans the wide range of the evolutionary process and stresses the meaning and nature of that process. Against that background of earned insight it surveys—at this mid-century and at community and national levels—the restrictive influence of organized religion in the areas of education, news dissemination, the theater, private life and public law. The relation of science to theology receives, and is here thought to merit, only brief or indirect discussion. From that long battle has come the evidence that the towering clash of our time is between advanced society and its infestations with institutionalized religions. Directly and specifically researched or examined here are the adverse intellectual and social consequences of organized religion in the world of today. This special and unique social survey could have full force and value even if it could be shown that theology correctly postulates a good and infinite God.

What is said thus briefly must again be said more informatively.

Go talk with them anywhere and you find that what people think and feel of themselves and surroundings—of their own origin and destiny—is clinched and colored by flying scraps from one or both of two sources: the teachings, past and present, of an organized religion *and/or* the teachings of science. Indeed, wherever the voice of science is more than a whisper, the thought of man shows a scale of wrestling compromise between the two. Worst of all, for most wrestlers, the key word of all science—evolution—is mangled and robbed of meaning by religious influence before it is granted a use in thought. So much for an ugly fact. How important this may be and what may be done about it are other matters. The teachings and goals of the many organized religions are perhaps too numerous to be either described or read by anyone. But such of their claims as violate science, and the avenues through which they smother the antitheological facts and insights of science, can be appraised within the space of a few chapters. Likewise the essentials of the modern and hard-won story of man's place in nature can also be told in a few chapters. Why not have a look at the two sides of the same shield in one and the same book? The almost incredible lack of such a direct approach to the two sources of modern man's self-appraisal and outlook would seem to warrant even an inadequate book.

Such a venture necessarily becomes at first a closer look at meaningful parts of the new revelation of science, with the intent of finding its full meaning. It next becomes a tour of the several areas—schools, news sources, churches, theaters, ordinance and law—where that meaning struggles against an entrenched and resolute opponent for a beachhead in institutions that reflect man's social goals and his thoughts about himself. This leash on thought is short and stout.

A few years ago, the Catholic Library Association was told in New York that Catholics may subscribe to a doctrine of "mitigated" evolution without violating the rules of the Church. Monsignor Joseph H. McMahon was further reported by the press to have said: "If science should prove that man descended from an animal, the Church would not refuse to recognize such proof, for Catholicism is founded on truth. But such acceptance would not affect the certitude that the soul of each and every man was the divine creation of God." In 1931, Dr. Howard H. Kelley, then a famous surgeon of Baltimore, was reported by the press to have said that man has been created in the image of God and that his progress has been downward and not upward. "I am a thoroughgoing believer in the special creation of man."

Rather similar *mitigations* of the meaning of the word "evolution" widely prevail in Protestant and other world religions; they are found in a flood of books for the reading public of all countries; they have even crept into several textbooks now very widely used in American schools and colleges. These "mitigations" everywhere rob the broader principle of evolution of its real meaning and of its vast ability to assist society to a level of sanity and warranted hope. They rob the race of its chance to build a genuinely modern society. Nor may we expect something short of comprehensive and direct challenge to check these sapping floods of undervaluation and distortion. It would seem that no story now needs to be more firmly written nor widely spread than that of *unmitigated* evolution.

The fact that state legislatures within the United States stopped the passage of antievolution laws not long after the Scopes trial in Tennessee, in 1925, has been much misunderstood by liberal-minded persons and especially by scientists in America. Those attacks in legislatures were blocked largely through the weak and erring plea that evolution and now current religion can exist in harmony together. And one may indeed admit that at that time probably no other argument would have prevented a very wide extension of such laws. But the basic issue of getting religion into state law—of undoing the separation of church and state, and of putting education at the service of religion—was left wholly unsolved by that poorly patched armistice. Moreover, the same general movement that antedated by several years those legislative battles still sweeps forward and now powerfully threatens all secular education at public-school and college levels. This movement now seems more dangerous to both the principle of separation of church and state and evolutionary thought than at any time during the past four decades—and it is not at all confined to the United States.

Many primitive peoples were and are inclined to accept higher animals as their kin, and even to endow them with many human qualities. Under the prolonged influence of Christian theology and the philosophy of Descartes, however, the Western world placed an unbridgeable chasm between man and animal. More than nine decades ago Darwin gave the thinking people of the world a new and enlightened way of acknowledging our animal origin and kinship. But at this date, what are the results? First, a durable *intellectual* revolution flowing from an acceptance by a small percentage of men of this very vital extension of the rule of natural law. Second, organized religion learning over the years how to make continuous and successful use of religion-born opinion to restrain—to suppress, mitigate and pull the teeth of—this and related evolutionary perspectives in all education, in most news sources, and in practically all public policy. Creeds operating their own schools, or openly directing those of the state, have most thoroughly accomplished their aims; but in the public schools of all lands the religions rarely and only temporarily lose a contest. Third, after ninety-four years of presumed opportunity to participate in the greatest of discoveries relating to ourselves, it appears that probably less than twenty-five per cent of American youth are impressed significantly by the principle of evolution. In a very few nations of Europe this percentage may be slightly exceeded, but in most countries of the entire world it lies perhaps between one and fifteen per cent. Fourth, throughout the world this quite effective smothering of the prime revelation concerning ourselves is little known and weakly opposed. Briefly, this is the cultural impasse of our generation. Roughly, this is the score that warrants the title of this book.

There can be no doubt that evolutionary thought has found a niche in which it will survive. All doubt relates to whether that survival may win its way to something of social worth; to whether it is to attain a stature capable of serving mankind or only a clan; to whether it is to remain a caged thought or become a leaven in public law and private life.

But if the battle for simple acceptance of the *fact* of man's animal origin has gone thus badly, there is yet to note that the thought-transforming and the society-lifting *implications* of that and related pregnant facts are *everywhere* educationally choked before the breath of birth. Though these same post-Darwinian decades have heavily reinforced biology, psychology, sociology and history with most meaningful principles (some of which are cited further along in this Foreword), it is a shameful, dangerous and intolerable fact that, because of influence clearly traceable to organized religion, the implications of these several superb triumphs of science are not being taught in the secondary schools of any Christian country. And they are being taught to no more than minute clusters of students in a part only of the colleges and universities of Christian lands.

Information obtained by the writer during several years as chairman of a Committee on the Teaching of Biology (appointed in 1936 by the Union of American Biological Societies) is especially valuable in fixing upon religion the responsibility for beheading and devitalizing the biology that is taught in the high schools of the United States. That information deserves a wide and a reactive audience; it finds a place here. Reflecting upon this and other information not difficult for an interested biologist to obtain, the writer recalls no vagary of the intellect of an age more amazing or depressing than this: in a day of alleged concern for the dignity of the individual and for sound social adjustment, most thinking people of the world are or seem unimpressed by the nearly obvious fact that, in all lands, every competent teacher of a cultural subject—from primary school to or through college—is now partly or fully restrained by religion-controlled public opinion from teaching the implications of a whole group of evolutionary principles that have high importance to the welfare and survival of the race. Even if this tragic miscarriage of thought and of elementary social freedom did not provide reason for the writing of this book, it would still be useful to record for history the dimensions of this blind spot of an allegedly aggressive era.

In earlier days, some parts of Christianity unquestionably contributed to the growth of democracy in the Western world. But the reader will here find convincing evidence that important parts of Christianity's power are now obstructing the birth of a socially vital outlook and an advanced level of society. Most unfortunately, Christianity today is joined with other religions in the suppression of the vital and the new. Must Western democracy fight its way—and gigantic conflict there is—without acknowledging and building upon the realities of human and social life as these are revealed by science? Is the Soviet competitor, which defies ethics in addition, to gain enough advantage from a restraint of religion to win the contest?

At this moment your neighbors and mine are wholly unprepared to give thought to the things that would flow from a widely accepted view of the natural origin of man, of his biological and social nature, of the animal and social, sources of morality, and of a world rid of the supernatural. On the other hand, the masses and their leaders everywhere appear to rest their entire case for *self*-help on things political and economic; and surely no reasonable person will question either the importance or the urgency of those problems. Again, entire nations are now splitting apart and dividing the world on questions of social and economic reorganization. Indeed, momentarily the very lives and freedoms of all Western peoples have been permitted to become endangered; and while this crisis lasts those peoples are perhaps necessarily facing the prime problem of personal survival. But whether or to what extent our existing society must reappraise its economic ideals seems, to this writer, a problem secondary to modern man's urgent need to see himself as a naked reality and to shape an entire social outlook in keeping with that reality. The veils and cobwebs of individual human minds are an immediate and an enduring threat to all that society does and plans. The mist-laden cobweb, however, clouds vision more than it resists touch; there should be some possibility of brushing parts of that fragile thing from the path of progress. And if man cannot find and accept basic reality regarding himself and the world in which he lives, his economic or other future lot is surely in doubt.

In this task of replacing the man of myth with the stark reality, it is clear that the word of the biologist is relevant. The writer here offers no apology for putting his voice on record.

Though this book is not directed solely to Americans, informed readers will recognize that the pale ghost of evolution now paraded grudgingly on the American scene serves as a gaunt reminder that a Huxley was not born here. In no part of the Western Hemisphere has there been a tempestuous moment in which the bishops have been chilled in biological debate. Partly in consequence, a battle on this continent that was thought by many to be wholly won must perhaps be rated as largely lost.

In any case, the much broader and still newer contributions to thought of all science today—far outranging the single matter of the animal origin and nature of man—now require full presentation in the forum of public opinion everywhere. That presentation of the case of naturalism against supernaturalism must continue to be the hard task of many men and the content of many books. But in one of those books—and therefore in this one—the public that is little trained in science should find an abridged but reasonably simple and accurate account of the humanly significant evolutionary processes and products in nature; and, very especially, of the bearing of this new knowledge on intelligent human thought, aspiration and behavior. Such an account—in Part I of the present volume—supplies the necessary background and prelude to a consideration of suitably documented evidence (Part II) that religious thought, practice and influence everywhere now block the avenues through which peoples may learn—for their surer and saner social survival they must learn—the sweeping implications of accomplishments such as these: the now established principle of organic evolution; the principle of integration and the flow of new properties from each new chemical union; the cogent indications of the origin of living matter under purely natural law; the oneness of the energy unit that is man—not *a* mind and *a* body; the wholly biological and cultural origin of ethics and values; the lifting power of social inheritance in mankind; and the chance-born basis for the biological *in*equality of all human beings.

Not all items of pertinent philosophy or of science are touched in these pages, but it is believed that the really significant ones are included. Moreover, the form in which these items are brought to the reader's attention in Part I—as sequences in actual history—should give the general reader a more satisfactory grasp of these matters than that obtained from a discussion less directly concerned with the story of science.

This book is the outcome of the author's growing convictions of the past twenty-five years regarding a group of related matters. He has become convinced that both the advance of science and the welfare of the race demand that the best of biology and the implications of all science be made much more available to popular thought; that, unfortunately, some of the more formidable foes of unmitigated evolutionary thought are those scientists who perhaps think they caress such thought but actually contrive to smother it; and that current insulation of our biological best from the stream of intellectual life and public use by organized religion must be directly attacked. Though not all our scientific colleagues can share these conclusions, there is reason to hope that some dissenters may respect them. Many a layman is little prepared to accept the seemingly drastic conclusions and indictments candidly reported here.

Further, at this moment liberal thought in general is immobilized by the dangerously mistaken assumption that socially harmful religious belief is waning, that fewer believe, that belief itself is softened, and that society is thus already protected from its harm by political gains or by other elements of modern life. Citizens, writers, scholars and scientists tend to adopt a complaisance that is in poor accord with facts assembled here. If today a liberal education implies any awareness of the forces most potent in society, a well-read chapter on the subversion of modern society by organized religion is indispensable. That chapter, however, can impress but little of its full force upon those who do not know the present range and reach of evolutionary thought.

Again, this book is born of the conviction that a biologist of this generation cannot fully serve his day and his science through either research or teaching alone, nor yet through both. In this swelling science, both exploration and drill are failing to counter and shift the older currents of thought. The potent warmth developed in these modern cloisters is being frozen by an unfriendly force. That force must be clearly recognized and resolutely combated if the vital contributions of a wholly unique century are to attain usefulness in society and law. It is the rather special shame of *biological* science that so few people give thought to the startling social cost of existing belief in the supernatural.

In sum, the scope and purposes of this book include three kindred areas of thought and action: first, the main and meaningful message of unmitigated evolutionary thought, particularly the basis on which it discredits a belief in soul and the supernatural; second, a presentation or an exhibit of now current worldwide restrictions on human freedom and purpose, on evolutionary insight and incentive, imposed by the cults of supernaturalism, and the formidable threat these restrictions and compulsions offer to society, freedom and democracy; and third, a cursory look at the good and the harm done by prevailing organized religions, which suggests that *such* religions are dispensable and expendable. These three areas relate thought to act, and, however brief and sketchy their treatment, they belong together in this book. Much experience, including a large correspondence here partly reproduced in Chapter 14, has convinced the writer that a volume with lesser range would prove inadequate for most readers and lessen the value of this effort. It seems evident and essential that thoughtful people should extend the area of their firm understanding and thereafter use their enlightenment and social maturity against trends identified as dangerous. Those two worthy performances are more effectively done when the entire need, aim and goal of action is seen clearly. Too, the basic need for increased and clearer personal grasp of actualities is exceeded only by that for effective social act.

Since a part, but not the whole, of the area of freedom of thought and expression is involved and considered in this book, some critics will suggest that this is important and unfortunate. That view is rejected. To write of the freedom of expression in general or of the issue of human freedom in the large would transgress the writer's competence and restrict further the broad survey that is indispensable here. Moreover, which human institutions have exceeded the organized religions in limiting overall human freedom and freedom of expression in general? And which areas of human endeavor potentially exceed science in power to free the hand, the foot, and the mind of man?

It was earlier hinted that Part II of this work surveys—in the world at this mid-century—several areas of everyday life in which the contributions of all science to logical thought are manhandled and perverted by prevailing religion. This portion of the book is in part a sociological research, but mainly it is what a reporter found. The report shows, however, that the case and the cause of science—of the life sciences in particular—are now in the rough hands of antagonistic community sentiment, and that the materials reported relate to the environment, the health, and the growth of science, in addition to the individual's thwarted freedom to think and to society's baffled obligation to adapt and advance. These materials, assembled with this partly in mind, are essentially filled-in blueprints of the residual and untoward social forces that attest John Dewey's verdict on current society—"the genuinely modern is still to be brought into existence." If Dewey himself had sought to document this most meaningful conclusion, these or similar materials could have provided his evidence.

Some technical topics have indeed not been wholly avoided in these pages, and the scientific subjects are usually so treated as to make them serviceable to most colleagues in science. Such inclusions seemed especially desirable since many scientists, including fellow biologists, have given little thought to the full meaning of some recent developments within the several sciences that deal with living matter. However, scientific colleagues assuredly familiar with both the facts and the implications of evolutionary science may profitably turn shortly to Parts II and III of this book. On several of the nonscience topics treated there, one can offer evidence but not logical demonstration or proof. The conclusions drawn from that evidence by reader and writer may be neither identical nor sound. Several hundred quotations, usually from recognized authorities, present the opinions of others. Presumably these quotations provide the reader with some assurance that this volume is not a home for lonely personal views.

"Organized religion" here receives much attention not because the writer claims unusual competence to discuss its aims, but because its sediment and current influence is found to cover and insulate, and to filch from now existing society, the best of the science that the writer has long served. In other words, because entrenched and organized religion now succeeds—in ways and to an extent little known to most citizens—in depriving populations of their right to know and ability to safeguard their own future; and because organized religion through its continuous and relentless smothering of evolutionary fact and insight has produced the cultural impasse that is the subject of this volume.

Usually the unqualified word "religion" has been avoided in these pages since its now current and twisted meanings (see Chapter 7) can lead to anything from swiftest condemnation to highest praise.

The over-all philosophy of this book is essentially that of John Dewey. It is the same as that of Freud's booklet of 1927, *The Future of an Illusion,* in which he showed that psychology must resist and defeat religion. And it is also the same as that of Reichenbach, whose book on *The Rise of Scientific Philosophy* appeared as the present work was nearing completion.

Man's chief *intellectual* problem up to our time is already resolved. Naturalism, not supernaturalism, sweepingly defines man's relation to the universe. Intellectually, the contest between theology and science is finished, though that between society and institutionalized religion may have only begun. That only a fraction of the people of any nation is aware of these developments is a most meaningful matter, but the hard core of the older intellectual problem no longer remains. What we have instead are practical and pressing educational, political and social challenges to understanding and adjustment. And the simple fact that live discussions of these pressing adjustments are so nearly absent from today's press, platform, book and rostrum is the one true marvel of our age and time. It is as if the celebrated Forum—forgetting warm debate on thought, on the state, on jurisprudence—had blandly turned to timely gossip on wine-making, war, and products of the kitchen.

This book is written for more than one kind of inquiring person. Three groups were especially in mind: students everywhere who in their early college years need a brief and nonexisting survey of the sources of reasonable thought about society, man, and themselves; colleagues in science, education, religion, and those others who have a common concern in the hard problems of liberal education and an earthly future for mankind; and, far from least, those citizens of any group or age who retain the fine curiosity of youth and do not fear to think and appraise. It is conceded that the task is overgreat; that much knowledge to replace the author's limitations is needed for success. It is granted, too, that peoples of this day—particularly our fellow Americans—may be unready to acknowledge and resolve the dangerous cultural impasse of their time.

PART I

WHAT EVOLUTIONARY THOUGHT IS

1

From Earth-Cloud to Man

New worlds are coming into existence; others are dying. . . . It is enough [for man] to know that the earth, life and man are still in the throes of creation.—*Sir Arthur Keith*

The larger viruses have a composition and properties which are characteristic, not of molecules, but of organisms. The viruses have certainly provided the link between the molecules of the chemist and the organisms of the biologist.
—*Wendell Meredith Stanley*

ONLY IN OUR DAY HAS THE MIND OF MAN GRASPED THE bold outlines of its own history—the origin and history of life on the earth. Today, too, the history of earth itself is being thrown backward into cosmic cloud. Triumphs of endless inquiry have so lifted man in the scale of living things that he may now meet and understand essential truth concerning his own nature and his place in nature. Indispensable fragments of this story of life rest securely in hundreds of volumes—the tributes of a dozen sciences. Can the outline and meaning of these many fragments be fitted into a few pages? This seems possible.

The whole of the known drama of life was performed in a narrow zone—quite near to the very surface of our own small planet. Even bacteria disappear in the upper reaches of the earth's atmosphere, while other life extends downward only to the ocean's floor. At no earlier time in earth's history has this been different. Though fossil fragments of once living things are found in coal and rock strata now a few thousand feet beneath the soil on which we walk, it is clear that these veins were land surfaces or ocean floors when they trapped the bodies of organisms. If, in an Arabian Nights' excursion, we could scan the earth from afar (say from a point 24,000 miles away, which is one tenth of the distance to the moon), we could rightly sense the narrow spread of life. On the great sphere that would largely fill our view to East or West we should then see all life imprisoned in a thin film—a living skin—tightly fitted to the very surface of that sphere. As we now know it, life sticks to the place where there is liquid water, with salts dissolved in it; where carbon, nitrogen and oxygen abound; where temperatures do not much outrun the meager range now obtaining in our lower atmosphere; and where surfaces can absorb sunlight for a continuous flow of free energy. On many planets or distant stars living matter probably finds a similarly cramped existence. That, however, is a thing unknown, and, if true, all such life probably arose only ages after the birth of the planet that is its home. Life thus presents itself as the most circumscribed and contingent thing in a world of things. It seems to be among the newer products of an ever-building universe.

Though life is most narrowly limited in space, the events that led to its origin (creation) fill an amazing lapse of time. Formation of earth and solar system was no affair of haste. And, thereafter, one or two billion years were necessary to form the compounds, and to attain the conditions of temperature and moisture on the earth's surface, that led lingeringly to the first and simplest living matter. The history of these two prolonged epochs is of surpassing interest. Though there are firm facts bearing upon the origin and earlier state of the earth, much is still unknown. It is not surprising, therefore, that the known facts are now summarized in two different ways.

An older view treats the earth as a fragment of the sun—its original gaseous materials as fiery hot as are those of the sun's surface today. It would follow that later stages of earth history were mainly an epoch of cooling and condensation, associated with a continuous formation of new and increasingly complex substances. The newer view, largely born of additional facts only recently obtained, traces both earth and sun—or at least the earth—from a cold cosmic cloud. This view derives our solar system, including the meteorites that fall upon the earth today, from an immense cloud of gaseous matter in which floated relatively smaller amounts of dust and solid particles. Probably the sun was formed from the much larger and perhaps more-or-less-detached portion of that immense cloud. The rotational and orbital speeds of remaining portions of the cloud were of rates that precluded the formation of an accompanying smaller sun and provided instead an elongated cloud circling the sun. That elongated and disc-shaped cloud later broke into many pieces which were destined to form the planets and the meteorites. Repeated fractures of the thinner edge of cloud—the part nearest the sun—gave origin to the smaller planets, including earth. Against the strong pull of the sun, or perhaps because of the limited gravitational fields of the gaseous particles, the lighter gases of the cloud could not be held by the smaller planets. For this reason, these planets were built mainly or solely from the cold meteoric dust and solids. Thus a variety of fairly complex molecules, already formed in the prehistory of the pregnant cloud, fell toward a common center and gave form and mass to an early and growing earth and to its midget twin, the moon.

How conditions favoring the eventual origin of life would arise under each of these two views can now be sketched. And first to be noticed is the course of events under the older view. When the earth-building material first separated from the sun, both its temperature and its gaseous state assured that several hundreds of millions of years must elapse before conditions favorable to the origin of life could exist on a sun-fragment of the size or mass of the earth. Following separation from the sun, the process of cooling and condensation of gases was speeded up. And though very few chemical compounds *could* exist in the superheated gases of either the sun or the newborn earth, many such compounds would be formed while the isolated earth slowly acquired lower temperatures. Under chemical laws new compounds *must* form in cooling gases.

The continued formation of a variety of new chemical substances was essential to the origin of life, and also to its maintenance after it had arisen. Some such stable and durable combinations (rocks, minerals) are still being formed from the release and cooling of heated gases within the hot volcanoes of the earth today. The building of solid-earth material from superheated gases continues to occur in our own day and is witnessed by our own eyes.

Some of the events and conditions that would attend the building of earth from cold dust and solids—the second view—are of highest interest because with each passing year it becomes more probable that they are parts of actual earth history.[1] First of all, such materials would provide the earth with what geolo-

[1] Harold C. Urey, *Science*, vol. 110, October 20, 1950; Wendell M. Latimer, *Science*, vol. 112, July 28, 1950; Harrison Brown, *Review of Modern Physics*, vol. 21, 1949; C. F. von Weizsacker, *Astrophysical Journal*, vol. 22, 1944; Gerard P. Kuiper, *Proceedings*, National Academy of Sciences, vol. 37, 1951; and Harold C. Urey, *The Planets: Their Origin and Development* (New Haven: Yale University Press, 1952).

gists regard as its original composition and structure—a central core of iron overlaid with a mantle of silicates of iron and magnesium and encased in an outer layer of basalt. Perhaps, however, iron and silica may have been fairly equally distributed within the early core of the solid earth[2] and may have attained their separation only in later phases of the planet's history. Among other things, this newer view also satisfactorily accounts for such facts as the presence of water—although the earth-cloud early lost its lighter gases—and for a newly found constant chemical (isotopic) composition of the matter of earth and of meteorites.

While the cloud was condensing under gravitational forces to form the earth, its gaseous material was lost and its particles of iron and silica were concentrated toward the center of the mass. The condensing cloud, by that time, was heated appreciably by the gravitational fall of particles and also from the exposure of its rotating surface to the sun. Heat was also being lost continuously both through outward radiation and by the escape of gaseous material. Warmth, but nothing resembling solar heat, should have prevailed in the earth-cloud at that period. At a still later stage, after condensation was largely accomplished, the more important internal source of heat came and still comes from radioactivity within the mantle and basalt layers of the earth. Uranium, thorium and other members of the radioactive series continue to sustain this powerful though now slowly decreasing fountain of heat.

This supply of heat from within and below the earth's crust was apparently capable of decomposing a part of the basalt layer into the granite that rose in the crust to form the continents and mountains. It also accounts for the presence of water and for the several gases of the atmosphere. Thus the heating of hydrated silicates and aluminates released steam, which rose to the surface as water. When carbides reacted with oxides of iron they formed carbon dioxide. Nitrides were hydrolyzed by steam to ammonia; and in hot regions ammonia decomposed into nitrogen and hydrogen. Finally, at a *later* epoch, chemical changes induced by light added our present supply of oxygen.

Most alluring is Kuiper's approximation to proof that not merely a few stars, as hitherto supposed, but literally billions of stars must have formed systems of planets as did our sun. This

[2] Urey, *The Planets, op. cit.*

greatly increases the number of places in the universe where conditions are suitable for separate and independent origins of life.

The events of later earth history, which prepared the way for the origin of life—and to careers for living things—may now be considered almost wholly apart from any special view of the way in which this planet acquired its substance, its temperature, and its atmosphere. Parts—though not all—of that fertile history can be outlined here.

When the earth became cool enough to permit its water vapor to condense and remain in liquid state on its surface, or when steam was released from hydrated silicates in its self-heating interior to form water on its surface, this planet of ours surrendered to the agent that would thereafter dominate and repeatedly remold its surface (and eventually assist in stopping its daily rotation by tidal action) and as a by-product build the cradle for life. Forever thereafter water would accumulate in smaller and larger amounts in oceans and billions of pools; and always thereafter water would leach chemical compounds from the diversified rocks, thus giving opportunity for endless interactions of these compounds in that unrivaled chemical laboratory—an aqueous solution. The purely natural forces that attended either a cooling or a slowly heating earth, and which later formed its first pool of water, decreed that innumerable new combinations of matter—each combination with new properties—would assemble in those waters.

The few gases that remained in a free state on a cooling earth, or the more abundant gases released by chemical action on a slowly heating globe, together with the sunlight, which now became an intermittent external source of light and energy, assured a still greater variety of new chemical transformations, because gases and light rays both actively invade water. Under the superlatively favorable conditions of an aqueous medium, these rays and gases early began their unending work on the water-held leachings of the rocks. In the older view, that primeval atmosphere probably contained very little nitrogen or carbon dioxide or oxygen in the free state, but ammonia, aqueous vapor, hydrocarbons (methane), and traces of carbon dioxide from active volcanoes—along with ammonia and oxides of nitrogen from the flame of lightning—were doubtless present. In the newer view, all the gases of our present atmosphere, like the earth's huge supply of water, are amply accounted for. The atmosphere of the newly contracted earth-cloud was probably rich in hydrogen, methane, ammonia and water, but without carbon dioxide. Along with the gradual loss of hydrogen, in the manner noted above and through later photochemical dissociation of water in the high atmosphere, the whole atmosphere shifted to the "oxidized" compounds—carbon dioxide, nitrogen, water and oxygen. And in either view, the nitrogen of ammonia, along with other nitrogen and oxygen earlier imprisoned in the great rock that is the earth, then awaited the rendezvous at the pool—the water crucible—for nuptials with compounds of carbon, sulfur, iron and molecular hydrogen, whose issue would exhibit the properties of life.

But who could be more handicapped in identifying the *simplest* manifestations of life and organism than the unschooled human being of today who views both of these things from present high attainments within himself—from products of a billion years of further growth of both life and organism? To men in general, an apprehension of even those large and up-to-date clumps of life called mushroom, clam, cow and elm is sufficiently vague and unsatisfactory. Nevertheless, the view that life and organism—two names, but *inseparable* in thought—in their crudest and *simplest* forms can be conceived to result from groupings and reactions of purely *inorganic* substance, may be maintained and well defended by the penetrant knowledge of today's specialist. "Life is the manifestations of organism, when manifestation is defined as anything perceivable by an observer, and organism is defined as a group of chemical systems in which free energy is released as part of the reactions and in which some of this free energy is used in the reactions of one or more of the remaining systems. Any pair of chemical reactions satisfying these conditions is an organism. Things as simple as two inorganic systems can meet the requirements."[3]

Here, however, the case for naturalism—a case on precisely that item which so halts and baffles thought in most men—really rests on a base much broader than a wealth of inorganic compounds. This follows from the fact that the materials at hand for the building of earliest life and organism were not limited on an unstable earth to *inorganic* substances. At that time and place organic materials too had been arising during perhaps a billion

[3] K. M. Madison, *Evolution*, vol. 7, 1953.

years. That process and several of the naturally arising organic molecules and colloidal particles now become additional subject matter in the development of life.

The very simplest pre-existing *organic* matter was simple hydrocarbon—the gift of air and stone—and this earliest union of carbon and hydrogen, under the action of photochemical processes (ultraviolet light, and lightning in the "reducing" atmosphere) [4] *must* have yielded a variety of substances such as aldehydes, alcohols, ketones and organic acids. These derived compounds can, in their turn, react with ammonia to form amides, amines, "ring" compounds such as porphyrins, and other nitrogenous substances. These later interactions of ammonia with the derivatives of hydrocarbon hurdled many steps in the long march to life. In the course of a thousand million years they could give to the waters of early earth a rich supply of organic material, including amino acids and other substances highly significant to living matter. Again, carbon dioxide as well as hydrocarbon could add to the store of organic substance in the primitive waters of our planet. Recently it was learned that radioactivity—then more forceful than now—acting on carbon dioxide and water results in the formation of formaldehyde and formic acid. Significantly, moreover, this reaction is hastened (catalyzed) by that abundant inorganic substance, iron sulfate.[5]

To return to the question of organism: one may observe a way in which even that simplest "inorganic organism" could develop a system of transport of free energy. If that organism (note its worthy definition above) so evolved as to obtain its free energy from the oxidation of (abundant) hydrogen sulfide and used that energy to reduce carbon dioxide to an organic compound such as formic acid, with both of these reactions catalyzed by then existing inorganic substances, it developed the ability to produce adenylic acid.[6] This substance, adenosine monophosphate, is of great importance to the living state. Organic phosphate compounds are present in all known persisting organisms where they serve to store free energy, transport it, and later release it to energy-using systems. Blum[7] has noted that "means of

<hr>

4 Harold C. Urey, *Proceedings,* National Academy of Sciences, vol. 38, 1952.

5 W. M. Garrison and others, *Science,* vol. 114, 1951.

6 Madison, *op. cit.*

7 H. F. Blum, *Time's Arrow and Evolution* (Princeton, N. J.: Princeton University Press, 1951).

free energy transport may have been this key factor in the transition from the nonliving to the living state."

In ways not to be recounted here, but recently stated precisely by Blum and by Madison, the arrival of adenosine monophosphate provided a mold for recasting the "inorganic organism" into one that was mainly organic. Thus additional carbon compounds in the environment became available for use by the primitive organism; new organic molecules were formed within the organism; a means of multiplying the chance linkages of amino acids to form peptides and proteins was established; the process by which adenosine was formed involved also the presence and use of purine rings; and "if the adenylic acid system introduced the purine ring pattern which later became incorporated into nucleic acid structure, it seems reasonable to suppose that this pattern was perpetuated in association with nucleoproteins."[8]

This stage in the development of organism and life, though equipped with nucleoproteins, was deficient in some properties those particular substances exhibit in the organisms familiar to us today. These properties relate especially to self-duplication, mutation and continuity on a genetic basis. A means of bridging this breach has been outlined by Madison. His proposal indicates how the now familiar nucleotides might be formed and a one-gene-one-enzyme system established. He bases his proposal on the fact that "a simple mechanism for genetic continuity of the organism and its progeny developed with the formation of the first nucleoprotein molecule"; that the chances for error in the duplication of molecules of nucleoproteins would be large in a mixture such as the primitive organism; and that when accompanied by the appropriate enzyme (catalytic) activity these molecular changes (mutations) would be transmitted to offspring, and genetic evolution would be thus initiated.

It is possible, however, that the most primitive or very earliest matrix of organism and life involved *organic* substances to a greater extent than is indicated in the preceding account. In any case, only the uninformed may doubt that a rich variety of organic substances—some quite complicated ones among them—preceded life in the waters of a bacteriologically sterile earth.

[8] *Ibid.*

Therefore we now take notice of the behaviors of certain organic compounds, and of colloids, which still other specialists consider significant in the origin of organism and life.

The origin of amino acids and proteins may be called a voyage, rather than a step, toward the formation of fragments of brevetted matter—matter alive or partly alive. The amino acids are the blocks from which protein is built, and all now living matter is built chiefly of protein. All these molecules expose several points at which they may make unions with each other and with still other compounds; under certain conditions they are thus capable of forming giant molecules of highly complex "colloidal" protein. Living organisms, now and in the past, put these building blocks together in an almost endless variety of ways, with each new way yielding a new and different polypeptide or protein—and each new protein has one or more properties possessed by no other grouping of matter. Though about two million living species are now known to exist, each with one or more proteins peculiar to itself, probably fewer than thirty different building blocks are used in the construction of these few million different proteins.

The *uniform* solution or diffusion of even high-molecular colloid compounds in the waters of our planet was not in itself a promise of a particle of living matter. But *mixtures* of two or more such colloidal compounds, whether hydrocarbon, lipoid, polypeptide or protein, cause the phenomenon of coacervation, a process somewhat akin to emulsification. In this process, myriads of minute droplets, or coacervates, are formed. Each droplet is a fluid mass, though the enclosed colloid particles hold less water than before their enclosure, and the whole droplet is sharply bounded by a membrane consisting of oriented molecules of water. Having such membranes, these droplets, or coacervates, possess surfaces upon and through which external substances may be adsorbed by the droplet. Moreover, some substances of the external medium can enter the droplets and form compounds with the constituent colloidal particles. Thus the small minority of coacervates that could maintain themselves (i.e., had dynamic stability) following a series of such absorptions from the external medium possessed also an extremely simple form of growth. Perhaps more notable still is the fact that under certain conditions the lesser particles within a coacervate partially assume a regular orientation in reference to each other; in other words, they acquire structure.[9] Through this localization and concentration of organic substance at definite points in space—through acquiring and expanding elements of structure— the coacervate assured that thereafter the behavior of its component *parts* would be governed not only by the simpler laws of organic chemistry but, in addition, by those of colloid chemistry. Occasional coacervate systems should have been capable of surviving as *wholes,* and thus of acquiring some properties that were subject to laws of a still higher order—biological laws.

Coacervates incorporating into themselves various substances from their fluid surroundings would in this way acquire "catalysts" and their "promoters." Very many simple substances that doubtless were present in waters of those early days are agents of these two types. Through these agents chemical and energy transformations, sometimes including growth, could be speeded up in the droplets. The changes should lead to increased stability in some droplets and to the dissolution and loss of identity in countless others. At this stage in the evolution of organic systems, high importance attached to the co-ordination of the speeds of these catalyzed reactions; and especially to co-ordinated growth-speeds in separate individual coacervates within which a more perfect (physicochemical) organization became of selective value.[10]

The traceable and spontaneous formation of adenosine monophosphate, which is the basis of an energy transport system, and the formation of nucleoproteins, which are the foundation of genetic continuity and change, have been already noted in connection with the "inorganic organism." At this point, therefore, we are prepared to include in one view the uniquely favorable environment that countless coacervate particles provided for the localization, development and use of those two life-generating substances. Fuller discussion of this particular item does not properly belong here; it is a personal view which has not been developed by others, but its possibilities seem both to widen markedly the road that led to organism and life, and to populate that path with greater numbers of independent origins of organism and life.

In any case, to whatever extent competition and differential survival became effective in advanced, self-sustaining coacervate

---

[9] A. I. Oparin, *The Origin of Life* (New York: The Macmillan Co., 1936).
[10] *Ibid.*

particles (organisms), a biological principle—natural selection—was thus born at the very dawn or predawn of the long history of life. Moreover, the entire population of coacervates that did not thus *graduate* into self-sustaining, living or semiliving particles—along with the huge reservoirs of both primitive and quite complex organic matter in aqueous solution—was destined to disappear into the swarming bodies of insatiable bacteria, which certainly evolved later and invaded the waters, the air, and the soils of earth.

Thus the bacteria, a simple form of life—though far from the earliest—could, in an ensuing epoch, soon sweep away all of the later steps of the very path that led to life. This sound reason for the absence of advanced, life-approaching structures and compounds from the bogs and waters of a parturient earth, as noted a billion years later by men of today, is now almost always overlooked. Now and again a man of science overlooks it, and even uses the fact that these intermediate substances do not now exist to emphasize a quality of severance and indeed of complete mystery in the living stuff. Actually, it seems both truer and more reasonable to say that present limitations on knowledge of the origin of life—apart from the facts that it could not be witnessed and recorded, nor could nature retain for our inspection the many materials and conditions from which she fashioned it—relate as much to our momentary inability to decide just which of certain now available structures are alive or partly alive and which are not as to our failure to comprehend the steps by which the simpler properties of organism and life arose. A hedged untruth now lurks in the timeworn statement that life is an "impenetrable mystery."

It is improbable that the foregoing history of the "inorganic organism" and of coacervate particles, their precursors, and their later acquisitions can ever be established as the course actually followed in the origin of the earliest living particles on this planet. But that survey does include a series of reliable items, and it sketches the probable sequence of some important steps in the origin of living matter.[11]

Only habit or ease of phrase warrants one's speaking of the

[11] See also B. De Jong, *Protoplasma*, 1951-52, four articles; J. L. Kavanaugh, *Philosophy of Science*, vol. 12, 1945; H. H. Horowitz, *Proceedings*, National Academy of Sciences, vol. 31, 1945 and vol. 35, 1950; G. W. Beadle, *The American Scientist*, vol. 36, 1948.

origin of life instead of the origins of life. On our own planet, living particles may have arisen thousands of times and throughout a period of thousands of years. Many such independent strains of a primitive living stuff may have persisted for variable periods and then disappeared from the living stream. Huge strides in evolution preceded the formation of those cellular mechanisms (mitosis, meiosis) which are now so widely associated with heredity and sexual reproduction. Within that older and probably more plastic phase of organic evolution, the conditions or factors for survival and for the formation of new types were probably not completely identical with the factors now found to rule in the far less primitive species of today. Though that view is widely current, it is quite improbable that persisting types of life—virus, bacteria, progeny of "Prophyta" and "Prozoa" —all descended from one single origin of living matter.

The viruses that cause such things as colds in humans and mosaic disease in tobacco and asters are extremely minute particles, each a molecule of protein but varying greatly in size; even the largest types can be seen only with an unusual kind of microscope. However, they are all probably pure proteins or nucleoproteins—and they possess some though not all of the usually accepted properties of living matter. Under certain conditions, the virus molecules can grow and so divide as very exactly to reproduce themselves. The virus can also mutate or change its character as does a fully living cell. Though the virus may be said to be only half alive it can be killed. Man, willow, worm, and bacteria, however, are killed even more easily.

As recently as 1932 the nature of viruses was a complete mystery. Viruses then contributed neither page nor phrase to the history of life. It was not then known whether they were inorganic, carbohydrate, fat, protein or organismal in nature. Within less than two decades the above and following facts relating to these links between the living and the nonliving world have been provided by the studies of biochemist Wendell Meredith Stanley his associates, and others. More than three hundred different viruses capable of causing disease in man, animals and plants have become known in two decades of rich discovery. The chemical changes involved in mutation—in change of type—in the viruses have been explored, and it is now known that either the gain or loss of one or more amino acids in the virus structure may accompany mutation. Of the method by which the virus particle is reproduced we as yet know little except that it requires contact with a specific type of living cell. The whole process, however, may require no more than thirteen to fifty minutes, and these potent particles seem able to use the *genes* of their particular host cells as aids to the duplication and reduplication of their own amazingly meager bodies.

More than a dozen different viruses have been obtained in highly purified form, and among these are found some of the smaller as well as the larger types. One exceptional and familiar virus, the bacteriophage, seems to include several molecules in its structure, though it still operates as a single individual. The smaller spherical or rod-shaped types of virus are apparently simple nucleoproteins whose chemical and physical properties tend to place them in the molecular world. In the words of Stanley, quoted at the head of this chapter, "The larger viruses have a composition and properties which are characteristic, not of molecules, but of organisms. The viruses have certainly provided the link between the molecules of the chemist and the organisms of the biologist."

Those similarly minute particles which in all living things bear the hereditary qualities and direct our development from egg to adult—the genes—are in several respects similar to virus molecules. Both operate only when associated with a cell. There the virus is sometimes an outlaw and possibly always a lone wolf, while the gene associates closely with other genes and works constructively. Probably both kinds of molecules are nucleoproteins: they are of about the same size; they both show a rare and peculiar form of instability, involving a change or changes of internal structure (mutation); and both seem able to reproduce themselves only when in contact with larger and more complex molecular aggregates that show more, or still other, of the properties of life. Probably no single gene standing alone is fully alive, and a few or many genes are packed closely together in all cells now generally credited with life. Too, since the various genes of a cell differ somewhat from each other, their very association or aggregation probably adds other elements of "liveness" to the aggregate.

In connection with gene and virus it is well to meet another molecule that has remarkable powers in the mixture usually associated with genes. The internal structure or chemistry of this particular molecule, unlike those just mentioned, is well known.[12] It is called heme (or iron protoporphyrin). It is several times larger than a molecule of benzine but very much smaller than one of protein or of virus. Its "ring" structure is such as to give it some rare properties. These include great stability, even resistance to destruction by acids and alkalis, and unusual power to absorb light. Furthermore, since all of the atoms of the ring lie in the same plane, it is a *flat* molecule. At the very center of this flat surface is an atom of iron. And both below and above the surface the iron atom is free to grasp one other atom or one other molecule. When the free "bond" below the plane surface attaches to the protein known as globin, the upper bond can attach to a molecule (two atoms) of oxygen, which it can also readily release when the entire ring structure moves into a region containing less oxygen. A compound of this type is called hemoglobin, the well-known oxygen carrier and red pigment of the blood. But when heme is similarly united with other protein than globin the union further acts as one or another of the respiratory enzymes which support the release of energy. Moreover, when the iron at the center of the porphyrin structure is replaced by magnesium and a molecule of colorless "phytol" is attached, a very modern molecule of chlorophyll is formed. Thus, besides its own special properties, the porphyrin molecule becomes a unique environment in which properties of iron (and magnesium) are intensified and applied to a variety of vital chores; altogether, a basis for respiration and energy release is provided through the two most characteristic pigments (light absorbers) of the plant and animal worlds.

Though they are not identical, "life" and "organism" are so related that they must be thought of together. What we call life is a grouping of properties really or apparently inseparable from organism. Organism is rightly regarded as an object—a thing. But life is not an object—it is a process involving a flow of properties from certain organizations or continuous reorganizations of matter.

The story of progressive change in present living things on this planet could indeed be largely told in terms of molecules—of chemical evolution. Some new molecules and myriad modifica-

12 Sam Granick, *The Harvey Lectures, 1948-1949* (Springfield, Ill.: Charles C. Thomas, Publisher).

tions and regroupings of old ones mark the steps taken by the higher organisms during much or most of a billion years. For the animal world that story has been written,[13] and the larger animal groups thus described by chemistry correspond well enough with those otherwise described in terms of anatomic structure. It is far simpler, however, to think of entire organisms under the structural names already familiar to everyone.

It was a long way—probably a billion years—from the earliest "living" particles to the man of a half-million years ago. But all was a continuing path of life; and since we now know the rules for the later half of that highway much of its course can be charted with fair accuracy and confidence.

In the great lapses of time, some genes within the simple, single particles or cells were able, as now, to undergo one or another change in structure—a mutation. In the laboratory, even X-rays, ultraviolet rays, temperature and chemicals have shared in causing genes to change. Here each mutated gene, not alone but when associated with slowly acquired rearrangements in the old store of genes, provided possibilities of a new type of cell—a new *species* —when the conditions into which it was born would permit it to persist and reproduce.

It is clear that rather early in the history of life, living cells of two or more distinctive types were formed. One type was adapted to capturing energy and was thus a promise of a plant world. The other type specialized in releasing energy, and this is an original animal quality. Now and again, however, these qualities probably were combined in one and the same cell, as we know they sometimes are in our own day. Microscopic *Euglena* moves itself with a whiplash thread of protoplasm, engulfs its food through a minute slitlike mouth, and digests this food as do the simplest animal cells (*protozoa*); but with its traces of green chlorophyll it also captures and stores energy like a plant. All cells of another distinct type, the bacteria, are not easily called either animals or plants, though they are usually classed with plants.

Highly promising were those cells which built within themselves some molecules of chlorophyll—or possibly, at first, a similar but simpler substance—that enabled them to use the sunlight and their supply of carbon dioxide for the capture of energy and

[13] Marcel Florkin, *Biochemical Evolution* (New York: Academic Press, 1949).

for the formation and storage of the sugars and starches. These stores, then as now, provided basic food for the animal world and also for that newer part of the plant world—the molds and fungi—that is, without chlorophyll. This single advance, accomplished in the morning of plant existence, did much toward forming the bud for an entire tree of life. It unleashed the cell of animal type—a cell specialized for the consumption and release of energy—into a world with an ever-replenishing food supply, a world in which that supply now rested largely upon sunlight and a rotating earth; a world in which even restrictions upon this food supply were destined to urge the formation of new types—new species—of organisms through the struggle for existence.

The well-established rule was for cells to grow, divide and separate. But occasionally all cells can undergo internal change in structure and capacity (mutation), and there must have been some cases of whole cells failing to separate completely after division, and of continuing this failure into later generations. The resulting cellular aggregates gave rise both to "colonial" forms and to extremely simple animals, or to similar plants, composed of several or many cells. Few events in the history of the earth and man are more notable than these unwitnessed cases of dividing cells remaining attached to each other, getting mutual benefit from it, and preserving the condition through all later generations. Perhaps not many primitive single-celled species have yet accomplished this, but those that did provided the possibility for flower and fish and man.

In the course of time these cell-communities could use some cells for one purpose, some for another, and eventually they thus built egg and sperm. Much of this progress was controlled by genes, but at least in the lowest forms of life we are now learning that the environment itself may take part in organismal change. A strain of bacteria readily killed by streptomycin can be made to mutate so that it will tolerate streptomycin and even depend upon it for existence. An equivalent action of the environment in the highest organisms, however, may not be assumed to follow in any forthright manner. Of immediate interest is the fact that the advent of egg and sperm multiplied the chances of starting out individuals with an unusual store of genes. Thereafter a new individual could arise not from one cell but from the union of two that had been part of two distinct parents. In more than one way, this provided additional chance for an offspring to begin life with a set of genes that differed from those of either parent. In fact, it was here that the method of human inheritance was born; and this event is worthy of one further word.

Within each cell of the cell-community, quite the same as in some though not all single-celled forms, the several genes align themselves into short threads, or rods, known as chromosomes—a few or several genes grouped very closely, in bead fashion, into a few or several threads. Normally, each chromosome (and gene) divides just before the cell itself divides, and one of the new sets of chromosomes passes to each of the two parts of the dividing cell. Incidentally, it is only in this way that any species may remain its relatively constant self. Under rare conditions, however, cells may acquire two, or three or more, complete sets of chromosomes. When this occurs following unions of egg and sperm of many-celled organisms—and it has occurred much more often in plants than in animals—the basis for a new species is fully provided.

Taking quick advantage of the simple conditions just noted, geneticists of the recent past have already added several new species of their own to those formed by nature. Just now the wit of man has added to the kinds of living things that grace or curse our planet. The species formed by nature, however, are the ones for further notice in this account. By whatever means marked changes were induced in genes, during the long past, a new species became at once a possibility. It became a reality in that fraction of cases in which suitable recombinations of genes occurred, and in which the descendants were variously fortunate— including birth into an environment in which, through an ever-present struggle or a fortunate co-operation, they could remain alive. The luckless bearers of the more numerous disadvantageous changes disappeared, leaving no descendants for the stream of life. Hundreds of millions of years provided floods of advantageous hereditary changes, some acquired slowly, others rapidly, and thousands of superior species of plants and animals overcame competition or found mutually helpful relationships and possessed the earth.

But an organism built of many cells is one that has met and solved continuously the special and oft-recurring problem of rebuilding itself from one or from two specialized cells. This is the egg-embryo-adult problem—the subject matter of embryology. For the "colonial" organisms this procedure is simple enough; but for oak and fish and man this story is not simple, and it is scarcely approachable in a paragraph. It may be remarked, however, that the sexuality that is expressed in egg and sperm was an invention of the single-celled ancestors; that the several forms attained by the embryo at least partially reflect ancestral adult stages of the organism; and that in a way not yet entirely clear the genes present in all cells largely direct the changing picture from egg to adult.

Within the animal series—but not in plants—the development of sensory and nervous elements of growing complexity and variety served to multiply—yet always to integrate—activity and behavior in animal bodies. The guiding fact in that book of facts is that different animals developed sensitivity to different elements of the special or of the total environment, and that their behavior is thus broadly accounted for. The group of five so-called special senses of humans is shared with some or many higher mammals. Numerous lower animals fashioned still other special senses. But nature alone—apart from study—equips no *man* to imagine what is afoot when fiddler crabs flee inland under the low barometric pressure that presages a hurricane; when bats avoid obstacles by supersonic-wave warnings; when male and female moths rendezvous through broadcasts of infrared light; when sound is perceived by structures far simpler than ears; or when radar—discovered by man only a few years ago—is now found to have modified behavior in certain fish during millions of years. Incidentally, indeed tragically, human behavior has been affected continuously by *lack* of special senses that blossomed unaware and bewilderingly in other species. That lack, to this moment, has added to the gullibility and superstition of practically every single span of normal human existence.

In the more complex animals the sensory elements were at once the germs and building blocks for types or grades of reactivity, self-consciousness and psychic life. It has become quite clear, too, that these countless grades of reactivity or "mentality" are *inseparable* from the complexity that is the organism; that none of the many items of "mentality" is an *object,* but only an activity or function—as is locomotion; that each organism is a single energy unit; that nowhere do we meet or deal with the development of *a* mind and *a* body. Everywhere and always, in plants or animals, we may find only wholly integrated organisms.

Now and again, during more than a half-billion years, splendid species, which had persisted for ages, totally disappeared from the living scene. Their fossil remains tell of such things as changing climates, harmful overspecialization, and new enemies— things to which, struggle as they might, these once living forms could not adapt themselves. But their traces and their bones, exhumed daily from sands and rocks of the successive ages of the earth's crust, continue to speak with assurance of the ascending series of living things that preceded us.[14]

Neither the ability to persist nor the capacity to change is at all equal in the many species of any one epoch. An odd way in which some organisms have asserted difference from other species is shown in their resistance to further change with passing time. Such organisms attain a level of structure and of life and thereafter stoutly maintain it. If this is somewhat questionable for the groups called species, it is nevertheless quite true for groups known as genera. About fifteen per cent of all the genera of invertebrate animals are said to have passed through two or more geologic ages without extinction and without apparent change. The family of Port Jackson sharks swam in Jurassic seas more than a hundred million years ago, and with little change they are still with us. The shellfish *Lingula* burrowed in oceanic mud of a half-billion years ago, and off the coast of Japan it does the same today. Probably some species of bacteria (e.g., *Nitroso monas*) have greatly surpassed the cases already cited, as they seem to have persisted practically unchanged since Archeozoic time. Stagnation and torpor are far from the rule in the living world, but in some cases they are associated with an amazing lengthening of the life of the species.

When a backbone was built a series of future rulers of the animal world was forecast for that teeming realm. Fish, reptile and mammal have taken their turns as king. Backbone pointed forward not merely to a bone-protected brain, but forward also to a mastery of sea, of land, and of air. That backbone, as first made of cartilage, was already built some four hundred million years ago—admitting, of course, that this measurement is not notably accurate. The first mammals date from somewhat less

---

14 William K. Gregory, *Evolution Emerging* (2 vols.; New York: The Macmillan Co., 1951). Here the reader will find a comprehensive survey of the animal world, and of fossil forms of animals and man in their relation to the geological record.

than two hundred million years ago. The tree shrews, or related and unspecialized tarsioids of about fifty million years ago, are usually recognized as the common ancestors of monkeys, apes and men. The hominoid group, including apes and men, probably arose much later and from a species other than the one that gave rise to doglike baboons. The unspecialized apes (hominoids) of the mid-Tertiary period, some twenty million years ago, provided two lines of descent. One of these led to specialized recent apes; the other, the hominids (men), led to fossil and modern men.[15]

Until recently, only scanty skeletal parts of a few species of fossil men—species largely filling the gap between us and our ape-like ancestor—had been found, studied and preserved. As yet, too, there has been only slight opportunity to do these three things. If made by nearly all men down to the narrow present, such finds would not have been studied and preserved but discarded or destroyed. Nevertheless, bones of some two to five species of now extinct fossil men—and of still other apemen—are already in the hands of scientists able to examine them. Among the examples are certain groups whose distinction from both modern man and the highest form of living apes is beyond question. Australopithecus was an apeman of South Africa who lived nearly a million years ago. Australopithecus had an erect posture joined to an ape-sized brain and to teeth intermediate or more apelike. *Pithecanthropus erectus,* of Java, was of rather similar age and type but with a somewhat larger brain. Sinanthropus, or "Peking Man," was also ancient, had a larger brain, heavy brow ridges, and showed large size differences between the sexes. Fossils of later, and of much later, human species have been recovered from western Europe. Among these are Neanderthal Man (or Men) and Swanscombe Man. Here may be found large men and weightier brains, smaller brow ridges, higher foreheads and developed chins. Perhaps Sinanthropus included more than one species and through this or other circumstances left two lines of descendants. One line, all Neanderthals, became wholly extinct; the other led perhaps through such forms as Swanscombe Man to modern man, *Homo sapiens.*

The little that we know of the performance of early members

[15] *Ibid.* See also W. E. LeGros Clark, *The History of the Primates: An Introduction to the Study of Fossil Men* (London: The British Museum of Natural History, 1949).

of our own species—which now seems less aged than was thought two decades ago—is not flattering to those early men. Surely, however, our understanding of ourselves would profit much by greater familiarity with them. And what an effect there might have been on the course of past and current thought and social structure, if only a small tribe of Peking or of Heidelberg men —intact and coarse—had persisted into this mid-century! Also of quite special meaning is the assurance that our distant forebears, like their still more remote human and apelike ancestors, were highly social and gregarious—readily expressing mutual aid and sympathy. This fortunate circumstance provides a purely natural basis for the origin and development of ethics and morals.

Early man took an unknown amount of time to learn to make a most simple tool—a flint for aid in obtaining food and for defense against animal enemies. But for ages thereafter he made little further progress. Once he had made himself a bit safer from dangerous animals, learned of fire, and somewhat extended his menu, man seems to have remained for hundreds of generations almost as nonprogressive as a population of black bears. If men of today could meet any of the early members of our own or of a still earlier human species, they would do so as superlative strangers. Tool-using and perhaps articulate man has been doing something on some parts of the earth's surface for as long as one-half million to one million years. Yet within probably less than fifteen thousand years he built his first city. A drab epoch of human futility! A brutal demonstration of primal limitations of raw, untutored man!

The limitations of even our more recent ancestors—those neoliths distant by no more than a thousand generations—are remarkable, and our link to them includes the emergence of a new and nonphysical factor elsewhere not met in this story of life. The man of today is, or seems, so different. How can we bridge the gap between modern man and his physically similar progenitor of thirty thousand years ago? That adjacent ancestor was probably equipped with nearly or quite as good heredity as are most men of today. Neither structural changes occurring in man, nor changes in the genes carried by him, adequately account for the changes in our species during that short term. Those changes, it is now known, came chiefly from *social* forces born of the maturing of language, with its gift of abstract thought, of the discovery and extension of agriculture, of the invention of alphabets and wheels, of the domestication of animals, and of the use of metals and harnessed energy. These external things—created partly by a clever few and shared by many—brought an outlook and new bits of security, plenty and leisure; these emergent social forces severed the tie to the typical animal mode of life. Thereafter a human being did not grow up as his mere primate self. From infancy onward his surroundings and senses were saturated with the cultural acomplishments of his race—a thing not possible to any other living species, and a thing little evident in primitive man during some hundreds of thousands of years.

The long trail of earth and life thus leads to modern, still-evolving man—the winner of an age-long race among many brothers. Like the sparrows, man arrives not as a single breed but in variety. Though the several races of the one surviving human species largely remain physically apart—only partially blended by intermarriage or by abode—they already begin acquaintance with each other. In numbers far greater than ever before, the warm bodies of man now caress the hemispheres. His art and effort have built cherished homes and laid a fruitful, checkered sward upon the continents. His growing mind has gripped the moon and a multitude of suns, has wrested speech from buried milestones of his own long path of a million years, has learned the outlines of the purely natural laws that created complexity, life and society on a small and partly chance-born earth. With the mantle of these triumphs about him, man now exalts both himself and those vast voids and wastes of distant universe that remain so aloof to his own creation, history and destiny.

2

First Principles

I do not profess to know what matter is in itself. . . . I wait for the men of science to tell me. . . . But whatever matter may be I call it matter boldly, as I call my acquaintances Smith and Jones without knowing their secrets.—*George Santayana*

With the acceptance of the doctrine of emergent evolution . . . the desires and aspirations of humanity are determiners in the operation of the universe [of man] on the same footing with physical determiners. What is to come in the future is not predictable from what has occurred in the past. The laws of nature are not immutable, in the sense that new laws shall not be exemplified as new conditions arise.—*Herbert Spencer Jennings*

A FEW LAWS AND INSIGHTS OF EVOLUTIONARY SCIENCE should get early attention if this book is to better serve some of those for whom it is written. Still other principles are more conveniently met in other chapters of Part I. Since some of the broadest generalizations of science and philosophy cannot be described solely in the language of everyday life, an occasional resort to technical words is here unavoidable. A pair of such items intrude themselves into the midst of the present chapter. For readers who have little liking for technical material, it is suggested that some pages near the middle of this chapter be scanned rather than read. Later and earlier paragraphs are less technical.

ON THE CONCEPT OF INTEGRATIVE LEVELS

The evolutionary process extends from the fragments of atoms through the known forms of energy and matter—the ninety-eight or more elements and their isotopes; through all nonliving combinations of these; through the cosmic bodies; the earth; organisms; and all human society. Roughly, these several states or categories of matter (levels of integration), exhibiting as they do the evolutionary accomplishment at progressively higher levels, provide a locale or a "home grounds" for the various sciences—physics, chemistry, astronomy, geology, biology, psychology and sociology. These several sciences, along with such overlapping ones as paleontology and anthropology—often with the aid of mathematics—have supplied the data that tell the entire story of evolution as we now know it. These sciences, together with those special considerations of relationships and meanings of phenomena called philosophy, seem to cover the universe of energy, matter, process and phenomena.

Since the appearance of Darwin's *The Origin of Species*, in 1859, the several sciences listed above have found that transformism—change with time, or evolution in its broad sense—is the principle of first importance to an understanding of that universe. Described otherwise and in terms of philosopher Dewey, the natural sciences discovered that the thing that is actually "universal" in their several fields is *process;* and though "this fact of recent science still remains in philosophy, as in popular opinion up to the present time, a technical matter . . . it is the most revolutionary discovery yet made."

This single, pervasive principle of change, or of process, placed the whole of the known universe within the territory ruled by natural law. It did not, however, give a complete account of the state, order and variety that exist in either the nonliving or the living world. Other principles are also involved. The living world most notably accommodates additional principles since, in this broad area, we sooner or later meet such dissimilar qualities or conditions as polarity, population problems, inconceivably complex organization, consciousness, society and civilization. Within this area several laws and principles that are perhaps less general in their application than the one that defines the method of the origin of species have been, and are now becoming, established. But, more notable still, a principle of even wider application than that relating to the process involved in the gradual appearance of new species seems to be necessary to a rational view of some biological and social things. Such a principle is that of levels of integration.

The concept of integration may be said to absorb and include much of the thought that became known as "emergent evolution";[1] for, as used in this book, the word "emergent" has no

---

[1] C. Lloyd Morgan, *The Emergence of Novelty* (London: Williams and Norgate, 1933). Herbert S. Jennings, *The Biological Basis of Human Nature* (New York: W. W. Norton and Co., 1930).

antimechanistic meaning. Again, the Darwinian and integration principles are complementary, not exclusive or contradictory. They mutually describe change, process or stage when, for example, they consider the origin of a many-celled animal (metazoan) from or through single-celled forms (infusoria). A new species obtained by a defined method of securing diversity is the point clarified by the one; the attainment of a higher order of unpredictable properties and possibilities is reported by the other.

From publications on the subject we here quote the following recent statement by Novikoff:[2]

The concept of integrative levels of organization is a general description of the evolution of matter through successive and higher orders of complexity and integration. It views the development of matter, from the cosmological changes resulting in the formation of the earth to the social changes in society, as continuous because it is never-ending, and as discontinuous because it passes through a series of different levels of organization—physical, chemical, biological and sociological.

In the continual evolution of matter, new levels of complexity are superimposed on the individual units by the organization and integration of these units into a single system. What were the wholes on one level become parts on a higher one. Each level of organization possesses unique properties of structure and behavior which, though dependent on the properties of the constituent elements, appear only when these elements are combined in the new system. . . . The concept . . . neither reduces phenomena of a higher level to those of a lower one, as in mechanism, nor describes the higher level in vague non-material terms which are but substitutes for understanding, as in vitalism. Unlike other "holistic" theories, it never leaves the firm ground of material reality.

Partly for the purpose of directing the reader's attention to an additional source of information on parts of the subject, a further quotation is taken from a book whose chapters were the work of twelve specialists. In a summary statement the editor of the book,[3] anthropologist Redfield, wrote as follows:

[2] Alex B. Novikoff, *Science,* vol. 101, 1945, p. 209. For earlier and more extended discussion see J. H. Woodger, *Biological Principles* (London: Kegan Paul, 1929); Joseph Needham, *The Modern Quarterly* (London), vol. 1, 1938; R. W. Gerard, *Philosophy of Science,* vol. 9, 1942.

[3] *Levels of Integration in Biological and Social Systems,* ed., Robert Redfield ("Biological Symposia," vol. 8 [Lancaster, Pa.: Jacques Cattell, 1942]).

What these papers seem to be saying, in most general terms, is this: The organism and the society are not merely analogues; they are varieties of something more general: the disposition, in many places in the history of life, for entities to undergo such modification of function and such adjustment to other similar entities as result in the development and persistence of larger entities inclusive of the smaller. "Fitness may mean cooperation for mutual benefit both between species and within integrated intraspecific populations as well as between parts of the organism." Departing from the language of science, one might say that the individual metazoan, the infusorian population, the ant colony, the flock of fowl, the tribe, and the world-economy, are all exemplifications of nature's grand strategy.

The idea of society developed by Hobbes and Spencer has now been abandoned by sociologists. Incontestable, however, are the facts that the larger entity, society, is composed of smaller entities, individuals; and that society has characteristics not identical with those of its individual members.

In addition to the easily seen structural complexity that arose step by step from single-celled animals to the near relatives of man (apes and monkeys), closer studies have shown also a series of functional trends. Three such trends that have much importance to the next higher integrative level—human society—deserve listing here: first, an increased effectiveness of the control of the many bodily functions by the organism as a whole instead of by the separate organs; second, an increased degree of independence of the animal from its environment, together with extensions of the range of the animal's activity; and third, an increased efficiency of the individual animal in accomplishing survival and reproduction, involving often a degree of ability to mold the environment. In fewer words, the many steps of animal ancestry led simultaneously to the structural complexity and to the platform of capacity that are united in individual man. And groups of individual men are the beginnings of human society.

SPECIAL ITEMS FROM PHYSICS AND PHILOSOPHY

The newer physics—nuclear physics—clearly requires notice here. Not merely organic evolution but the whole of the evolutionary process is the basis of evolutionary thought, and of naturalism. The inconceivably long and endless process of integration begins with the constituents of the atom. Certainly no appraisal of the question of the existence of the supernatural—of God—may overlook the very special position and creative powers of the atom and its several constituents.

No competent scientist now assumes, nor in any way asserts, that physical science can at this moment supply for the simplest living cell a description and analysis at all comparable with those it now provides for the atom. Nor can anything other than this be expected. The living state—organism—is much too involved to be fully accessible to mathematics, upon which physics relies. This topic has been discussed recently by one of our generation's foremost physicists. With slight change of sequence of his sentences the following is quoted from Schrödinger:[4]

The central problem: How can the events in *time and space* which take place within the spatial boundary of a living organism be accounted for by physics and chemistry? . . . The obvious inability of present-day physics and chemistry to account for such events is no reason at all for doubting that they can be accounted for by those sciences. . . . But the meaning is very much more positive, viz., that the inability, up to the present moment, is amply accounted for.

The "order-from-disorder" principle is a well travelled road in physics. . . . It appears there are two different "mechanisms" by which orderly events can be produced: The "statistical mechanism" which produces "order-from-disorder" and the new one, producing "order-from-order." . . . The new principle is not alien to physics. We must be prepared to find a new type of physical law to prevail in it [i.e., in living matter, a law that contrasts with] the "order-from-disorder" principle, which is actually followed in (non-living) nature and which alone conveys an understanding of the great line of natural events, in the first place of their irreversibility. But we cannot expect that the "laws of physics" derived from it suffice straightway to explain the behavior of living matter, whose most striking features are visibly based to a large extent on the "order-from-order" principle. You would not expect two entirely different mechanisms to bring about the same type of law—you would not expect your latch-key to open your neighbor's door as well.

One is here concerned wholly with the attainments and viewpoints of present-day science, not with those of even fifty or twenty-five years ago. Equally, the same should apply to philoso-

4 Erwin Schrödinger, *What Is Life: The Physical Aspect of the Living Cell* (New York: The Macmillan Co., 1946).

phy—and for identical reasons. Philosophy has recently found a way and a need to include the social sciences, and in doing so its former interest in metaphysics tends to disappear. A leader in that advance is John Dewey.[5] The philosophic background of the chief issues treated in the present book have been discussed by Dewey. From a paragraph of his Introduction to the two later editions (1948 and 1950) we quote:

> The supposed fact that morals demand immutable, extra-temporal principles, standards, norms, ends, as the only assured protection against moral chaos can, however, no longer appeal to natural science for its support nor expect to justify by science its exemption of morals (in practice and in theory) from considerations of time and place—that is, from processes of change. Emotional—or sentimental—reaction will doubtless continue to resist acknowledgment of this fact and refuse to use in morals the standpoint and outlook which have now made their way into natural science. But in any case, science and traditional morals have been at complete odds with one another as to the kinds of things which, according to one and the other, are immutable. . . . To the vested interests, maintenance of belief in the transcendance of space and time, and hence the derogation of what is "merely" human, is an indispensable prerequisite of their retention of an authority which in practice is translated into power to regulate human affairs throughout—from top to bottom.

This quotation deals with what is usually regarded as the central momentary conflict of natural science with theology and religion—and with ecclesiasticism, old and new. This fully matured thought of America's leading philosopher takes firm note of morals as a continuing area of basic conflict between natural science and the "vested institutional interests." It affirms that "the derogation of what is 'merely' human" by those vested interests "is translated into power to regulate human affairs throughout—from top to bottom." Better support in now current philosophic thought for the reality of the dangerous sweep and prevalence of those vested interests, for the philosophy of naturalism that pervades all these pages, and for need of the surveys, blueprints and specifications included in the present volume, will

[5] John Dewey, *Reconstruction in Philosophy* (New York: Henry Holt and Co., 1920. Enlarged ed.; Boston: The Beacon Press, 1948. Also a Mentor Book; New York: The New American Library, 1950). See also John Dewey, *Experience and Nature* (New York, 1925), and John Dewey and Arthur E. Bentley, *Knowing and the Known* (Boston: The Beacon Press, 1949).

not be sought or provided here. But after acknowledging the contributions of Dewey and his contemporaries,[6] one may now more properly regard this "morals" issue as being already resolved definitely in favor of science (despite the continued failure of many philosophers and some scientists to acknowledge that outcome) and the present conflict better described as one between society and theology or religion than—as was formerly true—between science and theology. For it is not additional scientific fact or philosophic effort, but the wider acceptance and use of an accomplishment, that is further involved. And that is a social matter.

It would seem that when scientist or philosopher now places "spiritual" and "higher" values outside or apart from the copious products of "material" change, he has failed to grasp the full sweep of evolutionary and integrative process. These values and desires have a fairly precise point of birth. Their cradle is farthest of all from the atom and from the immensities of the galaxies. It is indeed far, though less far, from the gene. It is vastly more localized than is greatly restricted and delimited life. It is still remote even from the fetus or the child of one year. These values are born always and only close to the skins of grown or growing men and groups of men forced by plenteous circumstance to long-continued life together.

To consider any human value any less valuable because its origins are comprehended or delimited is to do a wrong—it is to depreciate the actual stature and tasks of man.

A short reference may be made to the relation that exists between the real or absolute world and the impressions we obtain regarding the various items that compose it. Our concepts of these items are derived, primarily, through sensory impressions, and here the several senses (sight, hearing, touch, and so on) may each add information about accessible objects. Never, however, does our short list of senses, nor our measurements aided by the tools devised by science, present us with the complete, the absolute. In this field, science itself only approximates, however much it outstrips other human experience in its approach to reality. Again, we need not hedge in associating "objectivity" with such terms as "dog" and "tree," but adequate analysis shows that many abstract terms do not relate to objects, as has been

[6] Hans Reichenbach, *The Rise of Scientific Philosophy* (Berkeley and Los Angeles: University of California Press, 1951).

supposed, but to activities instead. This treatment of meanings greatly modifies a concept of both knowledge and nature.[7] Moreover, within the special new field of quantum phenomena many common-sense notions of "object" completely fail us, since permanence or stability—the essence of identity—is there lost. Indeed, a foremost physicist (P. W. Bridgman) has said that "the structure of nature in the direction of the very small is simply not the same as the structure of thought; and, being so, it is meaningless even to attempt to formulate what it is like."

The above brief reference to the lack of permanence and stability in the field of *quantum* phenomena provides further reason for venturing the following statement on the sort of physical state that has special significance to the processes of integration and evolution: We now know that matter, or mass, is interchangeable with energy; but it is *energy in the frozen form,* that is, atoms and molecules of *matter,* that provides a framework or basis for the several still higher integrative levels that are especially significant in the evolutionary process. In other words, it is *structure* (form), such as is exhibited in atoms of the ninety-eight or more elements and in molecules, that has been integrated and reintegrated and thus proved itself capable of self-inclusion in the ever-widening range of products of that process. The heat and light radiation that are thought to have dominated earliest stages of the exploded "monobloc"—from which the present phase of the universe perhaps arose—later became much less prominent. Frozen mass (atoms, molecules)—form, rather than energy or attenuated mass (light, electromagnetism)—is directly involved in the building of crystals and organisms. These statements, ventured as possible aids to comprehension, should suggest no barren role for the quantum in life processes. As one example, the food supply of animals is provided by plants; and the ability of quanta of absorbed light to be transformed in green plant cells into chemical energy is a steady sustainer of life.

Since the turn of the century the atom itself has become known as a nonliving and unstable type of organism. A score of elementary particles have been found to enter into its structure. Nuclear physics seems to have shown further that everything at all is simultaneously particle and wave field. The nature of matter nevertheless remains uncertain since, at the moment, physics

7 Dewey and Bentley, *op. cit.*

does not know how to combine these two truths about matter. Properties and new uncertainties relating to the atomic organism have transformed this area of physics, though they seem of relatively less consequence in biology (also in the part of physics that deals with larger masses) where much larger quantities of very complex matter are of primary importance. But the dominant mechanism that physics earlier provided—and now much qualifies through relativity, quantum phenomena, and the uncertainty principle—is apparently wholly lost at no level of integration. And an array of integrative levels—of examples of "order-from-order"—are interposed between the atom and society.

An over-all and usually overlooked result of advances in physics seems of the greatest possible consequence to all thought on the existence of God. That result indicates that it is impossible for human beings to transcend the human reference point. This superb achievement of science is thus stated by Harvard physicist Bridgman:[8]

Finally, I come to what it seems to me may well be from the long-range point of view the most revolutionary of the insights to be derived from our recent experiences in physics, more revolutionary than the insights afforded by the discoveries of Galileo and Newton or of Darwin. This is the insight that it is impossible to transcend the human reference point. The history of much of philosophy and most of religion has been the history of the attempt of the human race to transcend its own reference point by the invention of essences and absolutes and realities and existences. It should have been obvious enough, even without the experience of recent physics, that this was an impossible attempt. For even the mystic, convinced of direct communication between his soul and some supernatural external reality, would have had to admit that it was his soul, and therefore a human soul, that had the experience, and that the experience took place in his consciousness, and therefore a human consciousness. But considerations like these are so obvious that it is easy to overlook their significance. Recent experience in physics documents in another way the conclusion that it is impossible to transcend the human reference point, and by the emphasis of novelty may perhaps succeed in injecting this insight into the backbone of humanity.

[8] P. W. Bridgman, "Philosophical Implications of Physics," *Bulletin,* American Academy of Arts and Sciences, February, 1950.

ORGANIC EVOLUTION

More than two decades ago the following acute appraisal of the status and significance of the question of the *positive* identification of the factors concerned in organic evolution was written by biologist Haldane:[9]

No competent biologist doubts that evolution and natural selection are taking place, but we do not yet know whether natural selection alone, acting on chance variations (mutation) will account for the whole of evolution. If it will, we shall have made a big step towards understanding the world; if it will no more account for all evolution than, for example, gravitation will account for chemical affinity, as was once believed, then biologists have a bigger job before them than many of them think. But a decision on this question one way or the other will greatly affect our whole philosophy and probably our religious outlook.

Rather less than two decades later, through the researches of a number of now living biologists, this determinative decision was attained. The result is now a definite, magnificent, and largely neglected segment of the intellectual capital of the race. Two statements regarding the resolution of this central and dominating question—the factors concerned in organic evolution—will here be quoted from recognized authorities. From biologist Julian Huxley:[10]

With the reorientation made possible by modern genetics, evolution is seen to be a joint product of mutation, recombination and selection.

And from biologist H. J. Muller:[11]

The genetics of today traces the fact of evolution back to the existence of ultramicroscopic bodies, the genes, which not only reproduce themselves but, more important, reproduce their own changes, or mutation, and which can continue thus to change without losing the power of reproducing their changes, to an unlimited degree. At the same time there are mechanisms whereby

[9] J. B. S. Haldane, *Possible Worlds. A Scientist Looks at Science* (New York and London: Harper and Brothers, 1928).

[10] Julian Huxley, *Evolution, the Modern Synthesis* (New York and London: Harper and Brothers, 1942).

[11] H. J. Muller, *Genetics, Paleontology, and Evolution*, Part VIII (Princeton, N. J.: Princeton University Press, 1949).

the genes can become bound into an aggregate, and whereby the number in an aggregate can become increased ("duplication"). Following such increase in number the component genes can and eventually will undergo mutations different from one another, and thus not only the degree of compoundness but the complexity of the gene aggregate becomes increased, and with it its potentialities of forming a complex organism as a result of the interactions of the different genes of an aggregate with one another. In this way it has been possible for what we call living matter to advance from the stage of one or a very few kinds of genes, at first separate from one another and naked, to that of a large constellation, individually differentiated from one another within the group, which by their common activities result in the prodigious, stupendously intricate working system evident to us as the soma (body or phenotype). . . . That given phenotypes are maintained over long periods as a result of natural selection, and therefore must also have become established by such selection, is evident from the very fact of their long continued existence (and from still other facts).

In the preceding statement it is implicit that the later learned facts of Mendelism and mutation serve as refinements and extensions of Darwinian evolution. If the quoted remarks on the mechanism of evolution were expanded, they could emphasize the fact that *variation* (mutation) is the very stuff of change and that *selection* guides and gives direction to that change. Also that mere size of a population and the relative proportions of the genes carried within it have been found to affect the outcome of the interplay among mutation, selection and duplication.

Thus, with the principle of organic evolution finally and firmly established, the biology of today is itself in an exceptional position to affect and modify both philosophic and religious thought.

Social philosopher and lecturer John Fiske is said by many to have been the foremost apostle of Darwinism in America during the final quarter of the past century. Yet the pinnacle of his aims was to salvage religion from an overturn that he little understood. That, however, he did neatly and well. By merely surrendering the larger case of science—asserting that not only does science affirm the existence of God but this God of science is also the immanent God of a somewhat purged religion—he thus contended there was no conflict, and one need only hope for the noise to blow away. Fiske could have delivered no more lastingly welcome message to a growing body of religiously inclined Americans, nor could he better have camouflaged a conflict by use of error and a hope. Somewhat more realistically than by Fiske, the essentials of this contest were described, in 1897, by American historian-diplomat Andrew D. White, who in turn saw the essentials less clearly than did an English predecessor, scholar John William Draper.

ON SOME CLAIMS AND CONSEQUENCES OF RELIGION

For ages, theology and religion laid claim to a territory and asserted that science was excluded from it. Philosopher Dewey[12] first illuminated and then, with others, resolved this prolonged contest of natural science against its theological adversaries. Dewey traced the battle without victory, which was followed by an unquiet truce based on a division of fields and jurisdiction, wherein science "dealt only with lower material concerns and refrained from entering the higher spiritual 'realm' of Being." He demonstrated that "especially within the last generation, the settlement by division of territories and jurisdictions has completely broken down in practice." Its adversaries now charge science with breaking the truce, and also with "the present scene of disorder, insecurity and uncertainty, with the strife and anxiety that inevitably results." Dewey then summarized the case of science against this crucial claim of theology and religion:

Now the simple fact of the case is that any inquiry into what is deeply and inclusively human enters perforce into the specific area of morals. It does so whether it intends to and whether it is even aware of it or not. When "sociological" theory withdraws from consideration of the basic interests, concerns, the actively moving aims, of a human culture on the ground that "values" are involved and that inquiry as "scientific" has nothing to do with values, the inevitable consequence is that inquiry in the human area is confined to what is superficial and comparatively trivial, no matter what its parade of technical skills. But, on the other hand, if and when inquiry attempts to enter in critical fashion into that which is human in the full sense, it comes up against the body of prejudice, traditions and institutional customs that consolidated and hardened in a pre-scientific age. . . . Science as conducted, science in practice, has completely repudiated these separations and isolations.

12 *Reconstruction in Philosophy*, eds. of 1948 and 1950.

Only two or three decades ago did philosophy change and state its position on the vital question of the "location of authority" to the extent that it transferred to science the support it had accorded theology throughout the ages. This it did despite many dissenting philosophers. That quite recent reconstruction of philosophy now both deepens and highlights the chasm that runs through modern culture. Reorientation at this high level should have much and accelerating power—at least, wherever things of the intellect move men and where the will to believe is not overwhelming. It was in 1929 that Dewey showed that values relate to the contingencies of developing life, and not to the existence of some antecedent reality; and, further, that this had plunged Western society into a "genuine cultural crisis . . . a social crisis, historical and temporal in character."[13]

While gradually performing its many hard tasks, science has dehumanized the universe. It now turns out that this is perhaps its greatest accomplishment, a favor wrought largely because religious mysticism, through centuries of thought control, had forced the mirage of a humanized universe upon Western institutions. But to society—to men in general—the potential value of this foremost accomplishment of science can nevertheless be minimized or overwhelmed through a *continuance* of thought control by the institutionalized churches. If proof is provided that this source of thought control now operates effectively throughout the literate world, it will have been proved that here is an immense area of present conflict between advanced society and ecclesiasticism.

The subject matter of Part II of this book is neither science nor the methods nor the meanings of science. Central, there, is the evidence that society is now denied use of the essentials and best of evolutionary thought by a suppressive environment born of widespread organized religion. It is a consideration of means by which and the extent to which society is now denied existence on a new and higher plane. It is a survey of daily practices whose aggregate accomplishes our supreme social disgrace. ("The genuinely modern is still to be brought into existence," said Dewey.) Society, not science, is the prime and direct victim of this present power of the cults of supernaturalism. Otherwise the actual division, and a not uncommon indifference, among

13 John Dewey, *The Quest for Certainty* (New York: Minton, Balch and Company, 1929).

scientists would be unthinkable. The materials of Part II serve equally well as blueprints and specifications for more recent charges of John Dewey and others that ecclesiastical thought and practice inhibit and damage society, and as a unique and specific examination of the extent to which society is injured at this mid-century by those who teach and impose a trust in the supernatural.

The full story of the consequences and ramifications of religious thought and practice in the world of today could of course be reported only by many men through many books. In that crucial area of this book, this writer operates less as scientist than as observer and reporter. Preparation for this reportorial task has, perforce, included some personal and successful search for new information. Always, however, there was need to hear and find, to sift and weigh, and often to accept, the data and reports of others. In this field this reporter both notes and regrets his many limitations. Serious limitations certainly, though it could be specified that nearly six years of residence and travel outside the United States, on five continents, have permitted the extension of personal observations to a territory not highly restricted.

Other claims and methods of religion and theology require momentary examination in the light of the principles already reviewed in this chapter. First, the matter of *claims*. Whatever the pretensions and formal statements of the various religions, they seem to be basically concerned with the origin, the nature (moral and other), and the destiny of man. That their spheres thus overlap those of biology, psychology and sociology seems clear and certain; that conflicts between the two groupings may and do arise should be equally evident. It is undeniable that such conflicts were plentiful in the past. The eyes of unmitigated evolution, though not those of a protectively dwarfed and misty evolution, see the persistence of hot conflict down to our own day, in every quarter of the globe. Only the full retreat that is associated with a rejection of a supernatural element that exempts morals and values from considerations of time and place—from processes of change—and with a resulting redefinition of religion would provide escape from this conflict. This required redefinition involves religion's abandonment to science of its former dominions—the origin, nature and destiny of man.

The *methods* by which the thought and practice of prevailing organized religions were established and now operate are strikingly inconsistent with the procedures that have yielded our solid knowledge of nature and of man; this, although such religious thought, in actual practice, likewise presents or provides a sweeping interpretation of man and of nature. Here, in religion, observation and experiment give way preferably to revelation or to feeling; and the written word or legend, however involved and unascertainable its origin, is often made an object of reverence and a guide to personal behavior. In other words, what the method of science finds unacceptable is firmly embraced by religious thought. Again, it was observed above that, in science, chemistry, a higher organizational level than physics, attains its full meaning only in the light of physics; that sociology, a higher level than biology, attains its full meaning only in the light of biology. Religion, however, picked up a series of psychological and sociological fragments in the long-ago that, through the centuries, it has zealously pressed upon the minds of men; persistently it has either ignored or directly opposed the biological levels and tests on which the acceptability of those fragments must rest.

As long as religion or religions presume to provide an interpretation of man, of nature, or of conduct—or, again, as long as religion or religions have a supernatural content—nothing can be more absurd than the contention that religion and science are distinct areas and free from conflict. The extent to which that absurdity is propagated, even by many of those now entrusted with the advancement of science and learning, may be a major indictment of much of the intellectual leadership of our day.

IN RETROSPECT

Science. The more trivial discoveries of science are the things that nearly everybody thinks all scientific discovery is; namely, additions to the old store of knowledge. But, as recently remarked by English geneticist Darlington,[14] "This is not true of the fundamental discoveries, such as those of the laws of mechanics, of chemical combination, of evolution, on which scientific advance ultimately depends. These always entail the destruction of or disintegration of old knowledge *before the new can be*

14 C. D. Darlington, *The Conflict of Society and Science,* Conway Memorial Lecture (London: Watts and Co., 1948).

created" (italics mine). After noting that intellectual activity and scientific inquiry everywhere tend to become so grooved that such basic discoveries are delayed and endangered, Darlington continued: "We need a ministry of Disturbance; a regulated source of annoyance; a destroyer of routine; an underminer of complacency."

Philosophy. For some thousands of years one metaphysical system after another has tried its luck at solving the riddle of the universe, with the result that each was doomed in turn to collapse as soon as mechanistic science was able to test its contentions, J. Loeb points out. Dead speculative philosophies find no time or place for burial—"Great Books" and one or another religion hoist them to haunt further the horizons of humankind. A living, scientific philosophy contributes much to striving, questioning men. But what else in the realm of thought equates with the fact that what was mechanistic science, and now, as naturalism, is an upgrowth of it, has already provided an account of the probable way in which accessible parts of the universe and man himself evolved and operate? It is true, of course, that science and its method have not compassed the absolute; nor been able to determine both position and velocity of a particle at the same instant; nor to square the circle. But of these and related failures—all perhaps definite and permanent—may we not say: So what? They do not seem a serious threat to the world that science senses, tests, and increasingly discloses.

3

The Problem of Creation

The philosophy of Naturalism which wholeheartedly accepts scientific methods as the only reliable way of reaching truths about man, society, and nature, does not decree what may or may not exist. It does not rule out on *a priori* grounds the existence of supernatural entities and forces. The existence of God, immortality, disembodied souls or spirits, cosmic purpose or design, as these have customarily been interpreted by the great institutional religions, are denied by naturalists for the same generic reasons that they deny the existence of fairies, elves, leprechauns and an invisible satellite revolving between the earth and the moon. There is no plausible evidence to warrant belief in them or to justify a probable inference on the basis of partial evidence.—*Sidney Hook*

THE IDEA OF GOD AS A BLESSER, ONE WHO MAY BLESS AND who may be appeased, is among the most ancient—and certainly still the most widespread—of human conceptions of Deity. That conception, however, is of secondary interest in this chapter, because, like the question of fairies, it is difficult to discuss from the point of view of science, and because it is already so extensively discredited among learned persons. Even some leaders of liberal religion have abandoned it. Nevertheless, elements of the idea of God as a Blesser are wedded to the minds of most present followers of practically all existing religions; indeed, these thin-worn elements continue to sustain nearly all of organized religion. To most of the "faithful" everywhere, God is still both Creator and Blesser, and a prayer or a gift is a prudent investment. In the Roman Catholic branch of Christianity, creation, placation and veneration are more or less nicely apportioned among God, the Son of God (Jesus), and the latter's virgin mother (Mary).

THE BROAD SCENE

Of chief interest here is the question—asked by learned persons of our day—whether the universe exists and operates solely under its own inherent properties, or whether an additional agency or immanence—God—created the universe of energy and matter and in one or another way gives evidence of his existence. This learned group gives now, and has given in the past, conflicting answers to that question. And this diversity of view partly reflects a central fact in the matter, namely, that *positive* proof can be supplied neither for the existence nor for the nonexistence of God as a Creator. It seems certain, however, that in greater proportion than in any previous generation the living members of this learned group now reject ideas both of Blesser and Creator. Also, within this learned group, a greater proportion of leading biologists and psychologists than of lawyers and writers now accept a Godless universe. Finally, there is evidence (see Chapter 12) that the proportion of atheists in the total present population of certain civilized countries is much higher in some (France, Holland) than in others (Brazil, United States, Australia).

The last three statements are made here not as evidence on the problem before us, but as factual background for the general reader, and because they relate to an associated question: How do advances in science, and very close or less close personal contact with those advances, affect the thought of men in regard to God? And here the author ventures to suggest that, at one or another time in their lives, intelligent people generally have as much (or more) live interest in evidence relating to the existence of God as have the scientists who must contribute the crucial evidence on that question.

Fitting into the point last mentioned is the fact that wherever religion prevails this fine initial human interest in questions of ultimate origins—of the real nature of ourselves and surroundings—is rarely permitted to survive even to the tenth year of life. Before this age is reached, a soft and deadeningly unreasonable answer, closing the question and putting the inquiry to sleep, will have been provided by family, by associates, by church, or by school. Thus practically all those who later acquire the mental hardihood to review and revise this verdict, and to slough the imposed mental anaesthesia, must do so against the powerful "conditioning" influence of a view caressed through youth's tender years. Useful here are the observations of psychologist William James on the extent to which individual training may influence metaphysical speculation. These observations should rid us of surprise at finding that even some good scientists never succeed in making a thorough revision of the God of their youth.

The first chapter of this book scanned the long series of especially significant events and processes—the "creative" processes—as they relate to earth and man. The story began with earth-cloud substances which slowly achieved lodgment in a solid earth, and was continued through the building of the body and the consciousness of man. The field of knowledge thus drawn upon includes a fragment of astronomy, parts of physics and chemistry, geology, biology, psychology, and social science; in the matter of *time* the chronicle extends over an abyss of three to five billion years. Within that area, which dealt specifically with the origin of earth, life, man, and society, the "creative" processes were described as comprehensible in terms of inherent properties of matter, and nowhere was there found need for or evidence of God a Creator. The second chapter reviewed a series of principles and insights of science that seem to make all concepts of the supernatural unnecessary and even untenable.

External to the earth and sun, and hitherto largely disregarded in this survey, are the impressive spaces and the incalculable quantities of matter and energy of an incompletely observed universe. This expanse is surveyed by astronomy, mathematics and astrophysics. These sciences, through many triumphs of observation, measurement and analysis of the external universe, have made it evident that evolutionary changes—"creative" processes—are still in active progress there; but to many or most investigators—though not to all—those changes also seem comprehensible in terms of inherent properties of matter, and nowhere is God a Creator indicated or hinted.

This writer thinks that the principles and insights just mentioned above include the really important evidence relating to the existence or nonexistence of God a Creator. But he is under no illusion that discussion of the matter properly ends there. Any selected group of facts proves only what those facts prove, not what all facts taken together prove. Clearly, it is necessary to give further consideration to certain metaphysical and theological concepts relating to the existence of Deity. Still other facts of science also await attention.

Two Words—"Create" and "Evolve"

First, however, we define two words. The verb "to create" has a short staircase of meanings. To confuse these meanings is easy, but it would cheat all profitable discussion. In its relation to the Hebraic-Christian-Mohammedan Creator, the word "create" originally meant that He brought the world of things into being from nothing. Some, and only some, among these three large religious groups would *now* accept a lower step on the staircase of meanings, suggesting that energy and matter coexisted with the Creator and that even now He continues to "create" by merely molding and directing the forms still being assumed by energy and matter. This view is in full conflict with science, which marks out precisely this area for the reign of natural, not supernatural, law. Still more dilute meanings, unrelated to Deity, allow the word "create" to accommodate such things as the products of one's thoughts or imagination, or even an invention. The history of religious thought makes it misleading to apply the undefined word "create" to any succession of states of matter that occurs, or seems ever to have occurred, in the realm of nature (universe). "To change" is often the true and proper term. "To evolve" suitably applies to many broad areas of cumulative or formative change. The term "cosmic evolution" can cover the expansion of the material universe and the formation of the chemical elements, nebulae, galaxies, suns and planets; "organic evolution," or simply "evolution," covers the formation of all living things—plants, animals and man.

Now current evolution may have begun with an "explosive" expansion of inconceivably condensed materials of the universe ("monobloc") five billion years ago, and with the rapid building of the ninety-eight or more elements from elemental neutrons, protons and electrons. Granted too that the still earlier states or phases of the universe are now wholly unapproachable. Within the entire space-time area covered by the two terms "cosmic evolution" and "organic evolution" the word "create" really has no place whatever except as a well-emphasized concession to the widespread use of a demoted meaning of this particular word. The gulf between the two words—"create" and "evolve"—is best indicated by an attempt to push beyond the elemental neutrons, protons and electrons. To derive these or minor components of the atom from *nothing,* an uninhibited layman, and perhaps an undeterred theologian, will say, "Create"; and then forthwith also say or shout, "Creator"—all on a basis not better than imagination, myth, or wishful thinking. Nevertheless, it is precisely here that a Creator's job is to be found if there is a Creator. Making all the neutrons, protons and electrons from nothing—and thus "creating" in an undiluted, distinct, and unrivaled sense —is the stupendous and rather absurd task that believers place upon God a Creator. Here and here only is the kind of "creation" that might need or could clearly imply a Creator.

But nowhere does science attempt to deal with this impossible barrier. Nothing that we know or meet in any area or discipline permits the slightest move in that direction. Probably the topmost meaning of the word "create" actually applies nowhere in nature, but only to this wisp of mystical thought. And it should be so, since this meaning was born of mysticism or of theology, not of science. When this point is granted, but only then, the word "create" may thereafter—with due reservation—be applied to the very different and lesser order of change that is indeed easily found in nature and is there much more properly called "evolution." The entire sweep of science, except mathematics and a segment of physics, *falls within the path and performance of evolution;* and the exceptions never attempt the derivation of matter or energy from nothing (creation). The "earliest" state of the universe is, indeed, quite unknown. Rather more than suspicion puts all matter of the known universe in an inconceivably concentrated "monobloc" occupying a fixed point in space, with an "explosion" of the monobloc providing what our ideas of time call a beginning of galaxy formation, and with the final exhaustion of its radioactive stores providing an eventual end in darkness. Probably only pseudo-science now debates whether the widening space between the outrushing galaxies is being refilled by new galaxies continuously formed from matter in that widening gulf—all with no beginning and no end in time. New aggregations of lifeless matter, new organisms, and new endowments of organisms do of course arise. To all this natural process and result, accomplished under natural, not supernatural, law, the term "evolution" is rightly applied. And this clean and precise term would not now be widely misunderstood if church and theology, aided by the occasional scientist, had not, since Darwin's day, put their own several head-chopping *mitigations* on the true meaning of this word.

Within the sphere of living things—from virus or gene to bee, man and society—the evolving and integrative processes are met on a scale so tremendous that the trained observer—untainted with the thought of the mystic—sometimes ventures to use the word "create" in its reduced and casual sense. And, curiously and rather magnificently, to many performances of man himself the word "create"—thus naturalized and dwarfed—may also justly apply. Man never gets something from nothing, but his best is perhaps as good as that of an Olympian god. Man gets from matter (and energy) things that are neither inevitably nor predictably inherent in it; and, further, no organism not endowed with speech—and thus with abstract thinking—could bring them into existence. Thus man creates our knowledge of the principles of relativity and evolution; man creates a watch; man creates symphonies and the Taj Mahal; man creates alphabets and undying literatures. With zeal and fervor, and with an untainted logic, man may here salute the "creative" best in his own species.

Before approaching a matter of uncertainty—the question of the existence of God a Creator, for example—it makes good sense to come face to face with some certainties that probably bear more directly on that uncertain matter. It is quite probable that the visible material universe has existed during some four to ten billions of years. And practically certain that during that entire period the parts, or many parts, of that material universe have been undergoing change—change in position, in temperature, in combination or dismemberment, in organization or integration. Such "end results" as may be seen by man today—whether they be heavier atoms, galaxies, planetary systems, living stuff, man or society—are observed only *after* the parts of the material universe have been involved in evolving and integrative processes during a few billions of years. It is certain too that both law and chance shared in those processes, and that they continue to do so. When some people, including scientists, say, "I cannot believe it is all chance or accident and therefore I can admit a God," they merely build a straw man to assist them to a personally preferred conclusion. Nobody assumes that chance alone built anything that we observe in nature. Nobody conceives of accident as an alternative to God. First and most certain is the reign of law, natural laws; and firmly included here, regardless of chance, is the superior principle of the building process—each new combination of matter, each integration, giving birth to properties not included in the uniting parts. Again, chance does not play an identical role in the nonliving and the living world; not even identical at the bottom and the top of the living world. The one who fits God a Creator into his scheme of thought must do so by first refusing to accept the view that the simplified state of the universe existed without a Creator; next he must accept the view that a Creator built the universe from nothing, or toy with uncommon efforts to escape that gross conclusion; and finally he must accept the view that such a Creator is the thing that had no Creator.

SOME SPECIAL RELIGIOUS CONCEPTS AND PLEAS

According to Hopkins,[1] "The general doctrine that faith surpasses science is common to all mystics. Faith is real knowledge." If this view of the mystics is true, they and their successors may have obtained and may continue to obtain the most convincing evidence for the existence of a God of sorts—whether creative or not. This assertion of the mystics, however, can be received in only one way by the scientist—that is, by one familiar with methods of arriving at truth and able variously and repeatedly to prove the validity of those methods. Psychologist Leuba[2] has made a penetrating analysis of religious faith, ecstasy and inspiration. He has shown that much more than the raw datum of "immediate experience," which may not be denied, is usually very clearly involved in these cases. He further calls attention to the contradictory revelations—presumably from the same God—that continue to be received by Christians on such subjects as the trinity and predestination, with the result that the Roman Church has declared itself to be the final court of appeal regarding the truth of revelations obtained during mystical ecstasy.

The scientist will grant the "reality" of these special mental states only as he would grant equal reality—along with unequally

1 E. W. Hopkins, *Origin and Evolution of Religion* (New Haven: Yale University Press, 1923).

2 J. H. Leuba, *The Psychology of Religious Mysticism* (London: Kegan Paul, 1930). See also his *God or Man?* (New York: Henry Holt and Co., 1933).

"satisfactory" results—to any delusion whatever. To these ripples in the consciousness of a few or many mortals, however convincing they may be to the individuals experiencing them, the scientist can attach nothing of cosmic importance; they are born in human consciousness and cannot transcend it. Theology today still sets a supreme value on the activity called intuition, which, a biologist may note, has its known home and kin only in the mental processes of man, where in turn it is one of evolution's newer and less reliable productions.

The higher religions—Brahmanic, later Buddhistic, Hebraic, Christian, Islamic—all teach that faith is superior to reasoned or scientific knowledge; and that we can know God only by intuition, meditation and revelation, through which we must find the proof both for his existence and his goodness. In addition to what was said above, it may be granted that when the resulting personal exaltation or ecstasy of the meditator is coupled with suitably advanced moral codes—and when all this is shared with large groups of men—the product has high social importance; but it may be neither true nor socially advantageous. It is of interest to note further that this method of "knowing" God is historically very old. It was employed and apparently got its more conspicuous results—the various lengthily recorded revelations—many hundreds of years before man surmised or used the experimental method as a tool to truth.

This same group of higher religions, Judaism and Islam excepted, developed the intricate idea of "trinity," which seems to involve God a Creator and to give Him still other attributes through which perhaps His existence might be sensed or verified. Indeed, the three personalities, Krishna, Buddha in later times, and Jesus, are alleged by many to provide this evidence. Modern scholars may not be expected to accept any or all of these characters as evidence for the existence of a God of any type. Reason, history, and the higher criticism all contribute evidence that it was such things as legend, human credulity and illusion, invented miracles, virgin births, and resurrections, that later (or much later) induced followers to identify their leaders with Deity. Christian theology seems to have been, and even now to be, much preoccupied with the idea of trinity, although, according to E. W. Hopkins, the doctrine was apparently unknown (unmentioned, at least) to Jesus and Paul.

It is possible that an item of acceptable evidence for trinity, from the domain of religion, has been unintentionally over-looked in this review. The writer has not fully explored the jungle of mankind's Upanishads, where myth, legend, ecstasy, meditation and miracles fill volumes beyond the reach and patience of most truth-seeking men. The construction of a concept of trinity seems nearly impossible. But the fact that it has been done and *redone*—and even with quite *different* materials—compels both our admiration and liveliest curiosity. To a roving human intellect, it seems to say that theological architecture is whatever is wanted.

ALLEGED, BELIEVED, AND EXAMINED

At this point it is well to bring together the four things that many learned people now find most persuasive to a belief in the existence of God. First, reason and logic seem to them to require a Creator for the physical universe. That this is a false, naïve and unfounded view is made evident in both earlier and later pages of this chapter. Second, one's feelings or intuitions point to the existence of a Creator. The unsubstantial, uncritical and spurious nature of this argument has been noted immediately above. Third, our probings and scrutiny of the natural world provide us with evidence for the existence of the supernatural. Let the answer to this be the words used in 1926, at a conference on religion, by Yale philosopher Charles A. Bennett: "The fate of the older rationalistic attempts to infer the supernatural from the natural shows clearly enough that the undertaking is hopeless." Fourth, life, or living matter, is a thing so complex and mysterious that its existence and its superlative products point to a still greater, external Directive Agency. That this view should be taken by those who have not devoted long years of technical study to the specific problem—and the training and daily work of many biologists and medical men barely touch this study—is in no way surprising. But such a view is rejected by an increasing number of persons more familiar with the processes associated with life (organism); and it is inconsistent with now established principles that provide a purely natural basis for organic evolution. Conditions attending the origin of life were examined in Chapter 1. Thus not one of these four most commonly used props to what is termed an enlightened belief in God can find plausible support in the knowledge of our day.

Probably more than half of the persons met in the course of a day by a reader of this book think they are *reasoning* when they assert "a God *must* exist, because man and the universe are much too involved and complex to exist without a Creator." The reckless and shallow sophistry of this particular assertion should be evident, since one thus adds and forthrightly accepts a thing still more involved and complex than man and the universe as that which had no creator.

To believers in a "mitigated" evolution, any acceptance of the creation of life and of man through a succession of animal forms is usually coupled with what is regarded as an all-important gate of exit: "But the biologist does not *explain* what life is." This statement is hardly true, and again it is hardly necessary or philosophically important that anyone "explain" what life is. Biology and related sciences have provided evidence that certain organizations of matter show activities or properties called liveness or life; also that these properties seem to arise—as do those of all other compounds and integrative levels—from the elements and organization involved in the union. In other words, these properties arise under the unvarying rule, the *natural law*, that each new chemical combination or level of integration exhibits new properties. Moreover, the biologist has now identified and tagged the purely *natural* mechanisms by which living forms (species) arise and become diversified; and he has further clearly shown the important part played by the intangible conditions known as chance in the equipment of every human being with his several abilities and defects (Chapter 6). In a similar sense, the astronomer and philosopher do not "explain" what the universe is, but they possess more accurate and intimate knowledge of some distant parts of it than any human being had of the earth prior to five hundred years ago; the physicist does not "explain" what the electron is, but he can describe some of its varieties and properties; and the chemist does not yet fully "explain" the rather accessible thing called valence, though he now comes much nearer to doing so than was possible when this writer was a young student of chemistry. In a truer sense, however, the sciences do "explain" these things, since in larger or smaller measure they have reduced the multiplicity and variety of phenomena to simple and general laws. And science at its best merely completes that procedure.

Again, many theologians and other cultured people of our day *mitigate* their acceptance of the principles of organic evolution with the view that for certain special processes or stages, such as the origin of life or of self-consciousness, a nonphysical directive principle (vital force, prevision, purpose) enters into the creative process. Too, it is often postulated that this same vital force, or prevision, is expressed within and throughout the whole series of living things. This view—best known as vitalism in biology, but also a widespread and basic theological assumption—has tried persistently for a place in biology. But it is now more thoroughly discredited in biology than at any earlier hour in the history of that science. By way of evidence, a single but especially significant item from the field of genetics will be cited. Though the genetic adaptation of organisms is most meticulous and all-embracing, and though both individual and species survival generally rest upon this adaptation, "yet close scrutiny shows it [adaptation] most wanting in those respects which would have involved distant prevision rather than short-range trial and error."[3] And here factual observation looks directly into the very fountain of both persistence and novelty in the living world.

Momentary attention may now be given to the undeniable fact that the picture of reality obtained through scientific investigation is not full, complete or wholly exclusive. But does this mean that another *equally* valid view actually exists, or that such other view can neglect all that our senses can contact, survey or measure? Two decades ago a brilliant biochemist presented the rather lonely philosophical view that "science does not claim a *truer* account of the world than the organicism of the philosopher or the other interpretations, perhaps impossible to formulate, of the religious man and the poet or the artist." If one takes the latter part of this statement seriously, truth (and reality?) would appear to be whatever man, or perhaps groups of men, forcefully feel or think it to be. On such a basis, the snake religion of parts of our American scene—though poles apart from other religious thought—can match any other in its profound faith and in its ardent desire to show the domination of religious faith over a poisonous reptile. But here the making of a case for "equal truth" must be somewhat inconvenienced by the circumstance that now and again these intense believers are bitten and die.

Disregarding science and philosophy, the Western world,

[3] Herman J. Muller, *The Harvey Lectures, 1947-1948* (Springfield, Ill.: Charles C. Thomas, Publisher).

prompted by a logical paradox of Christian theology, has indulged in three centuries of rather barren discussion of "free will." Christian theology sought a way to justify the coexistence of two of its terrible doctrines; namely, that God has predetermined the fate of every man, and that He nevertheless holds every man responsible for his action, rewarding or punishing eternally. In more recent times, discussion has centered more specifically on whether volition (action, behavior) is completely (one hundred per cent) determined (mechanistic), or whether a small degree of indeterminateness is involved. Many have assumed the existence of a certain small amount of indeterminateness, and—herein lies the reason for these paragraphs—have further assumed that this "small amount" is an area for supernatural maneuver (God). These assumptions are hard to treat in a rational way. But the way in which the relatively new "uncertainty principle" of physics might apply in this case is suitably recorded in the words of physicist Tolman:[4]

I must caution you, however, that the opinion of one good physicist that the uncertainty principle brings free will and moral responsibility back into the world can hardly be regarded as sensible. As far as I know, moral responsibility has never left the world, and, indeed, could hardly be helped by a principle which makes physical happenings, to the extent that they are not determined, take place in accordance with the laws of pure chance.

The position of the life sciences on this question has been stated thus briefly by physiologist Smith:[5]

For the naturalists, free will was a countersense, a verbal contradiction. To "will" is to choose a course of action in which more than one course is potentially presented, and to choose one course of action as opposed to another requires not only knowledge of alternatives, but reason for the choice. Decision without reference to cause or consequence of that which is rejected or accepted could only refer to an act occurring in a referential vacuum, and if such could be conceived it could only be designated as an action issuing from nothing at all, *ab nihilo*, from absolute ignorance. Since willing can never be free of knowledge of either cause or consequence, it can never be free at all.

[4] Richard C. Tolman, *Science*, vol. 106, 1947.
[5] Homer W. Smith, *Man and His Gods* (Boston: Little, Brown and Co., 1952).

From metaphysics one may of course obtain other less binding, and even free-wheeling, evaluations of this overworked conundrum.

In theology and in much popular thinking, it is rather widely claimed that the order or orderliness that we all observe in nature, though involving natural processes, also involves a Directive Force that acts not dynamically but persistently. This tenuous and elusive halo can be neither successfully supported nor speedily dissipated. The alleged role of a Directive Force in this sphere is part of the broad theological assumption that the supernatural can be inferred from the natural world, a point dismissed by philosophy in the words of Bennett, quoted above. It is notable, however, that the immense and obvious amount of order in the universe is itself widely acclaimed by the world's clergy as good or sufficient evidence of God's existence. Of course, man and the living world arose in conformity with and in adjustment to all such things as the daily rotation of the earth, the seasons, and the usual yields of land and sea. The very existence of living things—like the earlier disappearance of many a noble group of species—is adequate evidence of the prevalence and necessity of this adjustment. When priest and minister endlessly cite these orderly relationships as considerate gifts of God to man, or as evidence for God's existence, they merely insult the information and intelligence of learned people.

Actually, our problem is to account for the order and for the seeming disorder displayed in nature. In this connection one may give a brief thought to the disorder present now or earlier in our well-observed solar system and in the living world. As bordering the disorderly, one may cite the movements of gases in the corona of the sun,[6] which (a) show both as irregular eruptive flashes and as long-continued storms, which (b) move 3,000 to 4,000 times more speedily than the winds of our own atmosphere, and which (c) may move outward to more than 250,000 miles—a distance exceeding that between the earth and the moon. Again, the repeated and irregular mountain-forming convulsions of an earlier earth and the earthquakes that still continue to shatter parts of our planet's slowly advancing growth are types of awesome disorder in the inorganic world.

Even more evident and more common is a variety of tolerable, though distressing, disorder in the living world. The daily

6 Donald H. Menzel, *Our Sun* (Philadelphia: Blakiston Co., 1949).

survival of many or most animals, including man, is based on strife and struggle with, and even the death of, hosts of other animals. Again, the very origin of most forms of life (species) is based partly, though only in part, on a struggle for existence in which this type of disorder abounds, and from which the noxious, the nuisance, and the degenerate sometimes survive along with the progressive, the majestic, and the mimic. "The evolution that Darwin discovered was process, not progress," declares H. W. Smith. The process by which man himself is reproduced assures that a low but tragic percentage of pregnancies and births will yield the defective or the deformed—physically or mentally—including the occasional ghastly clumps of flesh that must at once be put aside. Probably every one of the 208 bones of the normal human body has been repeatedly observed to be absent from babes born of women. The disorder involved in disease is mentionable but not measurable; the animal and plant kingdoms are its crowded home. Family, tribal, social, economic, national and international disorders are parts of a new, man-made world; and there, too, we must deal with both order and disorder.

Science does in fact examine successfully both order and disorder in nature and in society. It is done without the use or aid of the supernatural Directive Force to which theology appeals. Indeed, if only one single instance of the exercise of such a Directive Force could be established, that solitary *miracle*[7] would provide firm evidence for the existence of a Directive Force. That one case has never been established, though the trail of most religions is traced in the hidden and open frauds that are offered in its stead. In stating that science can deal comprehensively with both order and disorder, it is not implied that astronomy has solved its problems, that geology is a completed story, that biology has either asked or answered its endless questions, or that medicine has mastered the many ills of man. What is meant is that these and other sciences have proved that, with the method of science, they can continually increase the area of our practical, serviceable and solid understanding. In all fields this advance is made on the basis of uncontradicted evidence that phenomena, events and processes—whether of order or disorder—proceed under the rule of *natural* law.

[7] According to St. Thomas: "Miracle is a free interference of God; what has been done by God outside the regular course of Nature."

The miracles, and the sacred or healing relics, statues and monuments of religions, are rather clearly contributions to the history of human credulity, human yearning and desire—and sometimes doubtless baser human motives—but they are all false mirrors of cosmic law and reality. It requires some special consideration, however, to show that, along with a fragment of truth, an element of error or of fraud exists also in the "mental healing" practiced by both early and modern Christian priests and ministers, a practice asserted by some of them to prove Divine intervention, and therefore the existence of the Divine. There can be no doubt that the agents of various sects of the Christian and other religions have effected some "cures," as did medical (not priestly) agents in the Greek temple of Pergamon long before the Christian era, and as do *some* psychotherapists of today on a far better and broader scale than either of them. Here the element of fraud enters when priest or minister, all of whom should now know better, attribute the cure to a Divine source. The reader will find that psychologist Leuba[8] has treated this subject at length, and has indicated the purely psychological and physiological conditions that attend the valid instances of such cures.

One may not neglect the fact that a very few quite modern theologians and some less modern philosophers now speak of "divine immanence" with and without capitals. Here God is not put "outside" things in general but is postulated as being a process or presence within all parts of the universe; and sometimes even the quality of consciousness is not attributed to this presence.

This uncommon concept of God deserves notice here because of its modernity, and because His there alleged *intimacy* with the universe might, just might, at some time or in some nook, let Him become evident in science. Yet, to date, one could well apply to this view the exact words that Bertrand Russell applied to the concept of a supposedly highly intimate soul:[9]

One of the "grand" conceptions which have proved scientifically useless is the soul. I do not mean that there is positive evidence showing that men have no souls. I only mean that the soul, if it exists, plays no part in any discoverable law.

8 *God or Man?, op. cit.*

9 Bertrand Russell, *Unpopular Essays* (New York: Simon and Schuster, Inc., 1950).

Another argument commonly makes much of an "unseen world," a world said to be indicated by parts of our human consciousness, which, when followed up, provides a wholly different road from that of science to truth and reality. Within this "unseen world" one thus finds a basis for religion, and for a revealed God. This argument has been developed less by theologians than by some religious-minded physicists of our day. It provides a basis for much of the advanced thought of the moment on religious questions, and it will now be given further attention.

This argument was condensed and most ably presented as an address by astronomer-physicist Eddington[10] to the Society of Friends, London, in 1929. Its main features are to be found in the following quotation:

We have learnt that the exploration of the external world by methods of physical science leads not to a concrete reality but to a shadow world of symbols, beneath which those methods are unadapted for penetrating. Feeling that there must be more behind, we return to our starting point in human consciousness, the one centre where more might become known. There we find other stirrings, and other revelations (true or false) than those conditioned by the world of symbols. Are not these too of significance? . . .

Are we, in pursuing the mystical outlook, facing the hard facts of experience? Surely we are. I think that those who would wish to take cognizance of nothing but the measurements of the scientific world made by our sense organs are shirking one of the most immediate facts of experience, namely, that consciousness is not wholly nor even primarily, a device for receiving sense impressions. We may the more boldly insist that there is another outlook than the scientific one, because in practice a more transcendental outlook is almost universally admitted.

The first sentence of this quotation requires amendment. Our explorations of the external world are not in fact limited to "the methods of physical science"; or, if we are to infer that natural science—including biology, psychology, and sociology—is possibly included, the thing we call "common sense," as a first step in science, is certainly also involved in our arrival not at a "concrete reality" but in "a shadow world of symbols." And granting something technically short of concrete reality to all that we describe

---

[10] Arthur Stanley Eddington, in *The New York Times*, Sunday, June 16, 1929. See also his *The Nature of the Physical World* (New York: The Macmillan Co., 1928).

in science, the "shadow" we get in its stead has all the firmness (near-reality) of the existence of our friends and enemies, and also of our mates, who provide our chance to be physically represented in all future generations. Again, these friends, enemies and mates are and were similarly real, or near-real, for all the long line of animal kin that produced man. The maligned "shadow" does in fact represent the full kind of certainty, except that of arithmetic, which serves the common man to live out his years and to project himself and his species into the future. To attain these great ends man and his animal ancestors assuredly have not needed to go beyond common sense (or science) and search deep in consciousness for "the mystical outlook."

Amendment is necessary also in the sentence of the second paragraph that states that "those who would wish to take cognizance of nothing but the measurements of the scientific world made by our own sense organs are shirking one of the most immediate facts of experience, namely, that consciousness is not wholly, nor even primarily, a device for receiving sense impressions." Here the point for first notice and correction is that the contributions of biology and psychology to the scientific world are by no means limited to "measurements" and their associated symbols; they also include observations, tests and sequences that, though they do not measure, nevertheless illuminate and lead to comprehension. An example of this from the writer's own studies, described in Chapter 4, relates to observations and tests that show that a particular gene and the hormone prolactin are necessary— as is also the attainment of a particular kind of ripening of the organism—to the production of the parental instinct. And that with suitable use of these agents—not measurements—the experimenter himself has added *this element of consciousness* to rats and fowl and pigeons that never before had it. To assume that biological science is restricted to "measurements" and to "symbols" emerging from measurements is both to falsify and to degrade that science—the particular science, indeed, that accounts for the very consciousness to which this excellent astronomer appeals but also subverts. The second point for examination in the quoted sentence relates to the nature and content of consciousness. It is said "that consciousness is not wholly, nor even primarily, a device for receiving sense impressions." We may next note the broad and religiously unhelpful sense in which this is true.

The biologist and the psychologist must regard *conscious* or *self-conscious awareness* as a recent arrival among the mental activities that slowly developed as inseparable and essential parts of evolving organisms—from the low to the highest.[11] Somewhere within the broad and rather dark dome of the mental in man, a small illuminated spot, or window, could represent the minor part of the mental that is involved in conscious awareness (consciousness). Surely it is the larger, darker spread of dome—not the smaller spot—that deserves first and most attention, for from it the smaller spot was born. And, most meaningful of all, the entire dome was built of the successful organismic reactions and adjustments of an indescribably long ancestry. To say that the mental activities, conscious and unconscious, are fitted and suited to the organism would be to falsify utterly by saying far too little —they are, one and all, the very *parts* of organism.

Now the entire above-described dome—the mentality as a whole—may be called "a device for receiving sense impressions" and for very much besides, since in ape or man its participations, awake or asleep, are innumerable. But, however much this mentality—even in its brighter window of consciousness—may have developed its services, it is against all known biological history, expectation and principle to presume that mentality should evolve a segment that is not part and parcel of organism. Organism, in its long march to man and indeed in each instant of its billion years, has been forced inexorably to build itself in full conformity to the physical world—that is, to the "seen" world of science—and nothing would seem innately less probable than the production and cultivation there—in the whole of the subhuman world—of bases for a "mystical outlook" that would someday enable some members of *one* superior species to find equal or deeper truth in an "unseen world."

A mystical outlook can indeed be achieved by a man. It appears, however, to be rooted largely in the twilight area of the fluid emotions rather than in the brightest area of reason. It is certain enough that "stirrings" other than those of reason are present in all now known humans. They may take forms such as fasting, sweating, self-torture and hysterical dancing. But the practical value of the mystical (or religious) outlook as a way to get truth that is beyond scientific truth is also largely denied by the his-

[11] William A. White, "The Frontier of the Mind," *Journal of the Washington Academy of Sciences,* vol. 25, no. 1, 1935.

torical fact that enormous masses of men—not just individuals—have been led by it to a belief in God and in no God (early Buddhism, for example), and in wholly unlike varieties of God. Such facts make it evident that something quite *outside* those "other stirrings, and other revelations (true or false)" of consciousness to which Eddington appeals, is in fact largely responsible for his own view of "truth in the unseen world." The nature of that outside influence should not be hard to find in an astronomer *religiously* nurtured in a country where tradition is so powerful and respectable.

Eddington further states: "Religion does not depend on the substitution of the word 'God' for the word 'nature.' The crucial point for us is not a conviction of the existence of a supreme God but a conviction of the revelation of a supreme God. . . . I confine myself to the revelation implied in the indwelling of the divine spirit in the mind of man." This last sentence needs attention. If the "divine spirit" exists it would seem to be a "thing," and needs to find a home somewhere or everywhere. The "mind of man," however, is not a "thing" and the mentation referred to is an activity, not a repository. Moreover, as noted above, mentation, like locomotion, surely developed as a part of organism; and just as surely some of this activity may look like good or God, and some of it may look like the Devil.

Various views of God do indeed exist. One such—of a God *independent* of nature—admits that attempts to reason about Him or to prove that God is the Creator of the Universe are doomed to failure; but, it adds, in the words of Stace, "To ask for a proof of the existence of God is on a par with asking for a proof of the existence of beauty." Again, some current "advanced" theological thought confidently cites the support of one or more philosophers —mostly those of an earlier day though occasionally one adopted as a living authority. The authorities thus cited were, or are, those having a serious regard for such things as subjective philosophy, intuition, and essence—items discounted or abandoned by leaders in modern philosophic thought. Helpful here are the observations of philosopher Reichenbach:[12]

It has become a favorite argument of antiscientific philosophies that explanation must stop somewhere, that there remain

12 Hans Reichenbach, *The Rise of Scientific Philosophy* (Berkeley and Los Angeles: University of California Press, 1951).

unanswerable questions. But the questions so referred to are constructed by a misuse of words. Words meaningful in one combination may be meaningless in another. Could there be a father who never had a child? Everyone would ridicule a philosopher who regarded this question as a serious problem. The question of the cause of the first event, or of the cause of the universe as a whole, is not of a better type. The word "cause" denotes a relation between two things and is inapplicable if only one thing is concerned. The universe as a whole has no cause, since, by definition, there is no thing outside it that could be its cause. Questions of this type are empty verbalisms rather than philosophic arguments.

And modern philosopher John Dewey[13] wrote as follows:

The developments of the last century have gone so far that we are now aware of the shock and overturn in older beliefs. But the formation of a new, coherent view of nature and man based on facts consonant with science and actual social conditions is still to be had. . . . Faith in the divine authority in which Western civilization confided, inherited ideas of the soul and its destiny, of fixed revelations, of completely stable institutions, of automatic progress, have been made impossible for the cultivated mind of the Western world.

Individual stars and galaxies are indeed observed to be in various stages of organization—youth, age and death. Our own sun will slowly end its existence as an individual source of light and heat; and our earth is destined to cease its daily rotation and to disappear as earth. But such decay and death involve the dispersion and conversion, not the annihilation, of the ultimate particles of the atom. On this question modern physicists are the highest authority. They indicate that with reference to the constituent corpuscles of the atom neither creation nor annihilation is admissible. For him who asks: "What or how much preceded our cosmos of energy and mass?" there is no profitable answer; and the emptiness of the question has been indicated above.

Although the social God disappeared in the scientific advances of the previous century, some scientists and philosophers still continued, until the recent advent of the concepts of relativity, to think and speak of the Absolute or the Ultimate in capitals. Indeed, such is the power of the past that clarity sometimes requires a capital letter. Following the general acceptance (at least within

13 John Dewey, *Living Philosophies* (New York: Simon and Schuster, Inc., 1931).

the physical sciences) of the principles of relativity and quantum mechanics, our earlier concepts of such things as time, space, matter, energy, complete determinism and the absolute—even of absolute truth—have undergone radical change.

It is now held that all our observations, measurements and deductions are tinged with uncertain and ineradicable elements of our human sensations and consciousness. This may mean, for example, that you can say that eventually your neighbor, Mr. X, will probably die, not that he will die. Philosophy makes a point of the doubt. In the matter of cause and effect we still have possibilities of prediction, though only on a statistical basis. In this changed viewpoint of science, where indeed obscurities still attach to the basic concepts, some opportunists among the liberal clergy of various countries are busily engaged in making the obscurities the basis for the assertion that the conflict between science and theology is approaching reconciliation. The difficulties and the vicious error of this claim are evident to logical thought. They are partly indicated by mathematician Danzig:[14]

Is religion too prepared to sacrifice, waive, and renounce? Has it decided to give up its gospel of absolute truth? Renounce the absolute God? Substitute for its absolute world a floating and fleeting chaos? Announce that it has been the victim of anthropomorphic delusions? Or declare itself to be a mere emotional game of make-believe, played on human fears and hopes? Religion is not prepared to make such sacrifices; indeed it cannot make such sacrifices and survive.

And they are further covered by Harvard philosopher Perry:[15]

It would be presumptuous and foolish to assume that there is a kindly indulgence at the seat of cosmic control: presumptuous because unsupported by evidence; foolish because it would weaken man's reliance upon himself.

Nature would seem to be the nearest thing to a God that man is likely ever to know. Majesty and terrifying power are hers forever. The equality she grants is unquestionable; she makes no distinction between promising ape and noxious weed. And those creatures that understand or respect her are often rewarded with such things as safety and longer life.

[14] Tobias Danzig, *Aspects of Science* (New York: The Macmillan Co., 1937).
[15] Ralph Barton Perry, *Realms of Value* (Cambridge: Harvard University Press, 1954).

How account for the fact that many scientists and some philosophers believe in the supernatural and in God? By way of answer there is only evidence—certainly not logical proof—and parts of that evidence will be found elsewhere in this book. It is this writer's opinion that, for many of them, the core of the answer relates to their neglect of one or more of the outstanding insights attained by science. Chief of these insights, more fully quoted from physicist Bridgman in Chapter 2, is the newly reinforced and vital conclusion that it is impossible for human thought and feeling to transcend the human reference point. That summit of accomplishment has been ignored or rejected by at least many of the scientists who have written their beliefs into a library of books and addresses. And after ignoring such heights of present scientific attainment, these men seem to have swung freely into the blue, in a personally prompted pursuit of a "something further," a spirit, a "purpose," a God.

For an occasional scientist, it is certain, the case is much worse. Surely and strangely, some trained people are able to keep their science and their religion in two separate and nonleaking mental compartments. A famous professor of botany in one of our great American universities privately told of a brilliant nun, who accomplished her doctorate under his guidance. Before leaving the university, the Sister called to pay her respects. In the course of a pleasant conversation, she was asked how she harmonized all that she had learned with what her robes symbolized. In a matter-of-fact way she replied that she never let the two things mix.

It seems that just preceding the year 1600, the inventor of logarithms, the Scotsman John Napier, calculated that the Second Coming of Christ would occur between 1668 and 1700, and the great Sir Isaac Newton, though noncommittal as to date, declared his belief that the time was at hand. If history can prove anything, it proves that when science was less developed than it is now many good scientists accepted some or all of the most violently absurd theology of their particular age and place of birth.

Of course, science does not and cannot *prove* that God a Creator does not exist. Science cannot prove that a witch does not exist. And science cannot prove that a devil, or even a flock of devils, does not exist. But science can and does put natural law into many a dark corner of the universe and also deep into the crannies of consciousness. The nature of the world we are part of can be passably comprehended on the basis of natural law; "the eternal mystery of the world is its comprehensibility," says physicist Einstein. And from natural law it is clear enough that one can obtain a hint of neither God nor witch nor devil. Moreover, many things we do know are anything but innocent of bearing on the question of the existence of God a Creator, witches, and devils. These pages deal with a few such items that relate to God a Creator. Bearing directly on this question are the words of philosopher Hook, at the head of this chapter. They merit re-reading at this point.

One may not neglect the meaningful fact that the evolutionary process still goes on; that is, the "creation" ascribed by some to God continues through the present moment. For those who assume they can believe that the natural processes of evolution are God's chosen methods of creation, there are some most awkward facts that deprive Him of the attribute of mercy. Thus volcanoes and earthquakes are parts of the age-long evolution of the earth's crust; yet throughout a long past they have killed, in an indiscriminate and wholesale manner, all living things within the area of their effect, including large masses of babes, women and men. If a creative God operates through evolution, then an endless and frightful man-killing by volcanoes, earthquakes, famines, and like disasters are His work; and if He wishes also to be regarded as a *merciful* God, He absurdly blundered when He permitted man the fair intelligence that is now his. Or, in the words of mathematician-philosopher Bertrand Russell, "If indeed the world in which we live has been produced in accordance with a plan we shall have to reckon Nero a saint in comparison with the Author of the Plan."

Again, with reference to the religious type of Supreme Authority which often has not hesitated at breaking natural law—in order to slay the Philistines in a puny tribal war, to provide sixty-four kinds of miracles by Shiva, to convulse nature at the birth and death of Buddha, to provide loaves and fishes for an assembly that tarried beyond mealtime, to impregnate a virgin in an obscure village and epoch, and to bring a mountain to Mohammed—but which in our day refuses to rescue a man-filled world from tyrants who have devastated and killed by the millions, and who still choke the pulse of humanity while they threaten all civilization—*that* Authority is far too whimsical to be of the slightest use to struggling mankind.

A SOUVENIR FROM EARTH'S LONGEST TRAIL

It remains to point to a logical conclusion that may be drawn from one aspect of what seems to have been learned of the long trail of evolutionary processes. Phenomena, processes, and events —items of history—are, of course, usually better known and more easily studied in our immediate surroundings here on earth and in our solar system than in more distant parts of the universe. Within this restricted space, an enormous range of evolutionary change, which took place during three to five billion years, has been studied with considerable success. The *relative order* in which phenomena of lesser and greater molecular complexity, simple life, more and more complex life, and finally consciousness occurred or appeared within this long period supplies the basis for a meaningful conclusion.

Within this best-studied segment of time and space, the known facts indicate that, above the level of the atom, the primary *direction* of change has been—with rather minor or unclear exceptions (chiefly among Orders of animals)—*always from the simpler to the more complex.* Practically nowhere is the reverse of this observed if or when one properly rejects the quick, later, and transient processes of death, degeneration, disorder and decay which, essentially as by-products, inevitably accompany the central, expanding and enormously prolonged "creative" processes. In the inorganic sphere, the sequence is from split or single atoms to associated atoms of the same kind; from the simplest (hydrogen) to the more complex atoms; then to the simplest, and later to the more complex molecules; then to organic molecules of increasing complexity; then to the simplest living aggregates; then through less to more complicated living things—within which only the highest *animal* forms developed consciousness. Thus *thinking* comes as the highest and *most recent* development in the whole sweep of evolutionary change from the progenitors of the earth-cloud to man; and it then appears only in association with the most complex *animal* types—the most elaborate and intricate organizations of matter actually known in the universe.

To assume, or to propose, a Conscious Super-Thinker—God—

as a first or *earliest* entity and to characterize that Thinker as devoid of all *molecular organization* is to violate both the sense and the spirit of much of the best that man has learned about himself and about that part of the universe that is most accessible to study and test.

4

Evolution and Ethics

Besides love and sympathy, animals exhibit other qualities connected with social instincts, which in us would be called moral; and I agree with Agassiz that dogs possess something very like a conscience.—*Charles Darwin*

Instead of framing our philosophy around an abstract end, and reproaching the universe because it appears indifferent to the particular goal we have erected, it would be wiser to begin with the nature of life itself, and to observe at what point one good or another does in fact emerge from it.—*Lewis Mumford*

A DISTANT, UNREMEMBERED AND LARGELY UNRECORDED past would seem a sufficient reason for the little that is commonly known of the life of early man. Thousands of years separate us from primitive man's binding contacts with the wild and natural world. To that can be added our own wholly dissimilar way of life and our scholastic type of education. No one is now educated for life in a forest jointly shared with competing animals. Surely we have broken fully with the life of early man, and surely our blood link to him has been submerged in a deep gulf of incomprehension. Up to less than two hundred years ago even the wisest of philosophers and historians did not know that early man was a *social* being. Our sages usually conjectured early men as mere family units—as almost solitary hunters who, in the course of time, developed social tendencies. None of the "revelations" of the influential religions unrolled a vision of first men and their animal progenitors as highly social beings clinging to a tribe or a troop. Thus man's thought during all historic time, almost to our own day, left unexplored the consequences of our having prehuman and earliest human ancestors already endowed with sociality.

On this topic anthropologist Keith[1] wrote as follows:

> The brain which has fashioned civilization is that which was in existence before civilization began. Nothing could be more misleading than Hobbes's oft-repeated statement: "The life of man in nature was . . . solitary, poor, nasty, brutish, and short." Even Huxley described the life of natural man as unethical—as a reign of tooth and claw. Darwin knew better. He recognized that the bond which linked together the members of each local group was formed out of maternal love and family affection, that at the core of such groups was a primitive system of kindly pagan ethics, and that only the isolating crust was savage.

Without a social element in our distant forebears, it is improbable that any civilization would yet exist. In one of Darwin's several great contributions[2] to knowledge he presented the now well-confirmed view of the prevalence of sociality in man's nearest animal relatives, of its development from such things as sympathy and the parental instinct—both widely present in higher animals —and of the high survival and other value of sociality in subhumans and in man. According to Darwin, man owes his "immense superiority to his intellectual faculties, to his social habits, which lead him to aid and defend his fellows, and to his corporeal structure."

In an appraisal of the importance of sociality to man, later students may suggest that some of earliest man's "intellectual faculties" could never have been attained through descent from animals devoid of social instincts. The origin of some of those faculties seems to presuppose and require association with the troop or the herd. However that may be, the origin and consequences of sociality are extremely meaningful to the man that was, that now is, and that may be. The point of first interest in these pages is that we can now see that elements closely associated with sociality provide the key to the natural origin and basis of ethics in the individual human will. These elements are also involved in so much besides ethics, however, that the latter often seems submerged in the swollen stream of subhuman and very early human history and to be fused finally in an ocean of emerging civilization.

[1] Sir Arthur Keith, *Evolution and Ethics* (New York: G. P. Putnam's Sons, 1946).

[2] Charles Darwin, *The Descent of Man* (3rd ed. Rev.; New York and London: Merrill and Baker, 1871).

THE ORIGIN OF SOCIALITY

Though sociality has appeared at more than one level in the animal kingdom, it is the sociality of mammals—and particularly of the primates—that most illuminates the social foundations of man. In this connection, termites, ants, bees and wasps are mainly useful as reminders of the extremes to which sociality developed elsewhere in the many-branched tree of living things. The sociality of man shares its roots with mammalian kin.

Sympathy and sociality are quite different and distinct endowments. For the moment, however, it is convenient to think of these two and even other foundations of morality together.

SYMPATHY AND MUTUAL AID

The widespread presence of sympathy and mutual aid in mammals and birds is commonly known. Actual or seeming exceptions are also familiar; for example, a wounded animal may be expelled from the herd, or gored or worried to death. There is no clear explanation for this, which Darwin calls "almost the blackest fact in natural history," but possibly it arises from an instinct (or is it reason?) that the injured must be expelled lest beasts of prey, including man, be tempted to follow the group. In any case, this behavior is scarcely worse than the practice of certain North American Indians, who left their feeble comrades to perish on the plains, or of the Fijians, who buried their old or ill parents alive, or the rather extensive practice of infanticide by primitive peoples. And certainly these various human types cannot be presumed to be wholly devoid of sympathy. The feeding of blind (and fat) birds by companions in the flock has been repeatedly recorded. The sympathy and love of the dog for his master is proverbial. Darwin stated:

> I have myself seen a dog, who never passed a cat who lay sick in a basket, and was a great friend of his, without giving her a few licks with his tongue, the surest sign of kind feeling in a dog.

Acts indicative of sympathy and mutual aid in many species of apes and monkeys, both in the wild and captive state, are known to all those acquainted with these self-revealing cousins of man.[3]

The basis of "the all-important emotion of sympathy," which is distinct from that of love, may be sought partly in a lively memory of former states of pain or pleasure, the mere sight of suffering, independently of love, sufficing to call up vivid recollections and associations. Darwin noted that in animals this sympathy is directed essentially or solely toward the members of the same community, and therefore toward known and more or less beloved members—but not to all the individuals—of the same species. He further noted that this is not more surprising than that the *fears* of many animals should be directed solely against special enemies. Nonsocial animals, like lions and tigers, are thought to feel sympathy for the suffering of their own young though not for that of any other animal. Again, individuals and species that show sympathy and mutual aid would doubtless be favored by natural selection in the struggle for survival.

Further examples of mutual aid are the widespread use of sentinels and alarm calls among higher animals, hunting in packs by wolves and other beasts of prey, the common defense joined in by bull bisons (buffalo) after driving the cows and young to the center of the herd, and the united attack, particularly by the males, of troops of Hamadryas and of Geladas baboons, which even use stones against their enemies. But far more penetrating than these fragments of personal information, or the chance observations of travelers and earlier naturalists, are the results of recent research concerning the form and content of entire societies (troops) of apes and monkeys.

Intensive studies of individual and of group behavior, carried on among organized groups in their native haunts, have provided definite knowledge of societies of baboon (Zuckerman), chimpanzee (Yerkes; Nissen), and gorilla (Bingham). And Carpenter has provided similar knowledge for native and natural societies of

---

3 In 1901, the writer witnessed an unrecorded incident in an untraveled jungle of the Orinoco delta. In landing a small steam launch behind a thin tongue of land, our craft blocked the shoreward exit of a troop of monkeys, occupying a tree there. Much noise, confusion, and scampering to the treetop was soon followed by an impressive act of leadership, aid, intelligence and courage. The largest monkey of the troop jumped to a long, lower branch pointing shoreward, bent it greatly with his weight, and thus brought it nearer to a branch of an adjacent tree. That post he firmly held—though it brought him rather near to my reach—until all his twelve to fifteen companions had climbed over his rigid body and jumped to safety. Only then did he leap to safety himself.

several species of monkeys and apes. Thus troops of baboons in South Africa have been found to range in size from about 25 to more than 100 individuals. The troop consists of a number of harems, in each of which a dominant male is the central figure. Around each of these males clusters a group of females. Bachelor males may be only slightly tolerated by the "overlord" of the harem. Many more females than males actually remain within the groups.

Group organization within two other species can be similarly sketched. Troops of macaque or rhesus monkeys may vary from very few to more than 150 in number. One typical troop was found to be composed of four adult males, ten adult females, four of which carried infants, and eight juveniles. The males dominate all other individuals within these organized groups but not as completely as their counterparts among the baboons. Extra-group male macaques may live in temporary isolation, but much more frequently they live as a cluster of males. The gibbon, smallest of the anthropoid apes, displays a *family* pattern in its social group. The observed range of size of these groups was from two to six, a group consisting of an adult reproducing male and female together with their offspring, which ranged from infancy to early adulthood. Here the adult male and female seem about equally active in co-ordinating and controlling the family. Isolated or expelled individuals are found among both sexes, and doubtless some of these later find mates and establish new families.

Commenting on these societal groupings, psychologist Carpenter said:[4]

The *total number* of paired relationships and their various interactions result in a pattern of group organization. An individual within a group of considerable size, of the mosaic type, does not have free behavioral exchange with every other individual in his group because of established channels of relations and restrictions which structure the individual members of a large group into sub-groups. Each sub-grouping and each animal has its own particular status in the group. Processes of social integration and control strictly regulate the freedom of movement and limit the social contacts in organized groups of primates so as to produce definite group structures. The monkey or ape in its

4 C. R. Carpenter, "Societies of Monkeys and Apes," *Biological Symposia*, vol. 8, Lancaster, Pa., 1942.

natural group in the tropical forest has its freedom of movement strictly limited by the structure of its group.

These studies of primate societies further provide a rational account of the means by which attachments and antagonisms develop between individual primates. They also indicate that territorialism of groups—a defense curiously directed solely against members of their own species—is a primitive and basic characteristic of these subhuman groupings. Incidentally, this is analogous in several respects with primitive human behavior. Further, these studies definitely establish many facts concerning gestures and vocal efforts as means of communication in several species of higher primates.

These and other studies, mostly of nonprimates, show how individual dominance and subordination become established in the localized fragments of a species; and Allee has shown how the rank-order established in this way limits conflict among associated individuals and gives rise to relatively stable relationships that are truly social. On the still broader scene, studies by Wright, Fisher, and Haldane have shown that natural selection operates not only upon the individual but upon the entire population, the population being the seat or home of group relationships and activities like those heretofore described. Finally, newer evidence is supplied by Emerson and Gerard for the view that such elements as family, group or tribe, population, and society are all units with some of the properties of an organism and may be regarded as real co-operative superorganisms.

It is nevertheless notable that no one can yet give a complete account of the origin of sympathy. That lies deeply imbedded in the difficult field of animal behavior and psychology. But related facts stand out clearly enough. The bee is superbly social, yet without sympathy; in contrast, sympathy blooms in the chimpanzee where sociality is little developed and scantly structured in comparison with that of the bee and the ant. Of first importance to the present account, however, are the entirely clear indications that sympathy and mutual aid were fashioned in the subhuman world, and that these ready-made inclinations to sociality were passed on to earliest man.

Of the earlier species of man or of manlike beings—those that preceded our own *Homo sapiens*—we have only the findings of the past six decades. And most of these fragmentary findings—

skeletons or an occasional bone—now tell us something of the size and anatomy of those men, though usually very little of their social or other activities. Some less remote men left their records in favorable places such as caves, and our knowledge of them began somewhat earlier. Tools and other revealing objects exhumed with their bones uniformly give evidence of sociality among these men. This evidence is particularly convincing for Neanderthal man of one hundred thousand years ago, in Europe. The caves inhabited by those men, who obviously lived in groups, clearly show them as hunters who captured and killed such extremely large animals as the mammoth, woolly rhinoceros, giant deer, cave bear and reindeer. Their weapons included "throwing stones" and an early form of spears or darts fitted with flint points; but apparently they were without the bow and arrow, according to Osborne. It thus becomes apparent that these meager weapons must necessarily have been supplemented by a high grade of mutual and co-ordinated effort.

During the past two or three centuries, and in many parts of the world, skilled observers have been able to look into the lives of persisting uncivilized men, peoples of all present races of man nearly or wholly untouched by modern civilization. These studies, made within the past one hundred years, and indeed within every recent decade, have added amazing chapters to our knowledge of primitive man. One study of tribal life and organization over much of the world was published in 1916, and conclusions drawn from it recently reappeared in the form of a popular book by Sir Arthur Keith.[5] Notable are the conclusions that, while always seeking individual happiness, everywhere man tends to be gregarious and social; that the tribe is the really significant social unit; that the area occupied by the tribe is a separate ethical territory within which an ethical code—not savage, but usually of kindly pagan ethics—prevails; that at and beyond the tribal frontiers a wholly distinct, hostile and savage moral code comes into play.

The total of such studies permits us to know that locality, or territoriality, plays a prominent part in the social groupings of primitive men, much as it does among the higher primates. Here, too, blood relationship is recognized, and family and clan—the former with one or more mates—become established. Age and sex

[5] *Evolution and Ethics, op. cit.*

groups—partly or faintly reminiscent of the troops of higher primates—affect the social structure. In these primitive human cultures, local social patterns tend to become set in quite rigid frames. Into these frames large areas of conduct and morals are fitted and enforced by the tribe—largely through myth, magic or religion.

Myth, magic and religion are not identical—there is evolution even in illusion—and continuing search for borderlines in an effort to disentangle them engages many learned anthropologists. One of them, Malinowski,[6] recently wrote: "Myth as it exists in a savage community, that is, in its living primitive form, is not merely a story told but a reality lived. This myth is to the savage what, to a fully believing Christian, is the Biblical story of Creation, of the Fall, of the Redemption by Christ's Sacrifice on the Cross."

Moreover, this authority states that for primitive men "religion is a tribal affair." Whether called myth, magic or religion for them, it is more closely identified with tribe than with individual; doubt or denial are both unthinkable and intolerable among those whose grasp of *reality* is trusted so little. Here ethical code and taboo naturally ally themselves more with the area of acceptable myth than with the slender area of inconclusive experience. With tribally approved behavior (morality) thus anchored and enforced, myth, magic and religion could render an important service to the organization and survival of primitive, groping, hunted, social men, although to many tribes thoughout the ages they also gave or propagated the grisly practice of human sacrifice. This complex of traditional and emotional guidance often proved a worthy crutch to those ambling bipeds who took the early faltering steps toward understanding and civilization. By large segments of the peoples now charged with directing a highly complex civilization, a fairly full comprehension of *reality* has already been attained. What shall they do with their myth-built crutch?

In man, sympathy and love do indeed share a place with envy and hate; but, as in animals, all these feelings of untutored man are directed to certain members only of his species. And man alone, with his better brain, has had opportunity further to

6 Bronislaw Malinowski, *Magic, Science and Religion* (Boston: The Beacon Press, 1948).

restrict or to expand the scope of these emotions. We may therefore look for the ripest fruits of a revised and flexible sociality on the tree of man.

The marvelous development of articulate language has rested partly upon sociality, and it has shared heavily in the origin of values, conscience and morals. Darwin spoke of articulate language as evolving from man's powers of intellect, of the probability that language was man's greatest discovery, and as the thing —along with the mentality that produced it—upon which "his wonderful advancement has mainly depended." And Darwin patiently brought together the then known facts concerning the crude and simplest foundations (except those within the larynx itself) for speech and language as they exist in higher animals. The lowly origin and nature of this most human legacy has been more recently emphasized by historian Robinson:[7]

They [students] should learn early that language is not primarily a vehicle of ideas and information, but an emotional outlet, corresponding to various cooings, growlings, snarls, crowings, and brayings.

Though many early stages in the development of human speech are necessarily lost in antiquity, it is evident that—used for either emotional or informational purposes—language developed in relation to *associated men* as well as to self and to other things. From simple but adequate physical and social foundations, provided by his primate ancestors, earliest men slowly evolved the supreme tool of both emotion and thought—and abstract thought has gradually provided a torch to all tomorrows. This ever-enlarging fountain of speech has progressively flooded man's own world with values. The altered coo and growl, fitted to earliest man and given time, are the swollen and nourished seed from which later mankind has grown its increasingly fruitful trees of culture and its inexhaustible harvests of new human relationships.

Man is a well-known loser of instincts. But he has had plenty of time and fair opportunity to do this. During more than a million years his lineage has had a better-developed brain than any other specimen that survives for inspection and comparison. It would indeed be surprising if a "better brain," under prolonged

[7] James Harvey Robinson, *The Mind in the Making* (New York: Harper and Bros., 1921).

test, had proved unable to substitute its expanded reason, intelligence and abstract thought for some outworn blind-spots of instinct. Some elements definitely relating to sociality, however, were evidently *not* thus lost. Fortunately, neither man nor his primate ancestors were ever framed in a strait jacket of instincts at all similar to those found in ants, bees and other social insects. Sociality arose quite independently in these two groups. In the case of insects, this sociality embraced recognition and crude communication but not sympathy. In the primates sympathy was included. Stereotyped behavior was stamped upon the insect; a wider range of behavior was granted to every primate—and to man most of all.

Though modern man's heredity still includes an occasional social root, and though his present cultures usually wrap additional social garments about him, he seems to be engaged essentially in the pursuit of individual human happiness. That pursuit is on no straightaway; and ahead are tangled maze and dead end. In our day a few peoples have been asked to exist, like the social insects, chiefly for the state. Critics of this type of political organization say that only totalitarian ruin can attend such neglect of man's basic wish for individual happiness. The clue to the outcome of this controversy, and of others like it, doubtless rests in part upon what the last few millions of years have put deeply into man. But that is a wholly inadequate summation. Despite, or perhaps goaded by, an urge to individual happiness such things as fanaticism, hysteria and mob action—even on a national scale—sometimes establish political and social policy. And whether particular groups of men will long hug or soon discard their several self-made chains seems always a highly unpredictable matter. Clearly, however, their biological endowment, even as currently and partially reinforced by social garments, does not always protect them against wild extensions and misuse of their emotions. Clearly, too, only improvements in man's social heredity—his culture, education, tools, outlook and means to happiness—can provide an adequate defense against them.

Both apes and earlier species of man surely shared in endowing the person of today with the array of impulses and talents which set him apart from other organisms—which largely govern and often perplex him. Indeed, even his collective political pacts and performances are bent or bowed because, for him, the claims of sentiment are so often superior to those of reason. Any one of a large brood of emotions—prejudice, love, hate, mere inclination —may grasp the reins and plunge ahead. It is not surprising that many who have looked closely at modern man are unsure of his equipment for the harder, still unmet problems of politics and social life. Fortunately, most fortunately, the classics and history now vouch for the superiority of civilizations friendly to thought and reason. Eccentric Egypt sunk its concern and its wealth in care for the dead; and an ancient India trusted a blend of mysticism, contemplation and verbalism more than reason. But a later and glorious Greece was first to turn to thought and reason. Its light swept over segments of three continents for a thousand years, and after the eclipse of the Dark Ages its flame again glows in institutions of the West. Can a virile society of our time in any way or for any moment safely derogate the germs of rationality which, too, emerge in man?

Though possibly little related to the development of either sociality or morality in man, one must acknowledge the numerous indications that during some hundreds of thousands of years our primitive human forebears lived under conditions highly repugnant to modern man. Dirt, stink and lousiness probably pervaded that life. Some tribes that have remained in the savage state up to modern times illustrate this somber fact. The dens of some mammals are notably clean. But surviving primates—closest kin to man—are careless housekeepers. Dog and cat are easily housebroken; less so, monkey or chimp. Man is not a natural swimmer, and cleanliness was a virtue acquired quite late in a very long history. Apparently it was clusters of unclean humans who made so little progress during a half million years. In which ways, or to what extent, did these (to us) unpleasant elements of personal association affect that man's concept of ethics and his attitudes toward fellow humans? Mainly we know that within historic times, and in several lands, laws relating to the filthy or unclean were enacted; also that the few who learned to wash assumed an undemocratic superiority over the unwashed. It is thus evident that even the history of the truly human stretches to a near-eternity of shabbiness—or worse. For ages before man walked in a garden he rolled in dirt. The more urgent point here is that whoever wholly ignores this vast span of inelegant human existence—be he politician, theologian, student, or citizen—has only a myopic view of man.

MATERNAL OR PARENTAL LOVE

In the preceding section, the simplest of the elements of sociality —sympathy and mutual aid—were observed in higher animals, in earliest man, and in existing primitive peoples. Their role in the tribe, the truly human social unit, was considered together with indications of the far-reaching consequences of sociality to man. Though sympathy can be traced into the animal world— where it is lost—it was admitted that biological science (genetics) is as yet unable to give a full account of its origin. It is now profitable to discuss parental love, to note its distribution, its place of origin, its biological foundations—about which there is much that is quite new to report—and to indicate its large share in the development of sociality, human morality and conscience.

When Darwin wrote on precisely this subject he could venture only the following tentative remarks:

The feeling of pleasure from society is probably an extension of parental or filial affections, since the social instinct seems to be developed by the young remaining for a long time with their parents; and this *extension* [italics mine] may be attributed in part to habit, but chiefly to natural selection. . . . With respect to the origin of the parental and filial affections, which apparently lie at the base of the social instincts, we know not the steps by which they have been gained; but we may infer that it has been to a large extent through natural selection.

In the following pages it will be shown that biology is now prepared to put its tag upon some of these "steps"—unknown in Darwin's day—that were taken in the origin of the parental affections.

Whoever has seen a nesting bird, a broody hen fast on eggs or defending chicks, a python coiled over its slightly heated eggs, a perch guarding its eggs or nest, a cat caring for her young, or a mother caressing and crooning to her child, has witnessed the parental instinct[8] in action. All men in all times have been witnesses, but the instinct itself is vastly older than man. Indeed, if our species had sought to become truly *unique* among higher

[8] The word "instinct" (a better word is "drive") here means simply an unlearned reflex; a form of behavior not learned by the individual, but innate.

mammals, it could not have found a surer way than to arise *without* maternal love! Fortunately, man was built of accumulated unions—tissues and instincts—already formed in animal predecessors by processes that could continue to operate in him. And thus into the advanced mental equipment of man was anchored the deep well of parental love.

The similarity of expression of the maternal instinct in humans and in the higher primates—monkeys and apes—is worthy of recall. In certain kinds of monkeys, which the naturalist Brehm kept in confinement in North Africa, intense grief for the loss of their young invariably resulted in the death of the mothers. And among these species orphan monkeys were always adopted and carefully guarded by other monkeys, both male and female. A female baboon was observed to adopt not only young monkeys of other species, but to steal young dogs and cats which she continually carried about.[9] These and similar commonly observed displays of the parental instinct in higher mammals fully provide a background for accepting as truth the perhaps inadequately attested instances of very young children being adopted and reared by baboons (Africa) and by wolves[10] (India). Incidentally, if they are true, the published accounts of these very few animal-reared children beautifully illustrate the vast extent to which a civilized person is a product of human associations— that is, of social inheritance.

Much has been learned recently of the material basis for the origin and presence of the maternal or parental instinct, as this instinct is actually observed—temporarily and recurrently—in individual cases. It was long known that in some species the instinct is present in both sexes (parental), while in others it is found only in the females (maternal). The basis for even this single difference long remained quite unknown. This point and others recently became clear when several genetic studies were made on an especially suitable animal, the domestic fowl. Here a gene for "broodiness" was found, and it was learned that in some fowl this gene is carried by the females only; also, that this gene

9 "An adopted kitten scratched this affectionate baboon, who certainly had a fine intellect, for she was much astonished at being scratched, and immediately examined the kitten's feet, and without more ado bit off its claws." (From Darwin's account of Brehm's report.)

10 Arnold Gesell, *Wolf Child and Human Child* (New York: Harper and Bros., 1941). Lois M. Miller, "The Wolf Girls and the Baboon Boy," *Science News Letter,* July 13, 1940.

has been lost by most females of certain races of fowl (White Leghorns), and that such females, quite like their male mates, do not brood eggs or care for chicks. Since broodiness is merely another name for the parental instinct when found in birds and lower forms, it is evident that in this case biology has finally uncovered a specific hereditary factor—a gene—that is essential to the appearance or birth of an instinct; and indeed, in this case, an instinct that is at once an element of sociality and a seed of love. It is found, however, that the mere possession of this gene does not automatically provide its bearer with this instinct, but rather—as will now be shown—that in individuals that carry this gene the instinct, under a quite special set of conditions, will or can develop.

The other agent needed for the actual birth of the parental instinct was found, still more recently, to be a hormone. An approach to this discovery was made in England by Wiesner and Sheard,[11] who observed that something contained in crude extracts of anterior pituitary glands is able to develop or arouse the parental instinct in both male and female rats that have not yet produced young. The present writer and associates soon thereafter showed that the effective substance in these extracts is the hormone prolactin. Our studies further showed, among other things, that prolactin serves to arouse this instinct in male and female rats and doves, but in females only in fowl—and there only in those races or individuals which carry the genetic factor, or gene, mentioned above. Rats injected with prolactin began to build nests, within a few days. Both male and female rats, though they had never done this before, also became eager to carry various small, helpless, naked objects to their nests, where they cuddled and cared for them. Objects of this care included young rats or mice and also the newly hatched squabs of doves and pigeons. Though some substances other than prolactin were later found to have effects similar to it, it was further shown by special tests on doves[12] that these effective substances cause the animal's own pituitary gland to put greatly increased amounts of prolactin into its own blood stream. Since egg-laying, pregnancy, and birth are periods of an increased production of pro-

11 B. P. Wiesner and N. M. Sheard, *Maternal Behavior in the Rat* (Edinburgh and London: Oliver and Boyd, 1933).
12 O. Riddle and E. L. Lahr, *Endocrinology*, vol. 35, 1944. See also O. Riddle, *Proceedings of the American Philosophical Society*, vol. 75, 1935.

lactin, it is easy to understand the cyclical and temporary characteristics of the maternal instinct. This drive or instinct is not continuous, but it is thus automatically born and reborn to fit the reproductive periods of the species. Of interest, too, is the circumstance that, in mammals, the same hormone that is thus especially concerned in making the mother *love* her young also assists her to secrete milk with which to *feed* her young.

Exact knowledge of the present distribution of the gene that provides the possibility for the upsurge of parental love in humans is missing, as it is for practically all animals except domestic fowl. To the geneticist, however, this indicates the difficulties of investigation rather than expectations of our ultimately finding something really different in man. Certainly the situation disclosed in the studies already made on fowl, rats and doves is capable of accounting for the facts currently observed in man. But the uncertainty of facts for man deserves a further word.

Most women develop and express maternal love for their children. Whether some women are exceptional in this matter, and also what conditions may serve to weaken or suppress this instinct in some women, are not relevant questions here.[13] For human males, also, no positive statement will be ventured. These perplexing questions merit further study despite—or perhaps because of—the complications introduced by our family life, by our civilization, and by a special tendency to loss of instincts in our species. In at least some of man's closest relatives, this instinct is shared by both males and females; in the females of all apes and monkeys it is usually very strongly expressed. Reference was made earlier to Brehm's observation that, among certain species of monkeys, orphans are adopted and guarded by both males and females. Of some assistance to an understanding of the many restrictions on this instinct in humans is Tomlinson's evidence that certain monkeys cannot care for their young unless there is a carry-over of example—of learning—from older generations to younger.

The preceding account has shown that parental instinct (or love) is widely present in vertebrate animals and that it extends into our own species. In addition to nervous and muscular organ-

---

[13] In the case of pigeons, this instinct can often be quickly suppressed by so slight a procedure as changing the location of their nest and eggs by as little as one or two feet.

ization required for a use and display of the new parental activities, it is found that both a hormone and a gene are necessary for the existence of this instinct; and the particular hormone needed, prolactin, is known to be produced by precisely these same vertebrate animals—from fish to man. Again, even our imperfect knowledge of the gene involved in this instinct permits us to understand how one or both sexes of a species may or may not have this basis for the instinct, and also how it may happen that one or another individual of a sex thus usually equipped is without it. Also, indeed, how artificial breeding or selection may entirely remove this basis of the instinct from an entire race—a race whose eggs thereafter must be given to another race to hatch and rear. Above all, what we now know of this gene provides the evidence—wholly lacking in Darwin's day—that in at least some of the higher animals there is a material basis on which natural selection may work to propagate parental love and to favor populations that possess it. There can be no doubt that man obtained from subhuman forms both the gene and the hormone for this strong impulse to sociality and moral development.

These remarks on maternal care should encourage the reader to recall his own observations on the profound changes that the broody or maternal state brings to an animal. In many cases the entire personality seems transformed within a few hours or days.

THE ORIGIN OF ALTRUISM

Not long ago the view was maintained that there is in man no such natural thing as altruism, that all is egoism and satisfaction on a sensory level. This verdict on and indictment of human nature was upheld by many philosophers and psychologists, including Hobbes, Bentham, Bain, La Mettrie and Mandeville. Macaulay expressed a widely held opinion of his day when, in answer to the question, "What proposition is there respecting human nature which is absolutely and universally true?" he replied, "We know of only one; and that is not only true, but identical; that men always act from self-interest." It has now become clear that the one general statement regarding human nature that Macaulay felt certain about is wrong.

The case for at least occasional altruistic human acts is essentially proved by the consideration just given to the parental instinct. The acts associated with mother love in women are undeniably grounded in the same biology that produces comparable acts in fowl, doves and rats. And such a basis involves genes and hormones—neither of which may be viewed apart from a time scale or from a particular line of animal ancestry. The subhuman and human worlds are not filled solely with battle and egoism. Further direct proof for this point will not be discussed here. It seems necessary, however, to have a quick look at the broad area known as reproduction, within which altruism arose. There one meets still other aspects or outgrowths of reproduction—notably the perennial sex urge pointing to the institution of family life—which are closely linked in popular thought to questions of altruism and morality.

The Nature of Reproduction

Some who admit that the earliest form of overt altruism is found in parental care—and that it exists in the upper animal world—will ask a further question: Are we not compelled, within this upper animal crust, to derive altruism from antecedent egoistic activities? Once we adequately review the nature of reproduction—within whose borders altruism originated—it will be found that the answer cannot be an unqualified yes. A glance at the phenomena of reproduction in the lower animals should suggest the more accurate answer.

It can be neither overlooked nor denied that eating and reproducing are the main functions of living things, and that both of these are aspects of nutrition. Eating is indeed a typically egoistic function, and the one that leads inevitably to the immeasurable and endless conflict and carnage that truly characterizes much of the animal world. Reproduction, however, is a different and two-part story. It usually enlists an element of cooperation, and to one extent or another it involves sacrifices or loss of personal identity. In the second place, while altruistically preserving race instead of self, reproductive overproduction and fertility also actually doom to early death the vast majority of its own products. These facts, though not all of a kind or of identical meaning, indicate nevertheless that reproduction itself is to be regarded as the basic altruistic activity from which all similar activity has descended. With the qualification just noted, and while admitting the much-dwarfed status of altruism, one may say with biologist Holmes[14] that altruism and egoism are of one and the same age, each being approximately as old as life itself.

In the single-celled animals, eating leads to growth, and more eating to overgrowth and division of the one cell into two—and this is a large share of the total reproductive activity possessed by these animals. Something of self-identity, however, is here lost while a better chance for survival of a part of the earlier self is gained. Thereafter, each of two new individuals may occasionally meet, touch, and exchange traces of liquid with a member of its species, which it will probably never meet again. The degree of co-operation that entered this affair is not a very magnificent thing. Very simple reproductive activity nevertheless proved to be a basis from which several complex and quite dissimilar activities and states arose. In rare cases it yielded states both tragic and bizarre. For example: the young of whelks born in sealed capsules where they can only eat each other; the odd fly whose body must burst in order that her young be born; and the salmon that dies after laying her eggs. Much more widespread and of different nature were two other upgrowths of the simplest form of reproduction. These were the sexual drive and the parental instinct—already indicated as the more immediate sources of altruism and love.

Even the sex hunger that long antedated parental care became the parent of a brood of unlike things. The act of copulation is as truly egoistic as eating; this lust and the love that sometimes attends it can be confused only by self-delusion. But when, as in such higher mammals as monkeys and apes, seasonal sex hunger was replaced by *continuous* readiness to mate, it provided an incentive to family life—a thing of much value and significance to man. At the same time, of course, it opened an interminable and often unmanageable account in the lush ledger of human morality. This is illuminated by paleontologist Bradley[15] in the following way:

The form of marriage, to be sure, has been as fragile as glass. Man is the only animal that ever lacked automatic and unquestionable answers to the problems of sex because he is the only

14 S. J. Holmes, "The Reproductive Beginnings of Altruism," *The Psychological Review,* vol. 52, March, 1945.

15 John Hodgden Bradley, *Patterns of Survival,* (New York: The Macmillan Co., 1938).

animal that ever possessed an imagination and a conscience. . . .
From the beginning, men have sought the perfect compromise
between their sexual instinct and their honor, and from the
beginning they have failed to find it. But though the form of
the compromise has varied vastly, certain fundamentals of human
nature have remained essentially the same. When contemporary
alarmists assume that the collapse of the home and Victorian
morality is necessarily synonymous with the collapse of the human
family, they abandon history for hysteria. There are good reasons for believing that the family is as inherent in the nature of
men as the quack in the nature of ducks.

Animal societies everywhere are mutual-benefit associations,
says Holmes, and the altruistic instincts that make for social
solidarity and effective co-operation would be favored by natural
selection equally with variations that aid the individual animal.
Instincts leading to unselfish behavior therefore may be as basic
as are other elements of instinctual behavior.

A history that equipped man with a conscience did not fail to
provide him with very new and well-concealed problems. Some
events in his history are wholly peculiar to a few species. These
descended from the trees, mastered a new and greatly varied food
supply, and later became subject to social heredity, through
which they lost nature's means of ridding their species of the
unfit. Sooner or later, man must find a substitute means for that
discarded one of nature. And he must do this while he is still in
possession of a better conscience than exists in any other species.
In becoming free from earlier and deadly enemies, man tends
in several ways to become the enemy of himself. His species is the
only one that can deliberately promote its own deterioration.

By preserving the lives of some biologically unfit offspring, and
by granting the privilege of reproduction to some heavily defective adults, mankind is currently accomplishing a degradation of its stock of genes. And if this is not being done "deliberately," it is largely because a much-restricted education in biology
deprives him of the knowledge of what he is doing. A second
problem resides in the fact that, on the one hand, the world's
food supply is limited, and on the other, the prodigal womb
spawns two where only one may grow and live. Man's new and
ever-developing skills have sometimes overcome the clash that
inheres in this situation, and they will surely continue their
powerful if partial help. Perhaps at some future date complex
factory "farms" will be able to supply from one-celled algae much

of the protein that a crowded race would require. But present population pressures cannot wait for that possibility and distant time. Already many vast areas have too many people, though it was only yesterday that modern medicine, sanitation and housing began to prolong by many years the lives of most civilized persons. Unless births are restricted, an unfavorable density of population must arise everywhere at some future time. These two conscience-splitting problems have their origin in reproduction, and forever and everywhere ethics and social policy must deal with them.

The informed person facing natural death at the end of a normal span of life can know that even timely decease serves a social end. If the biologically unfit of any age—the present included—could live forever, there would be no hope for man. He who declares a "love" for *all* men has either slight contact with the species or great power of self-delusion.

APPLICATIONS AND REMARKS

The foregoing account of the origin of sociality and of morality in animals and primitive man is intended to do another desirable thing: to illustrate what is involved in the later evolution of the emotions. It has been elsewhere noted that the evolution of the mind (mentation) as a whole is inseparably bound up with the evolution of the body. They are not two agents but an indivisible unity. The present chapter has lightly touched at least some of the agents and conditions that have been effective in the purely natural origin of some features of the primitive human mind.[16] Still other features of mentation, such as the modest reasoning abilities of our near subhuman kin, are as capable of development and transmission to offspring as are the emotions of sympathy and maternal love. The machinery through which this is

16 "The dream has had a great influence in the building of the mind. Our ideas, especially our religious beliefs, would have had quite another history had men been dreamless. For it was not merely his shadow and his reflection in the water that led man to imagine souls and doubles, but pre-eminently the visions of the night. As his body lay quiet in sleep he found himself wandering in the distant places. Sometimes he was visited by the dead. So it was clear that the body had an inhabitant who was not necessarily bound to it, who could desert it from time to time during life, and who continued to exist and interest itself in human affairs after death. Whole civilizations and vast theological speculations have been dominated by this savage inference."
—James Harvey Robinson, *The Mind in the Making, op. cit.*

done is the same for these various talents. Surely all this is an exclusively biological matter, and here organic heredity rules.

But when one's concern turns from the biological origin or basis of ethics to the sources of *change* in ethics and morality, one must look further—to social heredity. That area, the subject of the following chapter, is partly at least a borderland of biology and sociology. In human societies fluctuating grades of understanding, emotion and will (volition) become a fruitful source of an ever-renewed and changing ethics. "No man can make a conscience for himself; he needs society to make it for him," says T. H. Green. In every case, however, ethics is wholly a man-to-man affair. Again, through both stages, or through every phase of its origin and of its changes, it is an affair of evolution.

The simple yet channeled way in which human intelligence and emotion readily united to build a moral code has been deftly described by physiologist Smith:[17]

As a fallen angel man would be most ludicrous. As an intelligent animal . . . he explores his world, and here is the first value that is uniquely his: he is more intelligent than any other creature, and from intelligence fired by curiosity comes knowledge, and from knowledge come power and the manifold satisfactions by which he surpasses all his fellow creatures. . . . But the need for knowledge has burdened him with the ethic of truth: to lie willingly to himself or others, to adhere to that which is suspect, however tentatively he holds to truth, is to forfeit his opportunity and jeopardize his dreams. This is the essence of all philosophy: to cherish truth for its uniquely human value, to search for it, to test and retest it by conscious effort, to communicate it, to be guided by it, to base upon it all purposes and plans.

But he who has purposes and plans must make a choice, no other can make it for him. A proper view of man finds no place for a priori "should" or "ought" or any categorical imperative, but only for this: that if a man so acts, that is *his* action, and his alone. This is the essence of all morality: a man is responsible for the consequences of whatever choice he makes. The degree to which he recognizes this and acts accordingly is a measure of his biological maturity.

Man is an animal for whom life is more than an experience to be passively endured. Below his bare perception he feels the resonances of the affections, joy, love, wonder, fear, anger, sorrow, which color every wish and vision until he can scarcely think but his thought is reinforced by feeling. . . .

[17] Homer W. Smith, *Man and His Gods* (Boston: Little, Brown and Co., 1952).

To neglect the creative dynamic of the emotions is to neglect the essence of human nature. Fear, anger and exhilaration move man as they move the denizens of forest, sky and stream, but the emotion that is uniquely his is pride: he will risk his life in combat rather than suffer loss of his self-esteem; and honor, jealousy and indignation contribute to the determination of his rights and duties, and elicit courtesy and consideration for the pride and privileges of others. He who is sensitive to shame will not be insensitive to the judgment of his fellows, careless of decorum, unappreciative of convention. He who through imagination can suffer another's pleasures and pains—his fear, anger, pride and even his prejudices and hatreds—will build a family, tribe and nation, and fabricate a moral code.

THE HANDLING OF VALUES

A secondary crest of this chapter is concerned with "values." Primary concern in these pages has centered in the wholly natural origin of ethics and morals. To minds sufficiently instructed and free from theological slant, it has long been possible to regard values likewise as products of existence at the human level. Near the end of Chapter 2 this matter was discussed briefly. Early in this century philosopher Dewey[18] pointed the way toward the building of a science of value on a behavioral basis. But for more than a century, while biology and sociology were encountering their first opportunity to take firm roots in thought, many scientists have sold their science short on this question. And they shamelessly and unscientifically still continue to do so. Their chant is that "science has nothing to do with values," and they sing it to the bishops. This represents nothing better than a bribe to churchmen to hedge on their opposition to science. At its worst it represents a cowardly capitulation of Western science. The Chinese were not misled by it.

It is to the field of social anthropology that one may best turn for a worthwhile account of the nature and handling of values. Since that subject is adjacent to this writer's special field of competence, the reader will be better served at this point by

18 John Dewey, "Theory of Valuation," *International Encyclopedia of Unified Science,* vol. 2, no. 4. See also H. A. Murray and C. D. Morgan, "A Clinical Study of Sentiments," *Genetic Psychology Monographs,* vol. 132, 1945.

the following lengthy quotation from Harvard anthropologist Kluckhohn:[19]

In human history one finds, broadly speaking, three types of assertions as to the source and sanction of values: divine revelation or command, tradition and custom, and human intelligence. Probably no culture has relied completely upon one to the entire exclusion of the other two. . . . The Greek philosophers and Confucius gave the world its first systematic conceptions of how human beings might derive their values in other than an authoritarian manner and with the possibility for change and growth. Central to the Greek conception of virtue was the congruence between behavior and conviction. For Plato critical intelligence is virtue, for Aristotle, it can be virtue. Both Plato and Aristotle brought values within the sphere of science, making virtue something discoverable and teachable rather than revealed or handed down and only preachable. The Stoics specifically proclaimed that to live in accordance with nature was the highest good. Cicero speaks as a Stoic when he says that "right is founded not in opinion, but in nature." The Confucian *jen* likewise considers nature the court of last resort.

In the Western world during the last century and a half a divorce between nature (as described and interpreted by science) and values has generally been accepted. This is expressed frequently in the writings of many scientists who make such utterances as the following, "Science only provides a car and a chauffeur for us. It does not directly, as science, tell us where to drive." . . . It is perhaps a not too gross over-simplification to suggest that the division of territory which ascribed to science the realm of "fact" and to religion and the humanities the realm of "value" was actually an attempted resolution of the so-called "conflict between science and religion" which plagued the nineteenth century. The forces of orthodoxy saw very clearly that new knowledge of the physical universe threatened credulity in the cosmogony of Genesis; that paleontology, biology, and archaeology had deprived much of the Old Testament of other than possible symbolic meaning. At the same time organized religion still had great power to block scientific teaching and research. In effect, the scientists were offered a compromise: "You may investigate the world of nature to your heart's content so long as you admit that problems of morality, of the aims and goals of human life, of ultimate values are, in principle, unanswerable by science."

Although this folklore has been verbally accepted by most

[19] Clyde Kluckhohn, "An Anthropological Approach to the Study of Values," *Bulletin of the American Academy of Arts and Sciences,* March, 1951.

scientists up to the present day, at least four intellectual movements of the present century have contributed enormously to the deflation of all notions of value. . . .

Equally incorrect are the views that "science has nothing to do with values" and the moral nihilism inherent in the vulgarizations of the psychoanalytic, Marxian, and the older anthropological standpoints. There is an alternative between dogmatism and anarchy. Ethical relativism correctly saw the diversity of actual moral codes among different peoples and quite rightly pointed to the scientific and logical flaws in metaphysical and theological ethics but quite wrongly concluded that there were no pan-human values and that no code was worth defending. . . .

These *universal values* have not yet been examined by social scientists in the same detailed way in which the gamut of cultural variability has been explored. We too often forget the extent of consensus as to the satisfactions for individuals which any good social order ought to make possible or provide. . . . As Lundberg has reminded us: "There is general agreement by the masses of men on the large and broad goals of life as evidenced by man's behavior. Everywhere he tries to keep alive as best he knows how, he tries to enjoy association with his fellow creatures, and he tries to achieve communion with them and with his universe, including his own imaginative creations. The sharp differences of opinion arise about the *means,* the *costs,* and the consequences of different possible courses of action." . . .

Some values are as much "givens" in human life as the fact that bodies of certain densities fall under specified conditions. These are founded, in part, upon the fundamental biological similarities of all human beings. They arise also out of the circumstance that human existence is invariably a social existence. . . .

In all societies the individual whose actions are completely unpredictable is necessarily incarcerated (in jail or asylum) or executed. No society has ever approved suffering as a good thing in itself. As a means to an end (purification or self-discipline), yes. As punishment, as a means to the ends of society, yes. But for itself, no. No culture fails to put a negative valuation upon killing, or indiscriminate lying, and stealing within the in-group. . . . But the core notion of the desirable and non-desirable is constant across all cultures. Nor need we dispute the universality of the conception that rape or any achievement of sexuality by violent means is disapproved. This is a fact of observation as much as the fact that different materials have different specific gravities.

To the preceding statements perhaps another may be added. To study people's values, as one does in studying the Hopi

Indians or the values of Middletown, is not to use science to find out clearly or directly what my values, or other people's values, *ought* to be. Nor is it clear or certain that these last-named goals are fully reached by even a most competent search for "universal" values. Can it be determined fully, by others—even by myself—and by any means whatever, what *all* my values *ought* to be? Are attempts at such completeness in that area sensible? Are the huge and countless biological inequalities of men described in Chapter 6 without consequence to the values of unequal individuals and peoples? The world's scholars find, describe and admit an imperfect universe. On which basis could they or others expect perfection in its most involved and evolved segments—the will and values?

At this point, and in the light of items discussed in this chapter, it is well to consider a point of view recently defended by educator-chemist Conant.[20] This author recognizes that many values fall well within the sphere of science, but attaches a supreme importance to certain *other* values, which, he thinks, fall into a "universe of spirit." He states:

As to the unifying materialistic World Hypothesis, my doubt stems from its manifest inadequacy. As a conceptual scheme attempting to account for everything in the whole universe, it seems to me unsatisfactory because it is incomplete. It fails to provide for the altruistic and idealistic side of human nature. It fails to accommodate what I regard as highly significant facts, not facts of science but facts of human history. These are the unselfish ways in which human beings often act with compassion, love, friendliness, self-sacrifice, the desire to mitigate human suffering. In short, it is the problem of "good," not "evil," that requires some other formulation of human personality than that provided by the usual naturalistic moralist.

Readers of the preceding pages may judge whether the naturalistic view "fails to provide for the altruistic and idealistic side of human nature." They should be in even better position to resolve that and related questions after an examination—in the following chapter—of the reach and strength of social inheritance.

But just here we are so close to the basis on which many scientists—physical scientists particularly—concede a separate realm of

[20] James B. Conant, *Modern Science and Modern Man* (Garden City, N. Y.: Doubleday and Co., 1953).

reality to religion or a "universe of spirit" that their reasoning should at once be further and sharply challenged. The fact that science is clearly no utterly *complete* guide for settling questions of what a human being *ought* to do—to training and directing conscience in the species that recently advanced the development of conscience and *also* automatically developed society in which "oughts" acquire prominence—is made the basis for *accepting* a kind of universe otherwise unnoted in evolutionary processes from atom to society. Those same scientists, within the field of the physical sciences, find that science can there give no *complete* account of the *weather*. But for that failure or default of science these scientists would stand aghast at any suggestion that this confessed failure of science called for the addition of even a single electron or quantum, or of other minute or unknown agency, to the familiar universe. Yet the two cases are parallel. In both cases the known or apprehended underlying facts amply account for the present partial impotence of science. Though both have been much accused of it, neither case offers the slightest evidence for a "universe of spirit."

These scientists then reach further to give a helping hand to the province they thus construct and call religion. Usually those from the Western world—and they only—refer gratefully to the Judaic-Christian religion. Yet at once they *reject* the vengeful and the personal God; the revealed story of creation; the doctrine of original sin; the need for an atoning Savior; the superhuman character of anybody; the manifold miracles. Then, from among the crumbled ruins, they bear away the persuasive admonitions to acts helpful to others rather than to self; the longings and trials of certain persons and peoples; and many memorable examples of self-searching that nearly all good literature provides. They ignore the admonitions and examples now clearly recognizable as horrible and harmful to society; and they seem to consider the immeasurable harm done by institutional religions as an earthy matter wholly separable from religion. Just why their personally resurrected residues of a shredded Book should bear the label "Judaic-Christian" is often unclear. Unintentional though it is, the main and larger effect of the endorsement these men give to "religion" is to support and perpetuate the "institutional religions" of this day. Is their own idea of what they *ought* to do at all inconvenienced by this circumstance?

At least in current advanced societies it would seem that the first and foremost moral obligation of the individual is to be intelligent and comprehending. And the obligation to breadth of comprehension may be greater than that to any other of its dimensions. A firm grasp of the natural origin of human feeling, aspiration and intelligence is the broadest plank in that platform—sympathy, love, sociality, educability, kinship and interdependence being surely found among the fibers that build its strength. A freshened grasp of recorded human experience in appraising and sorting values is another plank in that platform—justice, freedom, democracy, leadership, dignity, integrity and social worth belong there. Full comprehension, too, involves the fact that, to date, no society has been able to deal with many of these items either objectively, comprehensively or unobstructedly; that, to date, no individual having much insight has ever had the encouragement and approval of the whole society of which he was a part. Yet both the stature that intelligence adds to self—to one integer of exalted and interdependent human existence—and an assurance that the kindred attainments of others have usually nevertheless had lifting power can give great moral meaning to higher levels of unfettered intelligence.

A denial by theology or religion of the purely natural sources of morality and of values is a bald and crude pretension. It is clear, however, that the so-called religious impulse may, if it will, usefully contribute to the readiness and willingness of men to think and act within acceptable frames of morality and value. All human populations must persistently strive to make the moral also the acceptable and satisfying; and a good life must be both satisfying and productive. But, in advanced societies, a religion based on revelation, or indeed one that looks to any intervention of the supernatural in human affairs, mainly offers confusion to jobs that call for fact, clarity and unchallenged logic. In presenting an account of unmitigated evolution, the natural sources of morality and values—or at least their nonrequirement of a separate and "spiritual world"—must now be particularly emphasized because this point is so uniformly denied in religion, because it is so shamelessly surrendered by many scientists, and because the entire structure of evolution has neither consistency nor any deep significance without it.

5

Social Inheritance

People are always talking about originality; but what do they mean? As soon as we are born, the world begins to work upon us, and keeps on to the end. If I could give an account of what I owe to great predecessors and contemporaries, there would be but a small remainder.—*Goethe*

The problem is not how to produce great men, but how to produce great societies.—*Alfred North Whitehead*

FEW GIVE THOUGHT TO HOW MUCH THEY OWE TO HOW many. Much of the best in our lives comes to us from others who were here and are now gone: a gentled earth; a rich language; a literate parent; an ordered home among homes; a choice of pleasing tastes, sights and sounds; an arsenal against pain and disease; a prolonged chance to learn what others know and do; a frame of customs to guide behavior; a channel of law and government; an array of option to livelihood; a heavy list of unsolved problems; a granted bid to speed on water, earth, and air—all these are among the gifts of earlier men and societies to those born into the rich inheritance of this day. Here, if anywhere, is place for gratitude to many a fellow man. And who will fail to gain dignity in thus finding within himself a prized something that traces to Buddha, Plato, Artistotle, Jesus, da Vinci, Copernicus, Shakespeare, Newton, La Place, Goethe, Franklin, Darwin, and the rest?

The gifts cited above are an "inheritance" only in the sense that we inherit *property* from ancestors; custom, however, inclines to call them our social inheritance. This so-called inheritance was not won directly in the biological struggle for life and thus fastened firmly into our tissues in the form of genes as is the organic heredity discussed in earlier chapters. Organic heredity bears a stamp of seeming stability and permanence, while all of the items of social heredity listed in the preceding paragraph would drop as easily as a cloak from a child reared by bears or baboons, or even by truly primitive humans. It is

93

evident therefore that the thing we choose to call social inheritance is really much more akin to environment—and is in strong contrast to organic heredity. Social inheritance provides man with a new and highly potent part of his environment and is especially remarkable in being man-made and man-controlled. There may be truth, however, in anthropologist Kroeber's suggestion that the capacity, or an increased capacity, to develop culture may have occurred in earlier man through the change of a single gene.

The realm of the social widens and deepens under closer examination. Some students give it a broad (ecological) definition, which includes the "web of life"—an economy in which all living things have a part. In that view society truly becomes the summit of all integrative accomplishment in nature. But any further look at society in this broadest sense hardly belongs here. That subject was touched upon in Chapter 2, and it has been treated elsewhere by a group of specialists.[1] Animal species generally form associations, and many of them form societies—organizations in which relationships and performance are structured in a rich variety of ways. At the level of human society, however, a new mechanism is introduced that does not wait upon a modification of genes. Here communicated experience and abstract thought begin a controlled change of the human environment; they begin the shaping of customs, institutions and cultures. Here emerges a new factor in human advancement. In the terms of modern sociologists, society is made up of societies, but these components are called social groups. In the following pages the word "society" will sometimes be used almost interchangeably with culture.

The nature of social heredity, its qualities as an evolutionary emergent, and some discussion of its bearings on man's moral problems are the main items for notice in the present chapter. But no one should feel properly introduced to the word "society" until two points are clear: it is the topmost rung in evolution's ladder—the highest point reached in the series of integrations that began with atoms; and its creator is man himself—the supreme performance of collective man to date. Social inheritance is an evolutionary emergent—a phenomenon new in type and built on superb foundations of speech, gregariousness (sociality), brains

1 *Biological Symposia,* vol. 8, ed. Robert Redfield (Lancaster, Pa.: Jacques Cattell Press, 1942).

and hands, but giving birth to qualities and things quite unlike those foundations.

SOCIETY, CULTURE AND CIVILIZATION— EVOLUTIONARY EMERGENTS

For a moment let us look at hands, brains and sociality as they doubtless existed two or three million years ago, definitely *before* the advent of man. Let us, too, imagine that this look is taken by the detached "mind" of a modern biologist—though by one who has seen no man and no animal higher in the scale than those present on the earth in that faraway time. The then existing primates (early monkeys and apes), when compared with their ancestors back to fishes, would enable this mind readily to conceive of a still higher or more advanced type of primate. The hands and arms of the apes—limbs variously modified from fins in all higher vertebrate animals—would permit this mind to picture, for example, changes in thumb and fingers that would permit a more accurate management of minute objects—essentially as they now exist in man. Moreover, observing that the obviously higher species of primates tended toward erect posture, with less use of front legs for walking, this inspecting mind could picture a form fully erect with arms and hands wholly free for things other than walking and running. The knowledge that some of these apes were rather rapidly evolving toward both erect posture and larger brains might even provide clues to crude prediction. Again, the many advances of the brain and nervous system of such primates over those of reptiles—the vertebrates that began to build the cerebral cortex—would enable this roving mind to picture something still more efficient and complex. At least there could be no difficulty in forecasting for some animal of the future a further increase in the *number* of units of nerve action (neurones) in the brain. These, this mind would know, had proceeded from zero in early single-celled animals to perhaps three billions in the monkeys under examination. There are about ten billions of them in the brain of man today.

Finally, the social traits of that day, as partly noted in the previous chapter, must have been represented by the herd instinct, sympathy, mutual aid, the instinct of parental love, and the beginnings of conscience. Surely an extension of all these traits could be visualized by this prowling mind, and thus a more erect, more gregarious, more kindly, more intelligent, more social animal could be projected or conceived. It would also be conceivable that the union and integration of all these reinforced elements into one and the same species could yield an animal capable of dominating all others. All the above deductions would rest simply on the principle that new species arise through the slow heritable modification of existing species, and that when given time greater complexity of organization and performance arise from simpler states. The whole point of this odd mental excursion is to indicate that the chief qualities and traits of *earliest* man were readily conceivable, even though very imperfectly predictable, from knowledge of earlier stages of the living world that produced him.

Immediately it should be remarked that no similar predictability, by any type of mind whatever, could have been made on earliest and crudest man himself in regard to his present *cultural* powers and attainments, because these attainments do not rest directly on individual organic heredity but on a society, and a society, moreover, uniquely capable of endowing its new members with the cultural *accumulations* of its older or earlier members. These new, quite unpredictable, organically detached, and often intangible creations of man arise as emergents in society—a new and highest level of integration. They are products wholly unlike the foundations and ingredients that produced them. Those foundations were individual primitive men whose capacities included the following: a fair measure of bodily strength and dexterity; certain keen special senses; foundations of a conscience; a few helpful instincts leaning toward sociality; an approach to or actual articulate speech; reason and memory; emotions and will; near-ability to form concepts; and all of these somewhat or much above the standards of his nearest still-surviving animal relative.

The combination of these several qualities, when used by associated men and given plenty of time, has yielded the wholly dissimilar things—society in general and civilization in particular. Items of the latter include spoken and written language; a way of learning from a painting, a map or a printed page; long and numerous adventures in myth and religion; the recent and occasional substitution of knowledge and test for gratuitous assumption about the nature of objects, phenomena and self; mathematics and other fields of knowledge; clear and clean passage from the concrete to the abstract; development and revision of moral ideas; the extremely recent wish to progress; the metropolis; the ship and the airplane; the arts; and the skills of agriculture, manufacture, trade and government. In brief, these dissimilar yields are ideas, skills and institutions—the latter including kinship-structures, and economic, political and cultural institutions. Here, finally, we find ourselves dealing with an evolutionary emergent in terms both undeniably natural and very intimately related to our own lives. At the same time we are dealing with mankind—with human society—as a creator and director. Yes, here mankind extends itself—inevitably and usually unconsciously—into that self-enriching integration that is the climax of all known integration in nature.

MORAL AND SOCIAL RESPONSIBILITY

The opening paragraph of this chapter referred to several pleasing and satisfying things with which civilization clothes us all at birth. Not stressed in that list was a less pleasant thing. Birth into communities of our day also means baptism into a sea of pressing problems relating to life and action in a highly complicated social, moral, economic and political world. Because of their biological origin, some social and moral questions now require further notice. Among these are problems of behavior, morals and values that are largely frozen into custom; problems of organization and adjustment; and problems at once political and ethical upon which perhaps even the survival of nations or the race may depend. Some of these issues must be solved at the levels of the individual, the community and the nation—perhaps beyond the nation.

The burden of responsibility under civilization seems to be an increasing one. And certainly man is not equipped internally with instincts and genes to deal adequately with a type of environment that, as noted above, is a crusty, man-imposed emergent—not a thing well-met and conquered in the earlier and deeply registering struggle for survival that guided the building of the body-mind of man. This inescapable circumstance clearly points to man's need to find or to create, within his arsenal of emergents, suitable means for impressing men with their moral and social responsibilities. The agencies thus far developed and employed in this enduring task are all well known: example and discipline by parents and tribe—of highest importance; public approval—effective in varying degrees; education—till now usually in very light doses, highly variable in content and purpose; religion—in great variety, often in large supply, sometimes omitted; sports and play—significant but restricted; clubs and societies—varied but limited; ordinance and law—always available, often evaded, sometimes archaic and inflexible or worse; courts—usually impressive, occasionally abusive. Apparently it is within this broad area—or in some other unnamed segment—that men may look and must find practical and effective guides to behavior and responsibility.

Origins, and a free examination of the things implied by manner of origin, are the text and texture of Part I of this volume. Only these aspects of the elusive and much-discussed problems of social and moral responsibility will be even briefly touched upon in these pages. This account is especially concerned with the binding fact that our present social inheritance largely takes over, from the hard school of struggle and natural selection, the task of guiding the further progress of man; and next, with the social and moral implications of this meaningful fact. The solid rooting of this entire matter in biologic or natural law is as yet little understood by peoples and their leaders.

Just here one unavoidably meets a trinity of enduring dangers. The first danger is a public that is unaware of the evolutionary basis on which earliest man's capacities and progress exclusively rested, and is thus heedless of the continuing need to respect some parts of that pitiless mechanism. The second is a group of religions that, whatever their immediate service for good, continually immerse most of mankind in misty visions of "a friend in the sky," a Providence scrupulously guarding some or all of the interests of men; yes, a group of religions that in all advanced nations effectively blocks education from its obligation to carry to the public the full story—often even any part of the story—of the evolutionary origin of man and morals that recasts his thought and so largely conditions his future. The third danger is a crowded human race, now increasing in numbers faster than any known species, all largely engaged in seeking such things as happiness, food and liberty, highly unequal in both personal capacities and physical assets, and only relatively few of whom now have both capacity and opportunity to strive for the long-term welfare of the race. These menacing conditions are world-wide.

In momentary digression one might give thought to the foremost way in which the evolutionary insights discussed in these chapters can provide help for man in these present and durable difficulties. That help will come only as a result of a change in man's basic thought about himself—as a result of widespread and revolutionary acceptance of the *natural* origin of all that pertains to man. Organized religion is found to be the chief obstacle to that acceptance. Whether the help comes soon or late surely depends upon the speed at which the grip of the supernatural can be broken.

The extent to which the *improvement* of the race involves questions of morality and an enlightened social conscience will receive but little further notice here. One may recall that groups of citizens in various countries have formed organizations— eugenic societies—to forward knowledge and interest in the preservation and increase of desirable genes and traits now scattered through the race. This is a direct and useful response to an evident danger. Many observers think, however, that much more than can be done through organic heredity must and will be accomplished through social inheritance—through man's newly acquired dominion over his own environment. Both views have been fully presented by others.[2] Certainly both approaches can yield results. The larger question is: Will their sum be enough, and how can that sum be best and most expanded?

The related problem of overpopulation plainly involves pressing questions of ethics. But perhaps in no country, certainly not in the world at large, is there an enlightened public opinion to deal with it. On the contrary, there is widespread religious objection to already available means of a partial or a real solution. It is successfully asserted by dominant religious groups that life, even the dim life of the egg and embryo, is God-given and beyond the moral right of man to regulate. Meanwhile, overpopulation has become the problem of problems in Puerto Rico, Haiti, Java, Italy, Egypt, China, Japan, and in parts of India. There and elsewhere egg and embryo are preserved to points where they

---

[2] Herbert W. Conn, *Social Heredity and Social Evolution* (New York: The Abingdon Press, 1914). S. J. Holmes, *Human Genetics and Its Social Import* (New York: McGraw-Hill Book Co., Inc., 1936). L. C. Dunn and Th. Dobzhansky, *Heredity, Race and Society* (New York: Penguin Books, Inc., 1946).

have human sensitivity, consciousness and endless hunger—only to die after becoming objects of love, and after bringing further hunger and poverty to a family. On the other hand, uncontrolled use of contraceptives in some industrial nations may have become a threat to the political and economic dominance of countries that hitherto have been most productive in science and leaders of civilization.

Here, too, one recalls that the science and technology that produced the industrial age also made possible the present dense populations of some parts of the earth; and that in some countries, under conditions of the present and the recent past, such advances as the mercies of medicine and sanitation relentlessly contribute a threat of overpopulation and a severer struggle for the necessities of life. The new problems are inescapable and already in our laps. Their solution depends upon a wider spread of knowledge, a rational morality and self-discipline in men, and an unattained freedom to employ the measures found necessary.

It is clear that much social responsibility rests upon science and scientists. To some extent, scientists are now under attack on this score, and many of them are becoming conscious of this responsibility. Nevertheless, it is still frequently true that when or if a scientific investigator seriously attempts to meet this responsibility by venturing into the field of general education, he tends to lose caste with his colleagues. This book is the writer's personal effort to acknowledge such responsibility. Philosopher Dewey[3] has remarked:

Science through its physical technological consequences is now determining the relations which human beings, severally and in groups, sustain to one another. If it is incapable of developing moral techniques which will also determine these relations, the split in modern culture goes so deep that not only democracy but all civilized values are doomed.

Though there is no intention here of suggesting the extent to which science and technology may be expected to contribute to the solution of social and ethical problems, three authorities

[3] John Dewey, *Freedom and Culture* (New York: G. P. Putnam's Sons, 1939).

representing two fields of applied biology will be quoted on the nature of the problem. The Director-General of the World Health Organization recently spoke as follows:[4]

Most of us, by being civilized too early or too forcibly, have been driven to believe that our natural human urges are "bad," "not nice," "wicked," "sinful," or whatever the local equivalent may be. This is the dreadfully damaging concept of "original sin." . . . It appears that a system which imposes an early belief in one's sinfulness or unacceptability in one's natural state, with its consequent inferiority feelings and anxiety, must be harmful to interhuman relationships and to the ability of the human race to survive in the kind of world this has become.

Some unpleasant facts concerning the amazing amount of emotional stress and mental disorder in the population of the United States were recently supplied by Saul,[5] a preventive psychiatrist. He said:

Psychosomatic medicine is a whole new field devoted to exploring the role of emotions in physical disorders. . . . It is evident that the personal problems which psychiatry faces must be numbered, in this country alone, in the tens of millions. There are as many beds for psychotics alone as there are all other hospital beds put together—over half a million. . . . It is difficult to estimate the number of persons with classic neurotic symptoms severe enough to make help urgent. The figure derived from Selective Service examinations is five million. One child of every twenty born will spend time in a mental institution. One in ten will be incapacitated for some period by lesser breakdown. Even this is but a small fraction of the over-all problem. Physicians judge that from one-third to two-thirds of the practice of medicine deals with complaints which arise basically from emotional tensions, from the stress and strains of living. . . . This does not count the needless failures in career, the needless daily cruelties, and the emotional sufferings which never get into statistics. But it shows that psychiatry deals with such vast numbers that it must be basically *preventive*. There can never be enough psychiatrists to *treat* (cases). Only through prevention can psychiatry be effective. These figures show also that the psychiatrist cannot do the preventing alone.

4 George Brock Chisholm, *Science*, vol. 109, 1949.

5 Leon J. Saul, "Preventive Psychiatry," *Proceedings of the American Philosophical Society*, vol. 93, 1949.

With religions in mind, another psychiatrist[6] has thus expressed a first-rate difficulty in getting on with this job:

It is one of the graver lessons of our times that the new, the more liberal, the more effective, does not immediately succeed without our active assistance in driving out the old, the harmful, and the entrenched.

In our thought about these disquieting problems there should be no neglect of the fact that the various elements of our behavior—moral and other—are partly determined by the genes provided us by our fathers and mothers. Rarely, but sometimes, a cog necessary for moral behavior is left out of the new machine —a fellow being. When this occurs there slips into our problem the specter of personal moral irresponsibility. Now and again, in all communities, some who grow up with or among us will unroll one or another inherited defect that partly or fully exhibits this dark truth. Such cases of moral irresponsibility are not of great practical importance. They are much less numerous than cases of moral delinquency arising from the cultural environment; they are now incurable, and all societies have always had to deal with or endure them. But each case has real importance in illuminating the matter of the earthly origin of the moral sense. Evolutionary biology alone could put its finger on that source; Providence can no longer be accused of taking it away from some, nor can it be credited with bestowing it on others (see Chapter 6).

Ultimately, of course, it is the possession of genes that decides that squirrels will show squirrel behavior and that dogs will show dog behavior—various breeds differing markedly, despite all training. In man, the hereditary foundations are similar though less binding, because his control of the environment is unique and because the total environment markedly affects behavior. And notwithstanding many plaguing exceptions, men in common have comprehension, sociality, inventiveness and daring—all pointed to avenues of the new. Saints are rare or extinct, but most human beings are capable of adopting what seem to them the more important aspects of moral behavior. Though man's prospects for better moral adjustment are not wholly bright, they have not yet been proved to be wholly black.

6 D. Ewen Cameron, *Science,* vol. 107, 1948.

Acceptable personal behavior is best supported by a sensible social frame. A hothouse civilization calls for early and deep rooting in all now available human and physical resources. Most societies do not now utilize enough science to assure their survival, enough of the outdoors to protect their sanity, enough perspective to sense the direction they are traveling.

A further word on attitudes, on human orientation, belongs here. No reverence for the living stuff—for *all* that lives—is sensible or tolerable. Weeds and vipers must be destroyed in the same moment that we cherish loams and mountainsides. Understanding and common sense are often infinitely better than reverence, since endless and compelling integration operates apart from human need or convenience. Unlike and unequal men are necessarily appraised by each other. Must informed opinion and common sense there abdicate to reverence? Why? Are the choicest fruits of the integration process itself—wherever found—especially worthy of our highest acclaim? Are those fruits recognizable through other than a solid knowledge of nature?

HUMAN INEFFICIENCY: POSER OF MORAL PROBLEMS
AND CURB ON SOCIAL PROGRESS

Worthy of attention here are two accompaniments of human inefficiency. Within a human group the inefficiency of some of its members endlessly raises questions of moral obligation. That same inefficiency also curbs or limits many aspects of group attainment and social growth. Logically, one cannot place efficiency on a par with, for example, honesty or cheerfulness. Nature itself irrevocably restricts efficiency to certain periods of the life cycle; it does not similarly limit many other qualities of the individual. It is possible to overdraw or otherwise distort the role of inefficiency in human affairs. Nevertheless, some matters of social consequence can be discussed favorably under this title. This approach seems to have been neglected by those who have written on these subjects; the present statement therefore, though quite incomplete, is not unduly short. No documents or authorities are cited here. When the reader meets the three related generalizations stated below, he is especially invited to use his own personal experience as a test of their validity—and particularly that of the main thesis here set down as the first.

1. **Present human populations are made up of, relatively, a**

few *efficient*[7] individuals whose days and efforts are largely expended upon the care and training of the great masses of the *inefficient*.

This unusual cross section of nations, communities and families invites us to an unaccustomed view of ourselves and of human relations in general. It would seem that the prevalence of human inefficiency within all groups of mankind constitutes both a sturdy brake on the thing we call progress, and an early and still persisting source of a large part of ethics, morality and social adjustment. Inefficiency is a burden largely peculiar to humans, and a continuous challenge to conscience. Biological elements are dominant in some aspects of personal inefficiency, and that circumstance leads to this discussion of the problem. Training, or education, is the other and controllable element, and its relation to social advance is nowhere better observed than in its bearings on personal efficiency.

First of all, we ask: Who is it in the nation, community and family that overfills the ranks of the inefficient, and continuously maintains this group as the strongly predominant one? Rather high on that list are those millions in whom the scales of heredity are tipped adversely to one or another of the many components of normality (Chapter 6). For example, the definition of this word ("having and using the requisite knowledge, skill, and industry") excludes that considerable group in which even normal or supernormal ability to have and use knowledge and skill is linked, apparently hereditarily, with *sloth* rather than with "industry." Also very high on the list are those hosts with wholly adequate heredity whose social environment, lack of training or unfavorable training renders them inefficient. Clearly, the proportion of inefficients from this source varies much from era to era, from nation to nation, perhaps also from climate to climate. Even the most advanced nations, however, have a deep reservoir of such inefficients. Many are in this class solely because of their perturbed and distorted temperaments, which are derived from the badgering of mates, of kin, and of the cruel events of life. If such a distinction really existed, one could say that it is the mind rather than the body of mankind that is least at home in current civilization. It is notable, however, that wounded temperaments arise from environments, and from heredity as well.

[7] The usual meaning of the word is implied: "having and using the requisite knowledge, skill, and industry."

This list must enroll many others. There is a rather formidable connection of personal age with inefficiency. Of all animal species, the human offspring remains helpless longest; incidentally, these additional years permit the prolonged growth of the human brain. But this period of inefficiency—even in those who later attain efficiency—must be reckoned as many years. For the earth's population as a whole, life expectancy at birth is now perhaps less than forty years. Advanced age also often or usually becomes a period of inefficiency. Next, the ranks of the inefficient are everywhere increased by many of the malformations and crippling accidents of prenatal and later life. On a temporary basis, inefficiency prevails in adults during all periods of sickness and convalescence. The inclusion of immaturity, age and illness in these examples of the inefficient well shows that stigma does not necessarily attach to this word. The word covers situations both temporary and enduring, and no one can be efficient during the entire span of life. All who die early are denied a chance for even a moment of efficiency.

Again, whatever their earlier and later status in life, the inmates of jails and prisons, and prisoners of war, are at least temporary additions to the army of inefficients. Within the mixed crowds of the poor, the more impoverished and weakened cannot be otherwise than inefficient. Still others could be named. These, then, are the more notable groups whose sum provides the bulky battalions of inefficients who absorb the major efforts of the thinly recruited platoons of the efficient.

Next to be explored is the part of the thesis that asserts that the time and effort of the efficient few are largely expended upon the care and training of the inefficient many. First there is the commonplace fact that, at the family level, a very high percentage of all time and effort is directed to the care, feeding, clothing and training of its members—usually including the immature, the aged, and the sick. This work—this most arduous and persistent area of human effort—is indeed partly performed by inefficients. But many efficient adults also live their lives precisely within this area, and through their superior performance provide the better ingredients of a home and community life. The role of the efficient few at all higher levels of community and national life may be recalled but not discussed here.

2. Many past and future generations of mankind have been or will be likewise unbalanced toward inefficiency. This seems self-

evident. In the savage state, populations may have contained a smaller proportion of inefficients of certain types than do modern communities. Certainly the defective and the sick were removed from them by natural selection with a speed and effectiveness that is unusual in our day. In the savage state, too, the word "inefficient" would have a meaning other than in a modern industrial culture. The entire matter of education—of training—assumes very unequal relations to efficiency within those two cultures. Again, the extension of the efficient part of the total span of life—a thing now occurring—should perhaps tend to lessen the unbalance of inefficiency in future generations.

3. It is evident that elements of ethics and moral right must arise in connection with the dissimilar efficiencies present everywhere and always in family and community life. From early man onward the inefficiency of the immature has dictated not only that their needs must be supplied by others, but also that each child be taught a most meaningful lesson, namely, that he must *obey* parental and tribal commands. And this stream of experience, involving notions of submission to authority and *necessity* for obedience, provided at least a part of the foundations for both ethics and conscience.[8]

In our own day the inefficiency of the blind often makes these individuals both a family obligation and a proper charge of the state. Indeed, nearly or quite all groups of inefficients pose problems of ethics, economics, law and politics. There is momentary need for a revised ethics regarding the management of two wholly dissimilar, small, but far from negligible classes of helpless men —mortals who in fact can neither help themselves nor be helped by man. First, the bedded idiots and the grotesque monsters consigned to asylums, where a fishlike existence is often prolonged for many decades—at the heavy cost of efficient hands and state funds. Also, we should say, at the cost of a never-healing wound for parents and nearest kin. Second, the incurably ill who, facing months or years of pain and complete consumption of family resources, may not now ask and receive, under well-guarded controls, a merciful release through excess portions of morphine. Neither idiot nor incurable can be thus released

____

[8] Even in some animals certain wishes of parents and the herd are enforced upon the young. According to Darwin, the naturalist Brehm observed: "When the baboons in Abyssinia plunder a garden, they silently follow their leader; and if an imprudent young animal makes a noise, he receives a slap from the others to teach him silence and obedience."

legally from unwanted life. Most Christians now oppose these revisions of our ethical standards.

The point of view developed above poses a special set of questions concerning the advance of civilization during the past few thousands of years. It can be granted that the presence and efforts of the inefficients were doubtless often helpfully involved in that advance. Yet one may speculate on the result—the nature and appearance of man's world of today—had the thin ranks of the efficients merely been doubled during the last 2,500 years. Would this have enabled the Graeco-Roman civilization to survive? Would it have enabled the real but long-abandoned teaching of Buddha, with its lashings of caste and priest, to persist and shape to our day a course of liberal thought in India and the East?

Still untouched here is the question how religion affects the world supply of efficients and inefficients. It has been already noted that vast numbers of the inefficient result from lack of training or from still other features of the social environment. Since religion and political institutions are parts of this social environment they can affect—increase or depress—human efficiency. Both history and the current observations of any informed person clearly register effects. An advantageous effect of religion upon many people is obtained through its solace and personal encouragement. Anthropologist Marett observed that "religion is an art of self-encouragement in the face of the uncertainties of life." In itself such encouragement is unquestioned gain, since mental stability in multitudes of men requires many a supporting crutch. Even so, the best authorities do not permit us to forget that the concepts and practices of religion are themselves a prolific source of mental ills.

So, for man in general, we ask: Would equal, less or more of encouragement, mental balance and efficiency result from an alternative and enlightened view that swept from men's minds the fear of the unknown, the whimsical supernatural, and the forced gyrations of reason under threats of a hell or a degrading reincarnation? Again: How much of inefficiency directly results the world over from religious concepts such as propitiation, the Lord will provide, substitution of prayer for effort, rich rewards of contemplation, untouchability of meat, sanctity of life in insects and other disablers and killers of men, the relative unimportance of this life in relation to the next, the soul's obligation to journey to nearby and far-distant holy places, and the host of holy days on which productive effort is sinful and punishable? Is the cost to present society of these mere by-products of religious belief less than frightful?

From the array of specific cases included in the foregoing questions, this account selects a single one for further remark. During nearly three thousand years a notably large fraction of "civilized" mankind has lived in India; and most of this population has lived under the caste system for which the Brahmanical religion is responsible. Could a modern sociologist hope to sketch or invent another social arrangement that would unfailingly produce an equally and continuously high proportion of human inefficiency? With caste deciding daily tasks and type of performance, with number of performers in each caste decided by birth and death rates rather than by changing social need or by individual talents, and with other stagnating elements of a prevailing Hindu creed was built a society almost incapable of adjustment or social change—a place where progress might, with luck, have a head and a hand, but surely no legs! This is not to imply that there is no hope of a future for India, but rather to indicate that its future lies at the far end of success in the long, slow process of the transformation or the destruction of one of the world's most entrenched religions. That conclusion has been reached often by competent observers. The kind of problem faced by India, however, is quite the same as that of other nations. Only the staggering difficulties of the first few steps toward reality really distinguish the problem confronting the loose assemblage of languages, creeds and thwarted peoples in that subcontinent.

GENERAL REMARKS

Cervantes said: "Every man is as heaven made him, and sometimes is a great deal worse." In these pages a civilized man is seen to be an infirm combination of two men made in different worlds. In one of them, through age-long struggle in the hard battle of selection and survival, he became the moving crest on the wave of life. He it was whose earthly successes passed the magic point at which he could begin to plan and change his own course. In that unpredictable second world, still new and little tried, another man struggles toward maturity. Here all men are bathed with the new magic that should tame and inspire; but whatever the untamed residues in some individuals, much well-grooved behavior is exacted from all. This new self is often an unfinished one, and the fit to animalistic body may be neither good nor comfortable. Perhaps the one man made from two "sometimes is a great deal worse" than "heaven made him." However that may be, the behavior demanded by our present civilization is not obtainable from all members of the race. Both history and personal observation clearly show that the still deeper, firmer bases for man's primitive animal behavior are not negligible, that they, in addition to the many who become deficients or criminals through cultural distortion, are in fact a suspended threat to even the safety and lives of any one of us.

The lists of unethical things and opportunities that are born of civilization—of that second world of man's own creation—are surely a serious indictment of civilization if or where society fails continuously to struggle with them. Here and there over the earth these long lists include the exploitation of the labor of others, the unreasonable demands of classes and unions, and more or less of injustice in many of the transactions of government, trade, invention, communication and daily family life. It would seem that even a modest predominance of justice requires a most complete fusion of politics and ethics. In a democracy, the laggard citizen may become the stoutest foe of a manageable morality. And from this point of view it may be hard to find a people in condition to congratulate itself.

At any moment when these lines may be read it will be true that, despite bonds to society and boasts of One World, mankind still is a splintered panorama of savage and civilized, of lights and shadows, of grandeur and misery. It is the fragments of the species that cover the earth. The division, unfortunately, is still too deep and wide to enable everyone to be his brother's keeper. The sustaining hope of unlike and unequal men seems to lie in learning how to educate, in becoming utterly free to educate, and in educating endlessly. Perhaps one half of the individuals of our race are preoccupied with an enduring urge to obtain food; and for many of these, one should note with care, the lack of food arises far less from careless living brothers than from the toothless twins born with man—ignorance and inefficiency. Some aid for this array of men little prepared to lift themselves may and perhaps will come through education, technology and science. Thus it may be that the long arm of science—developed largely in Europe and America in an earlier century—will extend tomorrow to famished brothers in an India or a China. Hunger, however, is but one of an aggressive and heavy-fisted brood; disease, crime, hatred, domination and greed are widely scattered in torn humanity everywhere. All agree that these are *problems* for communities and nations; some further regard them as national or international *responsibilities*.

On this latter point many a political and ethical issue is joined. Decisions on such issues would seem to involve an appraisal of the history and nature of man—and, not least, the prevailing earthborn diversity of men. Those decisions, too, will recognize that science and technology, like other aspects of our social inheritance, create problems while solving problems. Here it becomes evident that at no time in the past has it been possible, and at no foreseeable time in the future is it likely to become possible, for a finished and closed code of morals to be written for the human race.

Civilization, with its obligatory panhuman contacts, must generate and follow rules peculiar to itself. It must do this even though it communicates also with primitive groups whose rules are rooted in another kind of world. Can the best of governments —that is, the best of united groups of diversified men—hope to exist without ethical transgressions in a world of multifarious men and cultures? Can a shattered mankind—still further separated by blinding prejudices and divisive beliefs—ever wholly escape its diversity and its sundered habitats on a scarred and uneven planet? Still above all this: Can the defensive and decisive task of endless education succeed while an esteemed and ever-present religious leadership denies the natural origin of morality, imposes alleged and discordant supernatural purposes upon men, and forbids mankind an acknowledgment of its need and duty to make its own purposes?

The above account has included a reference to certain responsibilities of science and of education to advanced cultures of the present day, and to that still higher culture that will arise only when or where supernatural purposes are supplanted by purely human purposes. Similar responsibilities certainly attach to writers—to literature. How may naturalistic thought or educational objective obtain a footing in lives before they are made attractive and familiar through literature? Who other than the writer is capable of portraying the world of things and relationships, of illuminating life and the daily scene, of focusing attention on the moral element, of guiding citizenship through its maze of responsibilities? Surely it is only a live, a brave and man-serving literature that will serve at all. A defeatist or cramped literature that avoids the full blast of our newly found, unique and magnificent human adventure is perhaps as sterile in one culture as in another. The most amazing story of the universe—and its earlier top secret—is that of our own slow but portentous arrival. The most meaningful truth concerning mankind is that associated human beings gradually create the world in which they live. Numerous writers nevertheless quaintly maintain that creation and existence, as we now know them, offer no new field or fresh incentive for a virile literature.

The boundaries of reality are indeed quite different from those of most bygone human imagination—of stretched and misty hope. Nevertheless, firm truth from biology and sociology now assures that two forms of near-immortality for man hover or hide on natural creation's crest: the same biological thread that saved for man the gains of a dim past ties also his own genes to a lineage that may end only with highest life on earth. And personal social worth puts its benign leverage upon the unborn tomorrows. Clearly, the countless blessings of our times point backward to uncounted personalities. A Lincoln or a Jefferson is gripped lightly by a grave, but firmly by a durably spreading society.

Man's critical explorations of thought and fact now vouch for a new kind of universe and for a sizable man. Both of these, some "conditioned" onlookers and writers nervously declare, seem to have shrunk. Actually, the gains to both are enormous, and the losses from each involve nothing more substantial than a halo. Though neither the whole universe nor any sheltered part in which a man lives may now be called "good," the ripening capacities of man himself—as yet never wholly free—are or may be socially useful, personally satisfying, and mostly admirable. Many a god created by man was of lesser stature.

6

The Biological Inequality of Men

It is comforting to appreciate the extent to which our behaviors
are determined by the genes, provided by our fathers and mothers,
since we can't do anything about it ourselves.
—*Richard Chace Tolman*

Nature never rhymes her children nor makes two men alike.
—*Ralph Waldo Emerson*

AT THIS POINT THE READER HIMSELF CAN BECOME AN OB-
ject of biological examination and analysis. Here light may be focused from outside us upon one or another point within ourselves.
Perhaps it is biology at its best that is thus concerned with our
intimate selves, and that here largely succeeds in picking personality to pieces. The biology of earlier chapters had the wide
sweep of time and of impersonality; here, and finally, it becomes
relatively personal. Moreover, by becoming personal while remaining related to all of the past, the science that deals with life
provides answers to some of the most searching questions that
human beings can ask about themselves.

The nature of those questions, and of their several answers,
should of course be clearer at the end than at the beginning of
this account. A few paragraphs, however, can supply at once a
bird's-eye view of the region in which the answers are to be
sought and found.

A PREVIEW

A first glance should scan the way in which the species, the
race and the individual actually share in a vast common store
of some ten to forty thousand *genes*—those agents that are present in every cell and which so largely determine the type of person that will issue from the egg. It is plain enough that the en-

tire human species as it exists today contains a greater variety of genes, and of genes combined in special ways, than does any single one of its races. It is equally clear that each race contains more genes and special combinations of genes than does any individual of that race. From this point of view one can see that an individual—a person—must develop from something progressively less than the total variety and arrangement of genes possessed by his two parents, by his clan, by his race, and by his species. An individual thus becomes a neat and shrunken pattern that almost everywhere incompletely fits the unfurled fabric that is his species; and the fit to his race is merely a little better. Moreover, an individual must develop from only one of the many possible arrangements of a reduced and limited number of genes carried by two parents. Everyone starts and ends his career with such a strait jacket of genes. Though this jacket usually permits considerable variety of outcome, it is always worn and is always persuasive.

A slight correction or explanation belongs here. Genes usually or often exist within the race in several slightly modified (mutated) forms called alleles, and where the word "gene" is herein used it would usually be more accurate to say allele. Each individual, however, will be provided with only one of the several forms (alleles) of the gene by each of his parents. He will thus be equipped with two alleles, and whether these two will be unlike or identical depends upon whether his father and mother gave him unlike or identical alleles of the same gene. When two unlike alleles are thus brought together one of them is commonly able to overcome, or cover up, the usual influence of the other. And one gene (allele) is sometimes, though apparently not always, able to affect more than one of the traits that later blossom in the man or woman.

Some outstanding facts relating to human individuality and biological inequality thus lie in the field now being previewed. First, the individual just discussed is provided with a special and unique assortment of genes that exists in no other person living or dead—except in cases of identical twins. Second, his physical and mental capacities, and a host of his defects, deformities and susceptibilities to disease, are either limited or definitely determined by the particular packet of genes from which he started his career as a fertilized egg. Third, the child's father and mother each carried some or many genes that could not be put into each sperm or each egg. These genes went instead into different sperms and different eggs, though only one sperm and one egg may be used to start any one person. Therefore, since this distribution of at least many genes to each egg and sperm is based on chance, and since unions of sperm and egg likewise occur on an unpredictable or chance basis, the personal characteristics, limitations and destiny of the individual arising from this union are based broadly and essentially on chance.

Just here, at the introduction of the word "chance," both the reader and the writer face a special obligation. The reader should now note well the point at which chance enters into this account. It enters through the *way* unlike eggs are produced by the mother, unlike sperm by the father; and also through the further and unrelated fact that the union of a particular sperm with a particular egg is an unpredictable or chance affair. However, any human egg uniting successfully with any human sperm will surely produce a human being, not an individual of another species; and if egg and sperm are of the same race the individual produced by the union will surely be of that race. It is *law* that governs these more basic matters; no unpredictability or chance relates to these points. But the genes for the personal or distinctive characteristics and endowments listed above are the precise items that whirl in the mills of chance. The writer's obligation is to make this point wholly clear, and also to note that the word "chance" requires a fuller definition, which can be given better later. Finally—and of first importance—since our talents, defects and personality traits rest upon chance, they do *not* rest on Providence—unless dice-throwing methods are used by that agency.

This preliminary view of the area included in this chapter sketches a most penetrating biological story, with one meaningful intellectual implication attached to it. This implication— the replacement of Providence by *chance* as the source of our very personal endowments and defects—has an educational value so pre-eminent that one can only marvel at the small margin by which it escapes complete elimination in the total education of everybody. The writer's experience leads him to estimate that this particular implication is being pointed out perhaps to about one or two per cent of college *graduates* in the United States. Because of religion-controlled public opinion, it is usually unmentionable at all lower levels of public and private education.

INDIVIDUAL AND RACE

Everyone has a valid, even if quite incomplete, understanding of what an individual is. But biologists and anthropologists need help, along with some of the liberty of arbitrary statement, when they attempt to define a "race" of mankind. Equal difficulties do not arise over whether mankind of the year 1950 represents more than one species; in other words, whether the few main types of surviving men should be called species rather than races. Practically all of the most competent opinion rejects the idea that these types have the rank of species. However, our look at the human family needs only to extend to the present records of fossil men to find some three to six true species of man.[1]

While the matter of species of mankind is momentarily before us, it is useful to consider the notable variation or diversity that is present within each of the now known species of man. Modern man, our own species, is by far the most variable wild species of animal with which we are acquainted. Associated with this fact, and partly in explanation of it, is the circumstance that man has a wider geographical range than any other animal species, with the possible exception of his parasites. Again, man's method of evolution, though leading to perhaps less extreme divergences than are found in some other species, has involved more intertwining or mixing of genes from all the extreme types that exist than is found in most other species. This, too, has resulted from the migratory nature of man. Over long periods of time the races of modern man have met and, in no small degree, have mixed. This has afforded unusual opportunities for highly diverse individuals to arise from the combination of genes carried in the most extreme types and races. It should be understood that the few primary races of modern man originated in a still earlier stage of prehistory, when isolation, not migration and mixing, was the rule.

When this question of the diversity of man is examined in some two to five fossil species of the human family, it is evident, first of all, that the whole of that unlikeness has not yet been

---

1 F. Weidenreich, *Apes, Giants and Man* (Chicago: University of Chicago Press, 1946). W. K. Gregory, *Evolution Emerging* (2 vols., New York: The Macmillan Co., 1951).

found and studied. It is nevertheless clear that these fossil men also were a diversified lot, probably much split into races or types. Indeed, some trained students of these ancient men also assign some of these species as wholes to a genus other than Homo, or modern man. In any case, the species of fossil men were probably rich in quite dissimilar individuals and races.

Returning now to the question of races within our own species, it is notable that the word "race" has been variously used and abused. Historically, this word was first used some two hundred years ago to designate the few and chief biological types of existing men. It would seem that confusion is the principal result of using the word "race" to mean something really different from that. To be sure, the Swedish naturalist Linnaeus (in 1738) used the word "varieties" instead of races for his four main types of man. And he tried also to use mental, or temperamental, rather than physical traits as a basis for his classification. A bit later Buffon—and also Blumenbach, the founder of anthropology—made skin color the basis of his classification, and so arrived at five races—Caucasian, Mongolian, Ethiopian, American (red), and Malayan.

In the light of present knowledge, skin color alone is too slight a matter to constitute a principal or an important biological type of man. Dark or black skin is widespread among the Hindus, who are otherwise characteristically Caucasian; and "White Indians"—with white skins but non-Caucasian—live in Panama. But associated questions arise: Are other and more significant evolutionary differences nevertheless tied to, or primitively associated with, skin color in some of these five color types? This seems probable. Are such significant evolutionary differences tied to something other than skin color? This is perhaps less probable, though the genes responsible for the so-called blood groups tell much concerning the lineage and relationships of the main types of mankind. When these and related questions are firmly answered, it will be possible to make a very short and more satisfactory list of races of modern man.

One need not doubt that human races really exist. The need is for further knowledge that will tell us which peoples best represent those races, and which peoples have resulted from the mixing of two or more true races or subraces. The human family, probably to its advantage, is already in no small degree a bulging realm of hybrids. And man now finds few questions more intricate than that concerned with his own not-too-recent ancestry.[2] Most anthropologists, utilizing color and several structural characteristics, now point to three primary races—White or Caucasoid, Negroid, and Mongoloid. They are also measuring and defining a much longer list of primary subraces—including Alpine, Mediterranean, Nordic, Negrito and Eskimoid—along with some blended hybrid groups called composite races and subraces. Knowledge and use of the "blood groups" has not led to a radically different classification of men. One authority (Weiner) arrives at the three above-named races, or at five; another authority (Boyd) finds indications of six primary races. Races are populations, originally isolated, that differ significantly in the frequencies of one or more of the genes they possess. Species differ from races in that the frequencies of a greater number of genes are usually involved; and also in the "reproductive isolation" of the population—for even when brought in contact two species rarely choose to interbreed. The number of pairs of genes that characteristically separate the Negro and White races has been estimated at from six to a few times that number. Man probably has ten to forty thousand genes.

Does race, as listed and defined above, have significance in relation to personality? Is race significant in relation to such physical items as strength, speed or endurance? Is it significant in the talents and skills of human handicrafts? Is it significant in the sharpness of one or another of the special senses? Is race of significance to things mental? Is it related to defects of development that arise along the path from egg to adult? Is race related to disease or to susceptibility to disease?

This list of questions may be left temporarily without answer. First of all, there is an obligation to point out that a difference—whether expressed in a race or in an individual—does not at once indicate whether this difference, or trait, is the result of genic difference (heredity, nature) or of some particular environment (nurture). Endowment with unlike genes may and frequently does lead to the development of biologically *unequal* men. Dissimilar physical environments (nurture), and also sociological elements of culture, may give rise to highly important *differences*

2 Earnest A. Hooten, *Up from the Ape* (New York: The Macmillan Co., 1947). L. C. Dunn and Th. Dobzhansky, *Race, Heredity and Society* (New York: Penguin Books, Inc., 1945). William C. Boyd, *Genetics and the Races of Man* (Boston: Little, Brown and Co., 1950).

in men. Races or individuals, however, may accumulate and monopolize a gene or a group of genes in a sense that neither of them can monopolize any ordinarily encountered element of environment or of culture. It is genetic inequality only that is of primary concern in this chapter.

The large extent to which a trait may be and usually is modified by a different or special environment is a large subject not discussed here. Nevertheless, present knowledge of the interrelations of nature and nurture, and especially our present ability to find or to trail a gene, frequently enables us to spot a difference that is hereditary. This task in humans, one must admit, is sometimes frustrated by limitations on the biological tests that may be performed on human beings. Fortunately, the more thoroughgoing tests already performed on our animal relatives often supply the essentials of this missing information. On the other hand, no species is comparable with man in the extent to which he has been routinely examined, treated and measured—medically and otherwise. Thus despite the relatively undeveloped state of human genetics, we have more than the beginnings of information on genetic inequality in racial and individual man.

With these considerations in mind, some answers to the above list of deferred questions can now be attempted; at least they may be approached. Does race have significance in personality? Two special and hitherto unmentioned items are involved in this question, and these far from negligible items must next be acknowledged. First, personality is surely not a single trait but a combination of very many traits—essentially it includes the entire organism. Some of these traits may therefore have an hereditary basis that is perhaps practically the same in all human beings. Second, the extent to which personality, as this is estimated by an onlooker, is a subjective rather than an objective matter is both confusing and important. To how great an extent does the "personality" of John Smith depend upon whether he is caught up by the eyes and ears of Mary Jones or by the coolie Chang Tan Wan? One may at least note that, if John is well bathed, Chang can smell him and Mary cannot.

To whatever extent personality is related to such items as stature, body build, skull and face form and proportions, skin color, hair form and distribution, and eye color, there is little question of its foundation upon genes whose frequencies vary from race to race. Indeed, these are among the principal items now used by anthropologists in the classification of races. With these remarks the writer may partly evade and longer delay an answer to a too complicated question. It is with the widely scattered components or fragments of personality that science is best prepared to deal, and items relating to these fragments will appear throughout this account.

The relation of race to such things as physical strength, speed and endurance is somewhat less complicated, although most of the data and measurements required to establish actual differences are not available. Physical strength is certainly much affected by stature, while stature in turn is largely controlled by genes. And measurements are scarcely needed to establish the fact that short and small subraces like the Negrito and Bushman have less physical strength than have much taller groups like the Nilotic Negro, the Nordic and the Apache. Racing speed is likewise affected by stature and bodily conformation, which are known to vary genetically in tribe and race. Endurance, however, is doubtless built of several or many traits. It is not clearly associated with either stature or with genes now adequately identified. This, however, is rather far from suggesting that endurance rests upon genes whose frequencies do not vary from race to race.

The extent to which race is significant in the talents and skills required in human handicrafts must be left largely unanswered. Here again not one but many separate traits are involved. For these several traits the relative roles of nature and nurture—the latter always includes the potency of practice, training and education—is unmeasured and unknown. It would be strange and unexpected, however, if the known and considerable structural and other differences of primary races failed to support some of these talents and to hinder others. And it is certain that several of these differences—structural and other—are based upon the special or characteristic share of each race in the total store of human genes.

Discussion of the extent to which the special senses differ in the different races will be limited to a single case. Several years ago it was found that a laboratory-made substance known as phenylthiocarbamide (PTC) is distinctly bitter to most people but quite tasteless to others. Family studies showed that the ability to taste this material is hereditary and due to a single gene. In America, about one person in four finds it tasteless, while three in four declare it is bitter. It was later found that chimpanzees can likewise be divided into tasters and nontasters of PTC. This substance is not found in nature, but, as is suggested in the preceding sentence, the gene that permits some men to detect it may be much older than man. Another substance was recently found that does occur widely in nature, particularly in cabbage and turnip, and it too is bitter to most people and tasteless to others. When several people were tested both with this substance and with PTC, it was found that every person reacted to the two substances in the same way—found them both either bitter or tasteless. The natural substance derived from cabbage acts as an antithyroid drug. In nature, therefore, the gene that gives ability to taste PTC is perhaps significant only as it directs the avoidance or the consumption of this other antithyroid drug.

In the present account, PTC and the antithyroid substance serve as a quite simple instance of biological inequality in men. The presence of a single gene, obtained from either parent, determines that the child will be a taster. It is further found that this gene is unequally distributed among various races of man. Whites show fewest tasters: Arabs, 63 per cent, and North Americans, 70 per cent; pure Nilotic Negroes, 96 per cent, and American Negroes, 91 per cent; pure American Indians, 94 per cent; Chinese, 94 per cent. It thus appears that both races and individuals are unequally provided with some of the sole means by which a human being may make contact with the external world and sense its properties. And this inequality of ability to learn clearly rests on chance, not on Providence.

It is easy to ask: Is race of significance to intelligence and mental qualities? But it is impossible to give an answer that is both informative and brief. In fact, what is known on and around this matter provides little more than the outline of an answer to this loose and composite question. A good grasp of that outline, however, would do more than provide a partial answer to this particular question. It would provide a sharper view of the extent to which both nature and nurture contribute to the bundle of things we call personality; it is that bundle—not really divisible into mind and body—that we are everywhere and always actually compelled to meet and study. Whether one wishes to examine the foundations of tallness or of ability to learn, those foundations will be found interlocked with a tangle of genes and with countless items of unrepeatable individual history—

growths, impressions, nutritions, movements, accidents, responses, and urges. From a unique store of genes, an unrepeatable history, and the never-ending contacts and impressions of life, each personality is built. To identify or partially to isolate the components of personality is no slight task. If some genes did not assert themselves clearly and forcefully the task would be nearly hopeless.

Much effort has been directed to the measurement of the "intelligence" of individuals, races, and various social and ethnic groups. As between individuals of the same family, community, school, or social group, it is clear enough that the tests used in such measurements have shown that some individuals are superior to others in some or several of the items that enter into intelligence. When the tests have been applied to various races and ethnic groups, the differences found were usually small. And almost always there has been a possibility or probability that the differences obtained could be assigned to environment or training, or to defects in the tests used in the measurement, and not to racial inequalities in genes. It seems that much improved tests, each directed toward one only of the several primary components of intelligence, according to Thurstone, are already being developed. Of the tests hitherto made, it is perhaps safe to say that actual racial differences averaging only five or ten per cent would not have been satisfactorily established by those tests.

Despite the difficulties and limitations that attend all such measurements, it seems clear that the *range* of many or all mental capacities is much the same in all of the three primary races and in nearly all of the subraces thus far studied. This is a very penetrating fact. It means that any and all of these ethnic groups (a very few peculiar subraces excepted) have a certain percentage of individuals—perhaps varying from a few per cent to fifty or sixty per cent—with a *higher* mental rating than has the lower fifty per cent of *every other* ethnic group. This, in turn, leads to the conclusion that the true home of mental inequality is the *individual,* not the race.

There is evidence for the existence of a gene that is essential to having or not having so important a thing as an instinct (Chapter 4); and some evidence also that this gene is being shuffled about in our own mills of chance. The parental or maternal drive, a probable progenitor of love and altruism, is the instinct here in mind. Whether this gene is distributed equally or unequally among the several races and the smaller ethnic groups, however, is quite uncertain; and even for the two sexes this point is far from clear. It is nevertheless probable that this gene is missing in some women and present in an occasional man.

Some of the above conclusions regarding mental qualities tend to agree with the firmly expressed opinion of Confucius: "Men's natures are alike; it is their habits that carry them far apart." The role assignable to habits—and to the somewhat broader and related thing known as culture—is an imposing one. Culture is of unusual significance in what we call mentality. And again —however loose and temporary this association may be—culture, habits and manners are items usually associated with ethnic group, nationality and race. Cultural differences and cultural achievements—such as attainment of a low or high state of civilization—are not based mainly on genetic differences. The explanation of such differences lies mainly in the history of the cultural experience that each group has undergone. Culture is at once the child, and often the long-term subruler or dictator, of a people, nation, ethnic group or race. So considerable is the weight of culture that it, and not the genes, probably often predominates and tips the balance for such composite and inclusive things as temperament, personality and character. If racial prejudices must persist, they may rest, with trace of justice, mainly on nothing more substantial than the alterable items of culture and manners. Yet the racial prejudice fostered chiefly in the Western world is now a threat to the security and welfare of all Western peoples.

The wholly surprising power of culture, as described above, is fully supported by the following statement by philosopher Edman:[3]

It is hard to believe, but as certain as it is incredible, that the modern professional and business man, moving freely amid the diverse contacts and complexities pictured in any casual newspaper, in a world of factories and parliaments and aeroplanes, is by nature no different from the superstitious savage hunting precarious food, living in caves, and finding every stranger an enemy.

The last two of our list of questions are concerned with the possible relation of race to defective development and to suscep-

[3] Irwin Edman, *Human Traits* (Boston: Houghton Mifflin Co., 1920).

tibility to disease. These two questions will be treated as one, and only a few of the known instances of this association will be cited.

Unlike other races, some Negroes carry a gene that, when obtained from one parent only, causes their red blood cells to assume irregular (sickle, holly-leaf) shapes when subjected to certain treatments. This condition is known as sicklemia and it occurs in about eight per cent of American Negroes. Such sickle cells have a normal span of life and apparently they do not inconvenience the individual. It seems, however, that when an individual obtains this gene from both parents, and thus has a double dose of it, these red cells have a very short span of life and they may assume the sickle or related shapes while normally circulating in the blood. This condition is known as sickle-cell anemia, and individuals showing it either die young or live only briefly beyond maturity. Nearly two (1.8) per one thousand American Negroes are born with this chronic and eventually fatal defect, and this proportion should be higher in the various African tribes, in some of whom this gene is known to be more than twice as frequent as in American Negroes.

A related though different hereditary defect of the red blood cells is found in the peoples who now or earlier lived around the Mediterranean Sea. In this disease the red cells are reduced in size and otherwise differ from normal cells; more especially they provide the blood with somewhat less than the normal amount of its oxygen-carrying coloring matter, hemoglobin. The disease is called Cooley's anemia, or better, since it exists in two quite unequal states of severity, Thalassemia minor and major. In southern Italian or Sicilian stock now living in New York, the minor type was found to occur about once in 25 births; the major type occurs about once in 2,400 births.

Another gene or group of genes affecting the quality of the blood, and which is sometimes responsible for the death of the unborn child, is associated with the Rh factor. This factor, an antigen, was first discovered in rhesus monkeys and has since been found in all races of men. When the father contributes the appropriate gene the developing embryo will produce the Rh antigen. When the mother lacks this antigen she may develop the corresponding antibody, which, in turn, may diffuse through the membranes into the blood of the embryo—sometimes with fatal results to the embryo. A first child of parents of this type is much more likely to survive than is a child from any later pregnancy. Only in very recent years has it been learned that, throughout the ages, unborn humans in notable numbers have been and now are being killed by this chance combination of genes—a chance combination regarding which the responsible parents could have no possible knowledge. About 85 per cent of North American Whites carry the gene for the Rh antigen. Considerably higher percentages are found among Indians, Chinese, and American Negroes.

Further brief reference will be made to one other series, but to one series only, of the "blood groups." These blood groups bring no defect or disadvantage to the individual, though they do become important to survival when transfusions of blood are made from one individual to another. The presence or absence of two antigens, called A and B, in your blood stream determines the blood group or type to which you belong—O, A, B or AB. People of Group O have neither of the two antigens, those of A and B have the corresponding antigens, while those of Group AB have both antigens. Three different genes provide the basis for the four blood groups. Your own blood group is thus genetically determined; it becomes established months before birth, and no known environmental circumstance—including age, disease or climate—can change it. The higher apes have four blood groups, which correspond to those in man. Human races vary greatly in the extent to which they share the four blood groups. According to Weiner, the Indians of Peru are 100 per cent O. Chinese are 30 per cent O, 25 per cent A, 35 per cent B, and 10 per cent AB. Negroes of the United States have corresponding percentages of 47, 28, 20 and 5. For North American Whites these values are 45, 41, 10 and 4. The significance of the blood groups in an interpretation of racial distinction and in human migrations has been earlier noted.

This account will be concluded with a list of some commonly known diseases that owe their presence to a gene or to genes. In most of these cases it is not known whether or to what extent the genes involved are shared unequally by the various races. Two such diseases, which relate to the use or disposition of food products, are sugar diabetes and gout. The nervous system thus suffers from a form of idiocy, from epilepsy, from ataxia, and from Huntington's chorea. The blood is thus subject to bleeder's disease, whose victims—because of the constant threat of the scratch that may prove fatal—can have neither an occupational nor an emotional life that is normal. The skin is exposed to hereditary urges resulting in rubber skin, in scaly skin, in purple birthmarks, and in the absence of sweat glands. One by one, the muscles of some boys at ages three to six may begin to fail and weaken, and in this response to a gene from the mother their life ends before their twentieth year. The skeleton may be predisposed to fragile bones, to legs and arms of disproportionate length, and to shortened fingers and toes. Much of deaf-mutism is hereditary. The eyes are thus subject to a three-way attack—by degeneration of the choroid and retina, by deformation of the lens, and by color blindness. And the body generally is gene-propelled to albinism, to dwarfism, and to allergy.

Races certainly share some diseases unequally. Negro children are relatively immune to scarlet fever; Whites are highly susceptible to yellow fever.

The foregoing excursion has scanned parts of the evidence for the definite fact that the total store of human genes is shared unequally by the various races and subraces of men. Also that individuals within each race or subrace share very unequally and often most tragically in that common store of genes. Again, both races and individuals share not only the desirable genes but also those that lead—effectively and practically—to crippling defect, disease and premature death. And distribution of these fateful genes that point to triumph or to tragedy is based on chance. Every human being is a monument to the sway and tides of chance.

CONSTITUTION AND PERSONALITY

There are two approaches to the problem of personality and the individual. The approach through genetics has been followed in the preceding pages. It remains to examine the other approach, which is through some segments of anatomy, psychology and physiology. It must not be inferred that these latter sciences are excluded from the preceding account, nor that genetics is excluded from the point of view that will next be presented. The facts are otherwise.

Anatomic and Psychologic Aspects

It is easy to understand what we are doing when we collect information on the "constitution" of each of a group of living men. To be sure, these men should wear no clothes while their physical traits are being observed and classified; and it is best to observe them with care and at some length—at work, relaxed, and at play—in order to assess the items of their temperament. But a child of ten years would have little difficulty in making a very crude assessment of constitution in each one of a small group of men. This child would label one as short, fat—also jovial and lazy; another as thin, big-fisted—also obstinate and courageous. Trained specialists can now set down many items for the structure and the temperament, and even for the peculiarities of the internal chemistry, of an individual. To study the pattern of these several items in an individual is to study constitution. The thing called personality is an actually achieved pattern. It is ordinarily understood and approximately described in terms of items of difference from other personalities.

Reasons for constitutional studies have varied among such aims as gaining a better knowledge of man as an animal species, the selection of the right man for the right job, and insight into the relation of constitution to disease. This latter relation was already partly recognized in the earliest medicine of India, Persia and Greece. And medicine—psychiatry in particular, at this moment—has provided much of the incentive to more and better investigation of constitutional factors. Finally, throughout the ages those disciplines which our universities call the humanities have recorded human inclinations, frailties, achievements and traits, thus mirroring the sort of creature that man was or has now become. History, literature and art all display the fascinating variety of the special qualities and disposition of particular persons and peoples.

There is now no generally accepted way of putting together, of classifying, or of describing the physical and temperamental items of the well-measured individual. Some of the classifications now in use relate especially to psychiatry or medicine rather than to human biology and psychology. The present brief consideration of this central topic will be based on the method of classification introduced (and still being developed) by Sheldon and his associates.[4] An outline of that method is as follows:

Group 1. A relative predominance of structure associated with digestion and assimilation—conserve energy and easily put on fat (called *endomorphy*). The temperamental or behavioral traits associated here are relaxation, conviviality, and gluttony for food, for company, and for affection or social support (called *visceratonia*).

Group 2. A relative predominance of bone, muscle, and connective tissues—tend toward massive strength and muscular development (called *mesomorphy*). The associated behavioral traits are bodily assertiveness and desire for muscular activity—dominating, love of action and power (called *somatotonia*).

Group 3. A relative predominance of the skin and its appendages, which include the nervous system—small mass in relation to surface, height large in relation to weight (called *ectomorphy*). The associated temperamental traits are inhibition of the traits under Groups 1 and 2; also overconsciousness, love of privacy, and avoidance of overstimulation (called *cerebrotonia*).

Of course, the descriptions provided above are sketchy and incomplete. They are meant to indicate with reasonable clearness the three major components of constitution within each of which several special structural and behavioral items naturally belong. It is not to be supposed that one has much chance of ever seeing a person built *wholly* of components of, for example, Group 1. Heredity is involved here; and in compounding an individual the versatile artilleries of chance will rarely or never use only shells of a single type. Larger or smaller amounts of the elements from all of the three groups will be found in every individual—often even within the head, torso or legs of an individual—and the extent to which this is true can be measured. This measurement is done by using for each group a scale, represented by the numbers 1 to 7. With only three numbers set in the following order, 5-5-1, the constitution of an individual can be rather fully described; and thereafter it is easy to find that many people of various races throughout the world fit this same description. In this grouping of numbers, the first figure 5 means

---

4 W. H. Sheldon, S. S. Stevens, and W. B. Tucker, *The Varieties of Human Physique* (1940). *The Varieties of Temperament* (1942). W. H. Sheldon, E. M. Hartl, and E. McDermott, *Varieties of Delinquent Youth* (1949). All published by Harper and Brothers, New York.

that the qualities of Group 1 are present in slightly more than average amount, since 4 is the average on a scale of 7. The second figure 5 has a similar meaning in regard to the qualities of Group 2. The final figure 1 indicates the lowest recordable amount of the qualities of Group 3. "These are squat, heavy organisms, built close to the ground. They don't put their necks out far."

Objective measurements can be made, of course, on any and every type of man. When these measurements, on both structure and temperament, have been classified and described according to this method, they bring into clear view many of the most remarkable facts now known about ourselves—about our constitution and personalities. Within each group it is found that structure and behavior are highly correlated. In this way the unity of the organism—the inseparability of body and mind—which elsewhere has been frequently observed, is again observed in man. On the other hand, components of personality that are seriously dissimilar and incongruous—as well as the more usual congruous components—are sometimes or often forced to integrate themselves into personalities. So marvelous are the powers of purely biological integration that both grit and grain are used to make this "bread of life"—at least, to make a two-legged type that can eat and walk and feel. But some of these products prove to be impossibilities at the social level; they swarm inside and outside our charitable and correctional institutions and our hospitals. Finally, these penetrating studies in constitutional psychology, though made with aims partly similar to those of Freud, serve to question or to replace methods used by Freud. Freud's method seeks the keys to mental disorder by trying to get a look at the subconscious self. For these same keys the Sheldon school looks at the entire organism, and directly to measurable items of body structure and temperament.

Quite other studies have shown that at chemical and physiological levels of cellular activity numerous gene-controlled differences and inequalities exist.[5] These may take such forms as addiction to drugs; absence of an enzyme or hormone; increased or decreased vitamin requirement of the cells; and perhaps a changed sensitivity of some tissues of the body to one or another

---

[5] Roger J. Williams, *Human Frontiers* (New York: Harcourt, Brace and Co., 1946). J. B. S. Haldane, *The Inequality of Man* (New York: Famous Books, Inc., 1938).

of its own hormones. In some strains of rats and mice it is now known that an appetite for alcohol is hereditary, and again that this appetite can be abolished by some unusual supplements to their diet. Here susceptible strains are ones in which parts of the cellular chemistry are so deranged as to require in their food extraordinary amounts of one or another vitamin, mineral or amino acid.

It was said above that some personalities that are passably welded together biologically nevertheless fail at the social level. Notable members of this or of a related group are those who lead an approximately normal social life for several years and thereafter develop a "split personality," or schizophrenia. Here the years of near-normality are those long years that humans spend in attaining maturity. And those same years manage to register much experience in forms commonly spoken of as conscious and subconscious. The person who becomes schizophrenic, however, is one who—among other things—does not readily or adequately accomplish maturity, and who ultimately also fails to deal suitably with his registered experience. His world too much resembles the dream world. It is clear that an integration, a suitable merging of parts, has failed at one or another level in the development of these individuals. The role of heredity in this failure has been very hard to establish; possibly that has now been done (Kallman). It is also probable that the functions of the suprarenal glands are deranged in this remarkable disorder. The psychiatrist's art can do much toward restoring normal personality to some but not to all of these victims. Considered as a disease, schizophrenia provides one fourth of all hospitalized patients in the United States, and of course in no country do all such cases enter a hospital. After prolonged study of this elusive constitutional disorder, in which one or another item in the welding process fails, physiologist Hoskins[6] has said: "It is doubtful that any mishap of existence causes in total so much human distress or entails so much unproductive expense."

Constitution and personality cannot be well understood without rather more than a passing glance at the system of involuntary (autonomic) nerves that presides, in belly and breast, over the main business of living and of procreation. All that occurs there, except for an occasional command of will to change the

[6] R. G. Hoskins, *The Biology of Schizophrenia* (New York: W. W. Norton and Co., Inc., 1946).

respiratory movements, is under the speedy and precise control of these nerves. Many of the keys to personality and to human inequalities are hidden in our deep internals—that very central part of our bodies that we may never have opportunity to see.

The ever-responsive glands of the body are touched and fingered by these autonomic nerves, as are also the deep-lying muscles within the body cavity and within the walls of the blood vessels everywhere. Even the muscle cells ruled by these nerves are wholly unlike other muscle cells; they are smooth or unstriped, while those large cells which are subject to the will are abundantly striped. These nerves communicate with a very special part of the brain—an old and primitive part bearing the names hypothalamus and thalamus. To and from this old brain flow the nerve impulses that regulate the basic and busy traffic of life; and usually we are quite unconscious of the tangles and the regulation of that endless traffic. The common and total effect of all these activities and precise adjustments is that we feel a state of well-being or that all is not well. Sometimes, however, the messages reaching the old brain are sufficiently intense to overflow and be carried somewhat indirectly by other nerve fibers to the newer brain, the cerebrum. There, in the cerebrum, such an overflow permits consciousness to intrude itself into the traffic problems, where it affects them and is affected by them. The importance of such overflows, with the resulting involvement of consciousness, lies in the fact that all this has an effect on personality and may even lead to self-imposed disease.

This item requires a word of further comment. Each individual has his own personal endowment of autonomic nerves; and this must have a genetic basis while exhibiting its many grades of inequality. At one extreme there is stability, the ordinary body functions proceeding calmly and automatically even when all sorts of exciting things are happening in the intellectual, emotional and physical spheres of the person's life. At the other extreme is found such a lability, or fluid reactibility, that autonomic functions are chaotically disturbed by every external pressure to which the individual is exposed. In the words of Draper,[7] "The functional quality of the autonomic nervous system, the speed of its reaction to stimulation, and its stamina in the face of long-continued strain are probably the basic factors in deter-

[7] George Draper, S. W. Dupertuis, and J. L. Caughy, *Human Constitution in Clinical Medicine* (New York: Paul B. Hoeber, 1944).

mining the physiological character of the individual." When an individual is described as having "guts," a really good autonomic nervous system is implied.

Within this area of nerve and brain two other items are worthy of notice. First, a word in regard to a home for the emotions in the old or primitive brain; and thereafter a note on the degree of fusion of older and newer parts of the brain. It seems to be fairly established (Cannon) that the emotions and instincts are rooted in the old or primitive brain—the hypothalamus and thalamus. And it has been noted above that these elements of the brain are closely tied to the autonomic nervous system. Evidently, therefore, the emotions are originated or modified by the impulses arriving from the deep-lying (visceral) organs, and those organs may in turn be affected directly by outgoing impulses, which the emotions themselves produce. It is on this basis that we may understand the fact that worry may lead to ulcer of the stomach, and that fear or rage will release adrenalin from suprarenal glands by way of preparing the person for fight or flight. And likewise we may understand that things like alcoholism and peptic ulcer are sometimes exorcised or cured by an unexpected emotional experience, such as "getting religion" or inheriting a fortune. The main point of present interest, however, centers in the meaningful fact that, in man, reason and emotion tend to dwell in separate segments of the brain. In what ways, or to what extent, are these two segments united? What are the consequences to men of an imperfect union of these two parts?

Full answers to these questions would relate to limitations on human performance, to the illumination of human behavior in all its diversity, and to many items of human inequality. Only partial answers and a further citation of some related facts belong here. Hitherto unnoted is the certainty that the newer brain, the cerebrum, is the seat of reason and the area of the more acute consciousness. This newer part has been added to the several parts of the earliest reptilian brain; and though a passable integration of new with all old parts of the brain has been accomplished—albeit by roundabout rather than by direct nerve-fiber connections in the case of the thalamus—human behavior can be and sometimes is dictated by the emotions, not by reason. In this fact lie enduring dangers to the race. "The voice of the intellect is soft," say Freud and others.

Nevertheless, it has recently been shown that some otherwise unassociated nerve fibers do pass directly from the thalamus to the cerebral cortex. And the discoverer (Walker) of these fibers interprets them as quite recent arrivals, and as perhaps indicative of nature's continuing evolutionary attempt to achieve in man a better association (integration) of emotion and reason. If this interpretation is correct, it provides an alluring—even if painfully distant—promise to the race. Moreover, some implications of this interpretation are intriguing. A recent origin, and a continuing increment, of these slender threads that expand the rule of reason in existing man provide a probability that our own species has thus advanced genetically over all earlier species of man. And they likewise suggest that the very unequal packets of genes carried by men now alive may include well-concealed inequalities of this hidden wave that slowly drifts toward reason.

HORMONAL AND SEXUAL ASPECTS

The hormones have much to contribute to a complete story of constitution and personality. No complete story, however, is being told here. The genetic inequalities of man—our central subject—are observed most readily either at their starting point, the genes, or near their terminal point, the fairly established traits of the going organism. To a large extent, the hormones play their parts at points between. The potent and wandering molecules of hormone that are released from certain special cells are mates and cousins to other molecules released in all cells under genic action—all involved in the orderly building and maintenance of the organism. However, in addition to or as a very special part of this widely shared task, these wandering molecules uniquely fraternize with the nervous system in making the many parts of the body into one, in the integration of the organism. It is perhaps largely because of the way our knowledge has developed that a few constitutional factors can now be best approached in terms of these intermediate agents, the hormones. The genes involved are in a dimmer background; the hormones earlier came within our view.

There is no uncertainty concerning the large extent to which most aspects of constitution, personality and behavior[8] are affected by absence or by excess of certain hormones. Even the circumstance that a male fetus must develop within a female body,

[8] Frank A. Beach, *Hormones and Behavior* (New York: Paul B. Hoeber, 1948).

and is thus subjected temporarily to the female sex hormone of its mother's blood, sometimes produces adverse effects in that male child. The female embryo runs no similar risk. The environment in general, as well as the genes, partly determines the amount of hormone that is actually produced. Absence of the thyroid hormone, thyroxin, from the developing embryo and child means that it will at best become a dwarf with an undeveloped mind (cretin). Another type of dwarf, with normal body proportions and mentality (midget), results from a deficiency of hormone produced by the pituitary gland. And an overproduction of hormone by this same gland in more mature years may lead to overgrowth and even to a deforming giantism. Many of these cases relate to abnormality, and their bearing on studies of normal human constitution is not too clear at the present moment.

On the other hand, it is clear enough that in the case of a boy the general growth of the body is promoted by one or more growth hormones of the pituitary, by thyroid hormone, and by male sex hormone (androgen). In a girl these hormones act similarly except that the androgen is largely replaced by female sex hormone (estrogen). It is possible that the special ability of estrogen to stop growth at the ends of long bones is one of the reasons for a shorter stature in sisters than in their brothers; perhaps another reason is the smaller amounts of androgen produced by girls. It is likewise clear that in boys the androgens assist the usual development of male genital organs, beard and change of voice; while in girls the estrogens similarly direct development of female genitals, uterus and breasts. There is fair evidence from humans for a gene that sometimes leads to incomplete sexuality (pseudohermaphroditism); for another that wholly prevents the development of an ovary in females; and for still another gene or genes controlling the time of onset of puberty.

The incomplete evidence for the three conditions last cited is greatly strengthened—and, in addition, proofs are provided for the existence in animals of a genetic basis for a wide variety of components of constitution—by tests that may be very briefly sketched. In a study conducted during a period of twenty-four years by the writer,[9] a strain or race of pigeons was permanently

9 Oscar Riddle, *Endocrines and Constitution in Doves and Pigeons* (Washington, D. C.: Publications of the Carnegie Institution of Washington, No. 572, 1947).

established in which both true and pseudo hermaphroditism occur in every generation among the birds that should become males; some of these individuals produce both ova and sperm. Likewise, through selection, a race of doves was formed in which the females become sexually mature—lay their first eggs— at six months, and another race that matures only at eight or nine months. The constitutional types thus established also included races with unusually large, others with unusually small, testes; races with large, others with small, thyroids; races with long, others with short, intestines; races whose tissues showed either a low or a high responsiveness to the hormone prolactin; and races with low, others with high, rates of heat production. This study was started with a careful selection of thirty-seven pairs of birds, for each of which the pedigree was known and, in some cases, certain "constitutional traits" were either known or suspected, and selection was continued during three generations. These birds were chosen from a large bird colony known to be mongrelized comparably with the human population of a large American city. It is entirely probable that exactly similar tests made on humans—granting the centuries and the many impractical items of the test—would lead to similar results. Certainly these briefly described tests on animals prove that in them genes really exist for all of the above-cited elements of constitutional inequality.

A concluding word on hormones should note that in extreme cases these agents may express themselves frankly in at least three large areas of constitutional study—the structural, the psychological, and the chemical. We meet, or perhaps there comes into a clinic, a woman whose arm span is greater than her height. That fact alone, as she enters the door, may be enough to tell the knowing student or the physician that here is a woman —and a body type called hypo-ovarian—in whom the ovary has for long been underdoing its work; in whom the gag reflex is diminished and the tendon reflexes are increased; and whose glands are now forming, and whose urine is now excreting, the sex-hormone products in reduced amounts.

Sex is an item of constitution so familiar that it would seem to need mention but not description. Nevertheless, it should be set in the same frame with other elements of constitution, and a further word should note its occasional incomplete development. Sex, too, is based on genes, but in an exceptional way. A certain group or block of genes occur singly in human males though they are paired in females. This results from the circumstance that two kinds of sperms—one with and another without this block of genes—are formed in equal numbers. When a sperm deprived of this group of genes unites with any egg, a male child results; the other type of sperm adds this block of genes to that present in every egg, and thus yields pairs of all genes—and a female child. It is wholly a matter of chance which type of sperm will unite with the rare egg that may be fertilized, just as it is chance that decides for all other genes which of four grandparents will be represented in a particular egg. Finally, sex, rather more than most elements of constitution, tends to get itself expressed very widely in the organism. Sex is a nearly wholesale endowment. There is little wonder that much of the subhuman world readily recognizes it.

Just as diversity of races is associated with some diversity of diseases, so the genes that decided your sex also apportioned you a larger or smaller share in certain diseases. Gout and the duodenal type of peptic ulcer belong mostly to males; while gallstones, osteomalacia and hyperthyroidism predominate in females. Even females of various types share unequally in some diseases. Dreaded pernicious anemia is frequent in florid, flabby blondes with rather large faces and blue eyes, but is rarely found in brunettes.

A unique thing about the abstract quality called sex is that it is observed and known only in terms of differences between two unlike types. In an all-male or an all-female world, our concept of sex could never arise. But once the idea and fact of sex difference was recognized by man, it became a dividing line between the two chief groups into which all humans are classified by themselves and by others. And here, indeed, there is no doubt of human biological inequality. To this contrast of bodily form and function, however, nature herself seems content to assign mainly unlike and unequal shares in reproduction and in a few diseases; meanwhile, and in addition, humans themselves have assigned to sex almost everything on earth and in heaven. In the numerous cultures of our own day—barbaric and civilized— there is scarcely a category of performance that is not apportioned on a basis of sex. To round out this absurdity, the apportionments of one culture contradict those of another.

Just as the idea of, and the wish for, progress is either unknown or quite new among men, a declared and purposeful search for sound and unexplored ways of developing and utilizing the special qualities of each of the two sexes—and especially those of the female sex—is yet to be attempted in any culture anywhere. It is true, of course, that bars and taboos are being lifted in many countries; but tradition, masculine vanity, law, custom and religion all compete for rule within this exceedingly broad area. The divisive aspect of sex usually still has right of way over any concerted effort to exploit the rich possibilities of what can be made of two rather similar genetic endowments that are so flavored as to yield two (often) superb and complementary personalities.

Let it be understood that the individual's huge endowment of genes, apart from the minute fragment that tips the scale for sexuality, does not under all climes and conditions rest on clear favor to either sex—ever-present chance is taking good care of that. And the traits that may develop from those genic impulses usually share nature's urge to full maturity and existence in both sexes. Some differences in such things as interests, physical strength, and averages for skills and reaction time—along with temperamental quirks in females that to males may look like those of an alien species—do indeed arise during early or later life. These differences are probably closely related to the dissimilar production of hormones in the two sexes. The bravery of a girl, however, is truly human bravery, whatever the tribe may call it. For the two sexes the overlap of traits, except the traits involved in sex itself, is almost complete and perfect; while for two distinct human races this overlap of traits, as noted above, is less complete, though even here the observed lack of fit seems to be of no great biological significance.

To those who have studied sexuality, that endowment has taken on hitherto unknown qualities during the past forty years. For much of a lifetime the writer himself has shared the privilege of joining his efforts to those of many others in a study of this intensely biological characteristic. It is now clear that, at or near its basis, sex is a rather fluid and a definitely reversible thing. But this innate reversibility of sex does not mean that it is everywhere practicable to change or reverse the sex—at any rate, not yet. And the group of mammals, including man, is precisely an area where this reversibility is not practical. The more intimate, intracellular and functional nature of sexuality—through or by which the sex genes do their work and on which the fluidity and reversibility of sex is probably based—is a large and incompletely explored subject which lies beyond the range of this account. But this brief reference to fluid and reversible aspects of sexuality, together with the enlightening fact that a basis for both male and female structures is laid down in all early human embryos, provides a background against which one should next scan a group of personality and sex problems of mankind.

In all human societies there are some boys who do not develop into "normal" males, and perhaps as many other boys who fear that they do not or may not measure up to normal maleness. Likewise among girls there are those who do not develop into "normal" females, and another number who have an unwarranted concern for their femininity. Among adults, both men and women, there are some who well know they have one or another sex abnormality or inadequacy; and there are some committed to homosexual practices or to countersex components of temperament, desire, voice or behavior that do much to color and complicate the whole business of living. Parts of this composite picture relate to biological inequality; parts of it relate to constitution; all of it relates to personality. A first remark about this area might well note that both medicine and education can now be of service in several of these cases. It is next notable that in any society the current social ideal of male and female usually has a great, and often a quite unjustifiable, influence on an individual's thinking about his or her own sexual deficiencies. This whimsical social ideal may be at variance with biological fact, and an awareness of this fiction may be all that is needed to throw off a fear or an illusion of inferiority. Among whiskerless tribes a whisker is not an accepted part of maleness; and perhaps the maleness of the fat man is better viewed in contrast to the femaleness of a fat woman than in comparison with the maleness of an athlete. A further word on the possible relations of constitution and sexuality may be helpful in presenting parts of a biological view of both sexual normality and abnormality.

If the bases of sexuality—viewed solely as an endowment of genes—can be fitted to male and female bodies as different as those of monkey, orang and chimp, these bases should also rather easily fit into the various human races and the several constitutional types—endomorph, mesomorph, ectomorph—and the many blends of these. To the truly genetic bases of maleness and femaleness there may have been no single addition or loss between early monkey and modern man. It is probable that occasionally a gene is displaced from the small group of genes that normally tips the balance for sex, and that this displacement provides the basis for some very rare cases of genetically based sex abnormality. In vastly greater numbers of cases, however, other sources probably account for both an unequal expression of a type of sexuality in different body regions and for sex deviations in the temperament of an individual. For example, the genetic trend to maleness in the skin and breast regions might not usually unfold itself as favorably in an endomorphic-viscerotonic as in a mesomorphic-somatotonic type of man. In other words, the prevailing local endowment or total constitution in a particular body region may make a difference in the degree to which a type of sexuality is expressed in that region. Sex is always a part only of the total constitution in any body region, and sex expression is a highly modifiable trait. Masculinity, in both sexes, is increased by hormones formed in adrenal tumors. Again, a lush variety of activities and introspections—fetal conditions, parental errors, early and protracted tribal comment, sex ideals, customs, taboos, nutrition and adventitious physical contacts—may adequately explain the great majority of the temperamental sex inversions and failures that actually occur. As already noted, environment is so potent in the mental and temperamental area that the usual role of genes in it is especially difficult to measure and declare. Mentionable, too, is the generally overlooked fact that at the end of the last century psychologists Krafft-Ebing and Freud studied and described cases of sex reversal of temperament long before biology was prepared to admit sex reversal in the bodily structures of vertebrate animals.

Of these temperamental sex inversions of mankind one further notes that certain groups within any current crop of victims may hopefully appeal to expensive psychiatry. Always, however, the fuller solution lies in prevention. Thus this huge and enduring task lies largely in the lap of education. The appropriate education can become fully effective, however, only when it is already a part of the training of parents and most other adults. And if this formidable educative task can be even started without teachers trained in, and permitted to teach, the best of evolutionary biology and some fragments of psychology, the race possesses some marvelous endowment that is entirely overlooked in this chapter and quite unknown to this writer.

THE IMPLICATIONS OF BIOLOGICAL INEQUALITY

The preceding account records several of the now known facts of human inequality. Facts do not easily lend themselves to dissent and disagreement. In contrast, however, any remarks on the implications of these facts may easily meet with resistance. Certainly a human factor, a thing that is not a part of the biological facts presented above, enters into any discussion of the implications of these facts. The present writer's outlook—his conditioning, his inhibitions, his prejudice, his frame of reference—may be wrong, timid or warped. In that case, what does the reader find to be implicit in the foregoing facts? Perhaps what the reader himself has found in them, and already accepted as usable in thought and practice, is all that matters. But the reader is not in position to set down his thought here, nor even to indicate whether his thought on these matters is satisfactory to him. To fail to attempt here some guidance to that casual reader who has patiently read this account, but is nevertheless uncertain of its several implications, seems a crime hardly less than treason to a partner in this adventure. Certainly, such a failure would represent an unpardonable reluctance to discuss, activate and disseminate a science—leaving it floating and free of useful application among men. The author therefore—hoping that his guards are up—readily takes this further look into and around the facts of human inequality.

One conclusion that follows an examination of the way genes are distributed to every individual has been indicated near the beginning of this chapter. The genes that so largely predispose or determine our small constellations of talents, defects, diseases, terms of life and personality traits are distributed at random— in such a way that it evidently is not Providence but chance that is involved in their distribution. The word "chance," if undefined, may not best convey the intended meaning to many persons. "Random" and "unpredictable" are words that help to give that meaning. Certainly Priestley's redefinition of chance as "the observation of events arising from unknown causes" may be clear enough to an electron-conscious few. But chance thus defined will not carry the intended meaning to most readers. To them the "how" of gene distribution—or, rather, of the distribution of certain particular genes—is far better pictured by the full, honest and thought-inviting statement that the events with which we are here concerned are unpredictable and random, and that the human mind finds it difficult or impossible to imagine a way of shuffling genes that would better exclude Providence or God from a share in it.

Though familiar human inequalities thus rest on chance, an incautious reader may not project this statement into a declaration that the human species likewise and equally rests on chance. The agents of species formation, of evolution, were elsewhere recorded as mutation, recombination, hybridization, genetic drift and natural selection—the last named and mutation being the key items. Whatever the part played by chance within a particular agent—and largely involved is chance plus selection—these agents, laws or principles direct the formation of species.

The true diversity of humankind, we now know, includes the diversity expressed in living men plus that much more imposing diversity that results from the inclusion of those several earlier and now extinct species (or genera?) of men. Within this total of human diversity there is, unquestionably, unfamiliar and staggering human inequality; this must spread from highest man to another species, which is or was more like ape than like earliest man. From this perspective one may wonder whether our account of the differences between existing races of men, as this was presented early in this chapter, does not somehow minimize those differences. On this point, and at this moment, the several now existing races should congratulate themselves on having presented so much evidence that their inequalities, though real, are neither shameful nor relatively very significant. For, in any case, the area of tragic inequality is the population within each race. The many households of any one race, through their sons and daughters, roll up the really impressive score of human diversity—inequality of interest, capacity and defect. And here, too, belong the more extreme inequalities associated with sex.

To the various peoples of this planet the many consequences of this human variation are, in some respects, similar. And the fact of human inequalities is, of course, partly or dimly recognized by all normal people. But it is extremely important that the reader now become fully aware of the very narrow and rather fruitless limits of that recognition. This general public, which speaks lightly of all inequality as "difference," is unaware of its sources and extent and innocent of most of its far-reaching implications. All this is truly reminiscent of the pre-Darwin day. Then, too, men looked at apes and dimly recognized many resemblances to man; but wholly and fatally missing from their understanding were the implications of those similarities—the firm link of ancestry, kinship and descent, which, once fully acknowledged, transformed the world of thought. From the whole field of biological science, only the principle of organic evolution has greater social significance than the principle of the biological inequality of men. No principle whatever has equal personal significance. This principle is basic in sociology, essential in psychology and ethics, and abrasive and subtractive in religion. In all these areas our momentary social and political progress depends partly upon widespread and full understanding and usage of this principle. Yet everywhere this supremely meaningful biological principle is almost avoided, and is skirted or touched with one finger in the halls of education. Is this restriction at all associated with the bearings of this and related biological fact on religion? Chapter 8 supplies the positive answer to this question. And the implications of this affirmative answer doubtless merit whatever action that answer may arouse in any reader seriously concerned with the welfare and future of the race.

While still fully conscious of the extreme range of human diversity, necessarily including a long history, one may digress for a moment—as one must in this book of dual purpose—to an item near absurdity. The so-called revelations of the several religions claim a high and soul-saving importance, and their proponents usually speak and think of them as given to *all* men. The perspective supplied by these chapters clearly outlines the fact that all of the now extant revelations were entirely withheld from all but *one* species of man! And from that one remaining species they were withheld during much more than 90 per cent of its history to date. Words that came from above fell chiefly upon the dust of men who, ages ago, had made the long and unaided fight. The few who thus late were shown a road to heaven had already built a wheel that rolled quite well on earth.

Once it becomes clear that genes bring into human society large amounts of mental disorder, physical abnormality, inefficiency and disease, many extremely important questions arise. Do the carriers of these genes reproduce themselves to a usual or to an unusual extent? How does substitution of civilization for natural selection affect the ratio of sound to unsound genes in the race's store of genes? Answers to these questions disclose impressive threats to future man that were briefly acknowledged in Chapter 5.

Within the United States there is a recognizable barrier, even though it be called a slight one, to getting the general public to give serious thought to human inequality. Certainly an appreciable amount of resistance to this concept is fostered by schoolbook familiarity with words of our Declaration of Independence —"that all men are created equal"—but by unfamiliarity with the intended sense and usage of these words. If an acceptance of reality provides any advantage to social survival of a people, it would appear that education may need to place unusual stress on this principle of biological inequality within the borders of the United States. This involves no reversal of early American aims and ideals.

Few human laws or institutions can be expected to persist successfully and indefinitely against biological reality. And men of good will wish that democracy may not be merely obtained but that it may also endure. The essence of democracy would seem to lie in continuous and effective efforts to increase the opportunities and capacities of all men everywhere. Those human capacities can be neither known nor fully nurtured apart from a recognition of the principle of human inequality; and sustained efforts directed to that end can be best supported by the truth. It is the Soviet Union that now denies and rejects outright the principle of genetic inequality of men. Democracies, if sufficiently educated and fortunate, may take truth as they find it; apparently they alone may hope to build their futures around it.

It is diversely unequal men and groups of men—as partly described in the preceding pages—that biology delivers to the attention, care and direction of sociology, economics, education, law and government. Within these latter areas—though not in biology—there is firsthand acquaintance with what men become as members of society and with the rules by which a human society passably maintains itself. If a biologist crosses these thresholds his first purpose should be to insist that biological man does not there become the forgotten man. In the often regrettable forms already pointed out, unequal men are in all of the woodpiles. As noted immediately above, one great government—perhaps one almost equally well called a religion—specifically denies the only type of man or race that biology knows. A doctrine of complete equality of all races or individuals, wherever it appears, is dogma rather than science—a yoke rather than a crutch.

Childhood conditioning and the chance-born pool of genes propel a lifetime to strengths and failures. Philosopher George Santayana was born and passed his early years inside a thick-walled city in religion-haunted Spain. His special kit of genes could build in him the keenest grasp of thought and the finest art of expression. But the total of those pressuring beginnings of youth and personality failed to form a man who could either like the modern world or find sympathy for the fate of contemporary men.

The many inequalities of men clearly mean—as this was pointed out by Davenport when genetic science was only ten years old—that there can be no *impersonal* science or art of medicine, hygiene or education, nor of any similar agency that deals with human beings. Only the genic jacket carried by the patient, plus the mental hazards he has encountered, will sometimes clarify the nature and the treatment of his disease. However we approach or are approached, personality clings to us as does our shadow. There are always questions of the practicability of obtaining this highly personal treatment in medicine and in education, but there is never a doubt about the need. In the handling of the criminal and the insane, civilized peoples already partially recognize this advance in our enlightenment. With respect to the extreme criminal, however, there is something further to remark. The worst of human beings may or may not be vastly worse than jackal or wolf. Their access to human companionship, however, certainly renders these human derelicts infinitely more harmful and dangerous to man—they can and do maliciously kill, torture and plunder. When society's disposition of these extreme derelicts is dictated by one or another religious sediment or sentiment, rather than by values born to clean biological fact and social health, the race is left with another wound to lick.

Without a further and specific statement about the things that may now or later be done to overcome the frequently unwelcome pressures of genes, the preceding picture is darker than the reality it should faithfully represent. Most, though not all, human traits are known to be somewhat plastic; some are extremely so. And the arsenals of organized knowledge have begun to bulge with instruments for counterpressures that transform disease to health; sometimes even abnormality to normality. The gene for diabetes is already partly countered with insulin; deforming mongolism of the embryo may be normalized by feeding the mother with vitamin A; pernicious anemia is successfully treated with vitamin B12; and the gene-slit harelip is glossed by the surgeon's skill. Again, as earlier noted, the exhibition of many capacities hinges merely on circumstance; thus culture fully bridges the gap between a savage and the professional man who happens to be your neighbor. To how many things will all these long arms of help extend in a later day? Surely the hand and brain of man are shaping tools that man can trust. And here as elsewhere it is not Providence but man himself that acknowledges nature and softens its many blows. That person who now only lamely trusts expanding knowledge and unaided human performance has indeed been robbed.

There remains for further remark only the forming, growing, inconstant thing called individuality or personality—a thread that runs through all of the preceding account. For this further look at personality it is more convenient to assign a name and dates to a fairly typical and entirely possible individual history. Thus we may see John Smith on a June day in 1950. He then seems tangible and definite enough as a person and personality. But the actually blurred and protean nature of the total personality of John Smith—in terms of 1870 or of 1900—will be grasped only when the simplest facts about him at these different dates are brought to mind. Early in 1870, he passed from egg cell to fishlike embryo, and in that autumn was a newborn infant—certainly with numerous elements of body and personality as yet unformed, and just as certainly he was then capable of assuming a variety of alternative physical and mental traits. In 1900, many of those traits had blossomed; an outstanding person was actively meeting the world. Here was a sensitive student easily, eagerly and rapidly adding to our knowledge of a difficult subject. Seeing him in 1950, we can find only a man who has already spent some years, in rebellion and violence, in a mental hospital. It seems clear enough that nothing of human inequality hitherto discussed in this chapter at all approaches the personal and private inequality that—from egg to end of life—is progressively unfolded by each of us, while acquiring and pursuing a human existence.

Personality is unrivaled in the extent to which it is a new and forming thing that is nevertheless incapable of remaining the same thing. In one of its ranges, only a step or an instant separates the comic from the tragic. Its beginnings—its foundations—rest on elements that may not be reckoned extremely simple, but they are certainly extremely unlike a personality. These beginnings, we have seen, trace to *two* quite small and quite independent bits of living matter, each bit—egg and sperm—enclosing several thousands of special kinds of molecules (genes) that in their entirety are either compatible with or complementary to the molecules (genes) enclosed by the other. And each of these bits of living matter, like everything else that is alive, somehow carries *time* as a fourth dimension. Time it is, too, that permits the merging of egg and sperm to become complete; it is time that permits further slow growth, along with the actual formation of new parts with new reactions; it is time that favors the origin of mechanisms for registering all, and for expressing much, that has occurred; it is time that permits also no end, during life, to these processes of registration and expression—that is, no end to changes in personality, no end to the molding and tempering of the individual.

At every point in all this sea of change the mind and body exist as a single unity, not as something separable. The two are a unified and common growth; "there are no mental states, only mental acts" (Herrick). Yet as late as 1900 some textbooks of physiology used in reputable American colleges spoke of mind and *soul* as interchangeable items; and, for purposes of certain creeds at scattered points on earth, it seems that some such texts are still in use. The early theology that gave man a "soul" also took from man his immense fourth dimension—time. While promising man a future, it filched from him a past that alone makes a present possible.

Surely those who today speak of a soul have need to provide it a home. If soul does not in some way represent personality, just what does it represent? If it *is*—or if it relates to—personality or individuality, to which end or cross section of that flowing panorama does the detachable soul attach? Can this be at the very beginning of the individual—the *two* things, the unmerged sperm and egg? If it enters at the instant of merging—according to recent Catholic dogma—who can possibly identify such an instant in an unbroken chain of life? To ask these questions is to indicate how thoroughly the concept of soul is now rejected both by fact and by reason. This entire survey of living matter and of personality has neither met that agency nor found a time or place for its entry or abode.

PART II

REINS HELD BY RELIGION

7

The Religious Impulse

I say that it touches a man that his blood is sea water and his
tears are salt, that the seed of his loins is scarcely different from
the same cells in a seaweed, and that of stuff like his bones are
coral made. I say that physical and biologic law lies down with
him, and wakes when a child stirs in the womb, and that . . .
these are facts of first importance to his mental conclusions, and
that a man who goes in no consciousness of them is a drifter and
a dreamer, without a home or any contact with reality.
                                        —*Donald Culross Peattie*

SURELY THOSE WHO PROFESS RELIGION HAVE FIRST OR HIGH
claim on the right to say what the word "religion" means to them
today. To many liberal-minded adults who profess religion, the
word at the moment has a meaning quite different from the
one it had in their childhood, when religion of a different color
and meaning began actually to mold and lastingly to "condition"
some parts of their personalities, preferences and outlook. Thus
even that minority of adults who can now define religion or the
religious impulse in an advanced way is unable to assert that
only that tint of religion actually influences and shapes their
lives. To play with religion at all, in childhood, is to make mental commitments and to train in a type of irrational response
that requires later reversal—but ends often in only illusory correction.

That detail, however, is an almost irrelevant and a quite minor
difficulty met at the outset of a search for an understanding of
the word "religion." The chief difficulty arises from recent shift-
ings of the concept and treatment of religion by a small but influential group of religious leaders in parts of the Western world.
In other words, in precisely those parts of the world where science
has acquired an audible voice, the word "religion" is now acquiring a full scale of meanings. This scale stretches from a mixed
fear and trust in a supernatural oversight and control of everything human, at one extreme, to—or almost to—a welcoming

acceptance of the ethical humanism that arises naturally and solely through man. That the word "religion" is thus being used to describe quite unlike things is a fact of much importance. It puts frustrating confusion into vital areas of human attitude and action. It tends to shrink the broad segment of social advance that is gained only over the path of intellect.

SOME DICTIONARY DEFINITIONS

Webster's International Dictionary (2d edition, 1936) offers the following pertinent definitions of religion:

1. The service and adoration of God or a god as expressed in forms of worship, in obedience to divine commands, especially as found in accepted sacred writings or as declared by recognized teachers and in the pursuit of a way of life regarded as incumbent on true believers.

3. An apprehension, awareness, or conviction of the existence of a supreme being, or more widely, of supernatural powers or influences controlling one's own, humanity's, or nature's destiny; also, such an apprehension, etc., accompanied by or arousing reverence, love, gratitude, the will to obey and serve, and the like.

From the *American College Dictionary* (1947):

1. The quest for the values of the ideal life, including three phases: the ideal, the practices for attaining the ideal, and the theology or world view relating the quest to the environing universe.

3. Recognition on the part of man of a controlling superhuman power entitled to obedience, reverence, and worship.

An additional meaning recorded as *obsolete* is "the practice of sacred rites, or observances." The reader was doubtless not expected to forget that many humans still regard "rites and observances" as the very kernel of religion.

From the *Shorter Oxford English Dictionary* (1936):

1. A state of life bound by monastic vows; the condition of one who is a member of a religious order; the religious life.

3. Action or conduct indicating a belief in, reverence for, and desire to please, a divine ruling power; the exercise or practice of rites or observances implying this.

5. Recognition on the part of man of some higher unseen power as having control of his destiny, and as being entitled to obedience, reverence, and worship; the general mental and moral attitude resulting from this belief, with reference to its effect upon the individual or the community.

It should be noted that standard French, German, Italian, and Spanish dictionaries provide rather similar definitions. None, however, of the several works consulted places as much emphasis upon the "ideal life" as does the *American College Dictionary* cited above.

TESTAMENTS FROM OUR OWN TIMES

Of much interest to the pursuit of this question is the comment offered some years ago by a select group of the American clergy on Dr. Albert Einstein's statement on religion. Dr. Einstein had expressed the view that "there are three stages of religious development—the first, that of primitive peoples: the religion of fear; the second, the religion which finds its source in the social feelings: the moral religion; and the third, the cosmic religious sense, which recognizes neither dogmas nor God made in man's image." In defining this "cosmic religious sense," to which Einstein himself could subscribe, he said that "the individual feels the vanity of human desires and aims, and the nobility and marvelous order which are revealed in nature and the world of thought. He feels the individual destiny as an imprisonment and seeks to experience the totality of existence as a unity full of significance. . . . A contemporary has rightly said that the only deeply religious people of our largely materialistic age are the earnest men of research." *The New York Times*[1] thereafter asked eight prominent clergymen of New York City to write for it their replies to the two following questions: "1. What is your definition of religion? 2. Do you believe that the highest form of religion is the cosmic religion, as defined by Einstein?"

The more significant statements included in those eight replies are reproduced here for what they may contribute to a now current definition of the word "religion."

The Reverend Robert Norwood, Episcopalian, said:

There is nothing above nature, for it reveals God in whom "we live and move and have our being." . . . It is only as we know that we can enter into eternal life. So Jesus taught . . . I can see no reason why anyone should quarrel with this Einstein statement. [He could not agree that men of research are the most religious] . . . The fact is, any kind of relationship in terms of one's devoted best is religious.

The Reverend Daniel A. Poling, editor of the *Christian Herald,* replied:

Another has said "Religion—any religion—is man's quest for God." Man's search for God is man's search for the answer to his being. What am I? What is my origin? What is my destiny? Where am I bound? . . . But does not the distinguished scientist fall into an error common to humbler mortals when he affirms that "the ethical behavior of man requires no support from religion" and that "the churches have always fought against science and persecuted its supporters"? Is not Professor Einstein too general to be accurate?

Rabbi Nathan Krass stated:

I have very little fault to find with Einstein's declaration. . . . But Einstein delimits religion. He excludes. I would include. He avers that the believers in the religion of fear, the devotees of the social-moral religion, the people of faith whose God is anthropomorphic and anthropopathic, dishing out destinies, rewards and punishments, are not in reality deeply religious. I think they are profoundly, mystically, most intimately, most naïvely, yet most genuinely religious.

1 Sunday, November 16, 1930.

The Reverend Ignatius W. Cox, S.J., expressed a Catholic view:

Historically, the term religion means the intellectual grasp by man of a bond between him and a Supreme Being of superlative excellence, upon whom man knows himself to be physically dependent. An emotional consciousness, called religious emotion, is a consequence of, but not antecedent to, the understanding of the above truth. Religion, therefore, is a moral bond supervening upon the physical bond of man's dependence upon God. . . . Cosmic religion is not religion because it does not admit the existence of a supreme personal God, distinct from the universe. . . . I think that cosmic religion is atheism, euphemistically called religion. . . . Indeed, in the words of the distinguished doctor, cosmic religion is "essentially" different from the first two. In that case it ceases to be religion. Things that essentially differ cannot be the same.

The Reverend S. Parkes Cadman, Congregationalist, answered as follows:

I find it difficult to concede so large a place to fear as that assigned by Professor Einstein. . . . Neither Judaism nor Christianity sustains his interpretation of their origin. They claim to have been founded upon a divine revelation which has generated humanity's virtues and repressed its vices. . . . Since the professor is here relating his experience, allow me to add mine, that religion is the very life of God in the souls of men.

The Reverend John Haynes Holmes, of the Community Church, replied:

Professor Einstein has given a noble interpretation of religion. . . . I have long since come to think of religion, in all its forms, as in essence man's ultimate reaction upon the cosmos. . . . Science deals with facts, religion with uses, poetry with the symbolic expression of the two.

The Reverend Harry F. Ward, Professor of Christian Ethics, Union Theological Seminary, approved some and criticized other parts of the Einstein statement:

To the scientist's deep faith in "the rationality of the world," they [the saints and prophets of an ethical religion] add an equally deep faith in the morality of the universe. . . . To separate them by claiming the support of a cosmic faith for one and denying it to the other is fatal.

The Reverend A. Edwin Keigwin, Presbyterian, expressed this view:

Vital religion, being something entirely beyond all our solutions, can be grasped only by faith. . . . A "cosmic religious experience" is considered [by Einstein], but no notice is taken of a personal religious experience, which is the kind of which the Bible mostly treats. . . . Vital religion is, beyond all else, an inner, conscious, personal relation to God.

The reader may be his own judge of the clarity, differences and definiteness of these several definitions of religion. Perhaps a trace, but not much, of these qualities has been lost by failure to print all of each author's words on the Einstein statement. On the other item—the attitude of this group toward the idea of "cosmic religion"—the replies range all the way from near (or is it near?) acceptance by the Reverend Mr. Norwood and the Reverend Mr. Holmes to complete rejection by the Reverend Mr. Cox, who frankly calls it atheism.

Whatever may be the lack of clarity and of agreement in the views of these eight eminent clergymen, one notes that they spoke for only a few of the many organized religious groups in New York and America. Reformed, but not Orthodox, Judaism was represented. Liberal rather than Fundamentalist Protestant Christianity was chiefly represented, though the latter branch enrolls a very high proportion of the total church membership. Other world religions were entirely unrepresented.

While the subject of Einstein's "cosmic religious sense" is before us it seems permissible for another scientist to record his emotional attitude, as this, too, was born in the pursuit of the new and the natural. The emotions of two who look into nature with equal confidence may not be the same. Mathematician and physicist Einstein apparently reacts emotionally in the way described by his words: "The individual feels the vanity of human desires and aims, and the nobility and marvelous order which are revealed in nature and in the world of thought. He feels the individual destiny as an imprisonment and seeks to experience the totality of existence as a unity full of significance." Now the present writer, a biologist, feels nothing whatever of the things expressed in the last sentence. On the contrary, he feels that his own perishable self and aspirations are things widely shared with a partly gentled species—a species much less ancient than the hills—whose recent and diligent mornings already have tamed and jeweled the earth, found fragments of freedom, and partly vaporized its own pain and fears; and he cherishes unstintingly these choicest prizes that a long and orderly course of nature has till now been able to bestow. Human effort, mutual effort included, has been thoroughly proved to be a path to human satisfaction and growth. Too, rather like his esteemed colleague in physics, this writer feels "the nobility and marvelous order revealed in nature and in the world of thought." But far from feeling "the vanity of human desires and aims," he greets the new years that lend added chance to sort out the fertile best of these—then grants them his effort, hope and confidence. If the crowded story of mankind does not vouch for such feats and hopes, does it bear any message whatever?

The point just treated is too central to be lightly put aside. To one who, at close range, has learned to chart high points on the course and cause of the stuff of life, there is more of promise than of perdition in the tribe that has learned to stand upright and sometimes to think. For most men and peoples of the Western world, the day of Jeremiah is long past. That outlook is dead. To vast crowds and communes, the value of existence may indeed have been doubtful in not-too-distant times and places: when liberty was yet a dream for the few, and for all a prize elusive and insecure; when subsistence was daily in persistent doubt; when a pope could choke a pregnant thought, and burn or shackle the bodies of Brunos and Galileos; when child and sweatshop met for their twelve-hour days; when the expected term of life bordered thirty years, years pursued by pain and lightly curbed disease. These, however, were but tragic tests put to man by abortive things in adolescent civilization.

Today triumph outshines tragedy. Though the brush that matches colors to our careers may not use all the rainbow, it need make little use of the black of night. It may spread the magic it chances to command to bring to sight the matchless strength and feelings born of love, of friendship, and of living to the full. And though those powers and feelings fail or dull with advanced age, so does all else that is part of life. At the full term of life, its best—slowly and little noted—has already ceased to be. Human existence is perhaps as variable as the stars—at its crest as gay as a parade, as meaningful as all history, though fragile as foam. And for us all—as against ancestors of a hundred thousand years after man became man—there *is,* unasked, daily deliverance and much hope.

Let it become clear that in those who take up the task of unraveling nature, the brood of emotions will vary from person to person. And perhaps these will vary among scientists according to the particular science pursued. Will all these variant attitudes and feelings find a home in "cosmic religion"? Which will be excluded? Is Father Cox right in saying that even a lesser variant of this type is not religion but atheism?

An informative glance at English thought on biblical questions was provided in 1928 by a volume *(New Commentary on the Holy Scriptures)* written by several devout and learned scholars of the Church of England. Some of their conclusions are useful here. Old Testament miracles, the flood, and Methuselah's longevity are discounted either largely or wholly; King Belshazzar did not exist; and, as Bishop Colenso had found and as Huxley had publicized four decades earlier, Moses did not write the Pentateuch. Certain Gospel miracles also are viewed skeptically. But the evidence for Christ's resurrection is regarded as overwhelming, and the narrative of the raising of Lazarus "is accepted with all its implications as the climax of all the miracles of healing."

Three years earlier Dean Inge, of St. Paul's Cathedral, London, had expressed disbelief in the physical ascension of Christ and recognized the close connection of this matter with the question of resurrection. Again, to Dean Inge "hell seems to have faded away" and "Darwinism has inflicted no injury on the Christian faith." On the contrary, professor of literature Clive Staples Lewis, of Oxford University, suggested in 1947 that the miraculous ("interference with Nature by supernatural power") not only can exist but has existed in human history; and Heaven has the reality of a Sunday-morning breakfast. Religious historian Dampier wrote in 1942: "The mystical and direct apprehension of God is clearly to some men as real as their consciousness of personality or their perception of the external world. It is this sense of communion with the Divine, and the awe and worship which it evokes, that constitute religion."

Following this short digression to views relating largely to the Christian religion, the more general topic will be considered in three additional statements. From an address by Professor

Sperry,[2] Dean of the Divinity School, Harvard University, the following is quoted:

It is only subsequently that the word [religion] achieves its present, common connotation, "Recognition by man of some higher unseen power having control of his destiny and as being entitled to obedience, reverence and worship." . . . In general the history of religion is that of a movement away from magic and toward mysticism. . . . In any case the word "religion" is a word like "companionship," "comradeship," "communion." It is a conscious experience of escaping from one's solitariness and of belonging to someone or something other than oneself.

Next, the New Testament, James 1:27:

Pure religion and undefiled before God and the Father is this, To visit the fatherless and widows in their affliction, and to keep himself unspotted from the world.

Do these words mean that not belief or worship but ethical conduct is pure religion? Is then the always unclear word "religion" a synonym for ethics?

Finally, we come to a Unitarian minister's definition of religion—perhaps of religion that is only in process of birth or of its complete absorption in ethics—which might enroll the writer of this book on the list of religious men. Some years ago Dr. A. Wakefield Slaten, minister of the West Side Unitarian Church, Cathedral Parkway, New York, spoke as follows:

The religion that is to be Christianity's successor is even now taking form among us. It is already the really important part of the religion of all church members, and it is the religion of all the decent portion of that fifty per cent of our population who are not church members. Everywhere you find men living it unostentatiously, even unconsciously.

Where the old religion made the supreme object God, the new makes it humanity; where the old controlled conduct by the assumed favor or disfavor of the Deity, the new makes the effect of one's conduct upon social well-being the controlling consideration; sociology takes the place of theology, and the world-hope of an improved social order replaces the belief in a blessed im-

2 Willard Learoyd Sperry, "The Present Outlook for Religion," before the American Academy of Arts and Sciences, December, 1948.

mortality. This is not paganism, or non-religion. All life has become religious, for it is realized that whatever is right is religious. The present life has been regarded as of relatively little import. It is now regarded as of supreme import. The great question has been, "Where will you spend eternity?" The great question now is, "Where and how will you spend the rest of your life?"

Now, somewhere along this trail of definitions something has gone wrong. For, according to the definition last given—perhaps the last two definitions—it is perfectly obvious to the present writer that, as regards his own attitude and beliefs relating to a good that is beyond self, he is as truly religious as atheist. And in our day this cannot make sense to the great majority of those who know and use either of these words in the English or in any other language. In memory of the good, clear, old and oft-forgotten word "ethics," it is suggested that it be applied where it belongs in two of the sentences of the able minister last quoted. Once the word "ethics" is used for the concept of ethics—with which atheism or secularism can have no disagreement—there remains the expected distinction between religion and atheism.

Physicist Langmuir once remarked that "no one has been authorized to make an exact definition of the atom." Perhaps that could not be said of the word "atheist." And since that term is used in these pages there is reason to state what it means to this writer. An atheist is a person without a God or gods. He may or he may not *deny* the existence of a force that is beyond or outside natural law, but he is beset with no doubts about giving it reverence. He believes that man is on his own, that man must lift himself. He rejects as false or insubstantial all so-called revelations and alleged personal contacts with an external pervasive consciousness or agency that is sympathetically concerned with human affairs. Usually he recognizes that some people, occasional scientists particularly, share both these beliefs and rejections but disclaim that they are atheists. Why they do so he does not know.

The partisans of religion were accorded the right of way in the preceding pages. The nonpartisan scholar will now be granted a mere page or two for his words on the nature and essence of religion. Philosopher Santayana's conclusion: "Religion is human experience interpreted by human imagination. . . . The idea that religion contains a literal, not a symbolic, representation of truth and life is simply an impossible idea. Whoever entertains it has not come within the region of profitable philosophizing on that subject." Psychologist Leuba[3] wrote: "So long as he seeks to discover God, man is a philosopher; he assumes the religious attitude when, having found God, he enters into social relation with him. That which is characteristic of religion is the way in which desires are satisfied, namely, by a social relation with one or several divinities." Clearly it is *means,* not *aims,* that distinguish religion from nonreligion. Leuba[4] later noted: "An institution must have some recognized way of seeking the realization of its purpose; and the specific way offered by our [Christian] churches, as well as by the churches of all the religions, is an appeal to, and a reliance upon, supernatural beings. It is the means then, and not the purpose that differentiates the established religions from all other institutions. That our churches also use natural means of action does not invalidate the preceding statement."

Of course, it does not follow that the religious element disappears through a retreat from "a God that thinks and feels and wills" if or when the person or institution confides instead in "an active ideal or spiritual universe"—with this joined or not joined to a particular ideal of life (the goal). The *means* remain as before—an appeal to something outside and beyond man himself as a source of good and hope for man. And the inclusion of a true or false view of the ideal life is an incidental matter that leaves the religious element untouched.

Again, social anthropology and history have supplied the objective and meaningful studies of religion that thoughtful people require. What are their findings? To primitive men, myth is religion,[5] and with them (Chapter 4) it is a tribal affair—a way such unlettered men must have to make the world acceptable, manageable and right. With their very slight hold on reality, primitive men grasped firmly at their own emotional needs and inventions for a consistent view of events—events that were otherwise merely uncontrolled confusion. "Faith in the supernatural is a desperate wager made by man at the lowest ebb of his fortunes,"

<hr>

3 James H. Leuba, *God or Man?* (New York: Henry Holt and Co., 1933).

4 James H. Leuba, *The Reformation of the Churches* (Boston: The Beacon Press, 1950).

5 Bronislaw Malinowski, *Magic, Science and Religion* (Boston: The Beacon Press, 1948).

says philosopher Santayana. Thus the mold in which religion was cast was a crusty and often cruel substitute for understanding.

Many moderns get great satisfaction from the false idea that the religious impulse is an instinct. As this term is used in science, instinct is a form of behavior that is not learned, not acquired by the individual, but innate. Clearly, however, man *learns* to think of and believe in supernatural beings, and he *learns* to worship them in one and another way. Indeed, both his god-conceptions and his forms of worship change—or disappear—as his knowledge grows.

From this survey it appears that a truly religious element is present wherever, in the mental processes of humans, values are located in the supernatural. The number and kind of values may stretch, contract and change from savage to sage, from child to adult; but always the religious element is retained. It cannot be established that—after rejecting the supernatural—a deep feeling of "the vanity of human desires and aims, and the nobility and marvelous order which are revealed in nature and the world of thought" is as well called "cosmic religion" as atheism.

Turning to affairs of the moment one finds that, at high cost to society, many responsible persons propagate a false meaning of the word "religion." A famous scientist says he believes in God but gives no clue to the type of god. Another endorses religion and leaves all in doubt whether he refers to ethics, the Christian scheme of salvation, or a heaven-born revelation to man. Widely scattered are community guides who strive to get the public to equate "religion" with an ideal or moral life and with man-to-man activities of the churches—two things foreign to *religion*. Here the more usual and apparent purpose of the deception is to increase the dominance of the church. For others, the purpose may be to avoid adverse consequences of a changed or lost faith. Here one meets the associated and unfortunate fact that state law and community sentiment still make a highly practical distinction between belief and unbelief. Most prominent among the persons caught in these nets are clergymen and academicians.

Nearly all the clergy are in a position to profit by a concept of religion that adds adherents to the church. And Christian churches are cast in the proselyting mold. But for a few pastors with advanced thought and much courage the matter is neither basely selfish nor simple. That part of the clergy whose thinking has broken with the creeds—and especially the part that has put the supernatural aside—is left with an obvious economic need to retain the word "religion." It was a church that assigned them to their posts; it is a church organization that provides their salaries; and if—in almost all the United States and in some other countries—ethics but not religion is in the organization's title, with few equating ethics and religion, its real estate becomes subject to taxation. This is one of the bizarre outcomes of exempting the holdings of "religious" organizations from taxation by the state.

In the academic area the thought that does not fit, or fits badly, abounds. But no observer of the academic scene can escape the fact that our universities had a monastic mother, and that the daughters—despite the Reformation and the newer and immensely greater revolution in the field of knowledge—have at no time or place broken sharply with her.[6] Among these halls and towers some find it more comfortable to let the view that, for them, so satisfactorily replaced religion go ahead if it likes under the same name—although the supernatural or the extrahuman has been dropped from thought. Here, too, one may look for one or another element of self-defense. Only in a few of our better universities will the instructor or professor who is frankly atheist find no disadvantage or inconvenience to flow from *his* views on religious subjects. In the great majority of American universities and colleges—as in those of other countries of the author's acquaintance—this disadvantage is usual; and sometimes it is total and terrible. Thus, whatever the value to thought of definiteness and clarity elsewhere, both the clergy and the academic world of our day provide their percentage of those who personally prefer an evasive and deforming elasticity in the word "religion."

But in this transitional age the harm done by the confusion of religion with ethics and morality is enormous and intellectually ruinous. This confusion is so widespread and high-centered that it constitutes an indictment of much of higher education almost everywhere, and particularly in the United States. This confusion writes itself into our technical journals and upon subjects of the

6 Worthy and early American approaches to such a break—the so-called complete secularization—were made by the University of Virginia, established in 1825, and by the University of Michigan, established in 1837.

highest national importance. As an example, witness the plea made in *School and Society* by Professor Williams[7] for the extensive teaching of religion in American public schools: "If we succeed in preserving democracy, it will be because we conduct an arduous and successful program of *religious* education" (italics his). Now Professor Williams could write this because he defines religion to suit himself, and in a way that suggests to many readers that "religious education" should be used almost or quite synonymously with "education in moral attitudes." But imagine, if you can, the thing that legislators and school boards would put into schools under the name of religion! If education in moral attitudes is what was wanted, it could not be mistaken under that name. And, as a reader (Dr. A. B. Wolfe) of the Williams article immediately replied: "Religion, as most people understand it and as it is preached every Sunday, is a man-to-God relation and outside the province of American schools." Morality, or ethics, is certainly concerned with man-to-*man* relations. The difference is clear and it separates things that are poles apart. It must be said that the prevalent academic assist in the portrayal of religion as the necessary foundation of morality is intellectual high scandal.

WHO IS A CHRISTIAN?

Who is a Christian? To this question the many followers of Jesus of Nazareth now give notably different answers, but always an answer that is unacceptable to many of those who acclaim him. This confusion doubtless rests on more than one thing, and not solely on the prime fact—unimpressive or unknown to many or most of his past and present followers—that probably no one can ever know just what Jesus did and did not teach. It would seem, however, that if the title of Christian is to have a religious meaning, and not merely an earthly ethical one, it would require belief or faith in one or more of the following doctrines—all of which are incompatible with sound knowledge of evolutionary science: the fall of man; the need for an atoning savior; virgin birth; resurrection and ascension of Jesus; Sunday as a holy day; heaven and hell; immortality; answer to prayer; the Second Coming of Christ; the Bible, or parts of it, as the inspired word of God; the perfection of Jesus as a guide to modern conduct. If ac-

[7] John Paul Williams, "Religious Education: Ignored but Basic to National Well-Being," *School and Society*, May 22, 1943.

ceptance of any single one of the above doctrines is a necessity for a Christian, the members of that faith should know that never again may they hope to count among their faithful any person who has acquired a sound knowledge of evolution.

This assertion does not in any way suggest that those who now best know the story of earth, of life, and of man and society are aloof to the history of ethical thought. These trained persons—unlike the mass of Christians, past and present—note and deplore the meager and unsure reporting of the life and teaching of one of earth's great men, the myth and mental confusion in which those reports were immersed, and the shrouds of sham, miracle, exploitation and intolerance that so effectively veil what possibly was quite largely—though by no means solely—a teaching of good works, of superior love and kindness, and of a way to look beyond self.

To qualified students of human thought and behavior, the foundations of Christianity have meaning if or when those foundations can be examined clearly and objectively. Now and then a competent scholar contributes something toward this desirable objectivity. Certain it is that Jesus was not responsible for the ignorance and supernaturalism of a small and detached Asiatic homeland in a prescientific age. And since the real thought of a thoughtful man is wanted, it may be counted civilization's misfortune that Jesus apparently carried the triple handicap of teaching for only a few years of a brief life, in a quite restricted and static area, and without reducing his thoughts to writing.

It is also clear enough that these several conditions provided a maximum of opportunity for others to ascribe to Jesus incredible and discreditable things that he may neither have thought nor taught—a virgin birth, a resurrection, the practice of the miraculous, ordering or permitting demons to enter the Gadarene swine, and an extraordinary kinship with God. To this same mangling process were subjected also the humane and atheistic teachings of Buddha (Gautama Siddhartha) whose thought—though likewise not put into script—perhaps had a better chance of comprehension by associates than that of Jesus. Again, surely not Jesus but church leadership was responsible for the religion-based outrages, killings and scourges of the Dark Ages of Europe. Likewise a now much-divided Christian clergy, not Jesus, must be held responsible for the active and passive repression of modern thought, for a slight or a heavy degree of harmful institutionalization of religion, and for a continuing world crusade to propagate in its entirety what is probably both the true and the quickly perverted thought of one who spoke briefly but forcefully to lowly neighbors little prepared for a stirring thought.

Here indeed is doubt. Did the use of an Eastern tongue—grown in symbolism and around religion—lend itself less to clarity than to allegory and emotion? Could deep feeling in illiterate listeners of that part of the Eastern world then fail to clothe itself in supernatural cloud? Would the true thought of Jesus on ethics, if known, now suffer greater change than that of Plato on the world of things? Was his voice more than all else a voice against the prevailing thought, and acts, and miseries of men? Has the message been so mangled that the teacher has been robbed? Could it be that—despite my own atheism and very low opinion of Paul and the church fathers—I, too, am now sharing in that teacher's tasks?

OTHER PRACTICAL RESULTS OF UNCLEAR DEFINITIONS

Largely in pursuit of its own self-defense, the religiously inclined mind during recent decades has so tortured and changed the meanings of the words "God," "religion," "spiritual," "natural law," and "evolution" that, in a company of five or more persons, no one may now use any of these words without further explanation, if he hopes to be reasonably well understood. The preceding pages give some of the evidence that this is true for the word "religion." Similar evidence for the term "natural law" may be given in the words of astronomer Eddington, whose friendly attitude toward religion has already been reviewed in Chapter 3: "I am aware that many religious writers have felt no objection to, and even welcomed, the intrusion of natural law into the spiritual domain. Probably, however, they are using the term 'natural law' in a more *elastic sense* (italics mine) than that in which the materialist understands it. . . . But I am sure that those who take this view have never understood and faced the meaning of the ideal scheme of scientific law. What they would welcome is not science, but pseudo-science." Finally, it can hardly have escaped notice that the several caricatures of "evolution" currently masquerading under that name made it necessary to write the six preceding chapters of this book.

There are immediate and disastrous consequences of the Western world's present inability to use the unqualified word "religion" without being more misunderstood than informed. The word now has, or is tending to acquire, a spread of meanings that wholly prevent applying either statistics, condemnation or praise to it. Though it should and more commonly does refer to a social relation between man and the supernatural, through localized twisting it tends to become too elusive to be characterized with assurance as this or that. Therefore, not "religion" but "organized religion" is recognized in these pages as the obstructionist influence now producing the cultural impasse of our time—the impasse itself being the central theme of this book. The "organized" groups do indeed provide the footprints of their written creeds; they have histories of effort, aims and performance; they distill themselves into present community sentiment, practice and law. It is this still-unfolding record that can be examined.

That record, as partly reviewed in the next following chapters of this book, leaves no doubt that organized religion everywhere —at this mid-century as surely as in the past—opposes the new, distorts the true, and insulates most teeming multitudes of men from the culturally vital facts and guidance of evolutionary science and modern thought. Specifically, organized religion now effectively blocks all of earth's civilized peoples from a bulk understanding of man's slow and natural origin, from the resulting deep sense of opportunity and obligation to lift themselves by their own efforts, and from the urgent societal need to substitute clear-eyed human purposes for outworn and myth-born supernatural purposes.

8

School: The Sepulcher of the Intellect

For we shall have shut the door in the faces of their minds; we shall have pinched the young fingers of thought; we shall have checked the ardours of response.—*Henry Chester Tracy*

Education is only the image, the reflection of society. Education imitates society and reproduces it in abridged form, but it does not create it. Education is healthy when the nation itself is in a healthy state, but not having the power of self-modification, it becomes corrupted when the nation decays. . . . Thus education cannot reform itself unless society is reformed. And in order to do that we must go to the cause of the malady from which it suffers.—*Emile Durkheim*

DURKHEIM'S STATEMENT, QUOTED ABOVE, POINTS A DIRECtion that must be followed to a definite end if a profitable discussion of the disposition—the utilization or the entombment— of evolutionary biology and modern thought by schools is attempted. Granting the biological illiteracy of peoples and the truth of the statement that the current thought of society transfers itself generally into education, may we expect the *schools* to break that rule in the special case of evolutionary thought? On the contrary, since for nearly a century biology has been the one subject pregnant with disclosures either suspect or anathema to every new and old religion, may we not expect to find religious community thought directed rather especially *against* that subject in all the schools within the state? And who will doubt the past and present militancy of religion in the thought of communities and nations?

The schools of a truly free society could, of course, do much more with intellectually awakening discovery than wall it in. But who—despite much declaiming—can point to one such society? To most Americans the burial of thought in primary and secondary schools is as little understood—or resented—as the burst of a "nova" star some millions of light-years ago.

Certainly the French social scientist whose words are quoted had no illusions on these questions, and a further reference to

166

French experience is useful here. The French society and education that Durkheim studied was that of the end of the previous and the beginning of the present century. At that period of French history the Jesuits—the Teaching Order of the Roman Church—were in exile. But it was clear to every Frenchman that Jesuit influence on French education had not been and could not be wholly exiled. Their permission to return to France in 1920—granted in payment for papist intercession with the rebellious Catholic populations of newly acquired Alsace-Lorraine—well illustrates the tie-in of national or international politics with both religion and education.

Thus the question at hand becomes at once the reach of religion in schools. An accurate estimate of religious repression of the teaching of evolutionary thought in the schools of the world today is far beyond the ability of any one living person. This repression is almost as pervasive as the sky—equally beyond touch and grasp. When institutions entrenched in society also become a sepulcher for thought killed in birth, a firm focus on the performer—and on the intangible relics—is not made easy. There are, nevertheless, the best of reasons for examining parts of this overspread tragedy of suppressed and buried thought. In perhaps no country of the entire world are more than a few per cent of the population aware that their schools deprive them of anything. Yet if that repression were removed—even for only a single generation—most countries with advanced educational systems would emerge with changed and new societies. And the freedom thus acquired could thereafter be defended as a basic human right and as a vast leaven to social progress. All this would follow because a generation in which authority, leadership and education could freely accept a short list of truths—the universal rule of natural law, the animal origin and nature of man, the human sources of morality, and the biological inequality of men exacted through the rules of chance—*is*, quite clearly is, an entirely new society. In that society the institution most displaced or inconvenienced is organized religion. Religion therefore is now the foremost resister. We may agree with Durkheim that "education cannot reform itself unless society is reformed." We may see why John Dewey's verdict is true, that "the genuinely modern is still to be brought into existence."

It is the vital best of biology and related science, not the wide stretches of descriptive details, that is directly or intentionally

entombed and silenced in education by religious thought. It is
the awakening of the intellect of youth, the release of those minds
from an elemental fear, and the launching of personalities into
a world in which they face the clear problem and the fine opportunity of control of their own destiny that religion resists. These
superb stimulants could brace high purpose in all education and
put life and vigor into the eviscerated husk of biological or other
science that is now permitted to be taught. Such insights and
goals could soon transform biology from a shunned to a sought
subject in all secondary education, and could deepen the meaning and force of every cultural subject taught there. Thus the
same agency that openly robs biology of its best, indirectly steals
also from other millions of youth practically all contact with
truly liberating thought. And what is stolen relates directly to
their own bodies and talents, to the foundations of social thought
and act, and to the broadest aspects of their own origin, destiny
and responsibility. Through constant choking of evolutionary
thought in schools everywhere, entrenched religion thwarts the
birth of a genuinely modern, saner and freer society.

The most discouraging thing that this writer has learned during a lifetime given chiefly to biological research is that so little
of precisely this revealing knowledge of ourselves, with its clear
invitation to build a truly free and "genuinely modern" society,
is given a chance to enter the consciousness and life of modern
youth.

It is convenient to divide the present subject matter into four
parts. World-wide restrictions are first viewed as these arise
through the influence of the Catholic Church. Thereafter the
similar influence of other religions is examined. A third section
reports the informative results of a special study of religious influence on the teaching of biology in the high schools of the
United States. Finally, an over-all summary of the entire subject
is ventured.

RESTRICTIONS THROUGH THE CHURCH OF ROME

An Associated Press dispatch from Vatican City, dated October
3, 1949, quoted a Vatican announcement to the effect that the
Catholic population of the world then stood at 432,000,000. Also
that since 1920 that number represents an increase of 119,000,000.
Further, within this same period world population jumped 600,-

000,000—an increase of one sixth—while in comparison Catholic Church membership increased by one third. Nominal and actual church memberships, of course, differ appreciably. But in view of the totals cited, and more especially in view of the superb and wholly unmatched organization of the Catholic Church and of its intensely militant role in education, there can be no doubt that its present influence on Western education is far greater than that of any other existing organization, religious or secular. Moreover, its membership is centered where it is educationally, socially and politically most effective—in Europe and the Americas. Within the United States its membership resides principally in the largest cities, and these cities largely dominate the entire pattern of life and thought of the nation. Its nominal world membership is greater than the combined populations of the United States, England and Russia.

During many centuries this powerful organization has dictated an acceptance by Western civilization of a whole philosophy that is basically counterbiological and antirevolutionary. That very ancient teaching survives in the laws of many nations, in the present thought of peoples, in the music and arts of centuries, and in the educational programs of numerous large populations. It is well to list some specific items of that contrabiological philosophy: the special creation of man, or at least of the "soul" of men; the descent of all men from a single pair; the inheritance of original sin; the virgin birth of Jesus; a multitude of old, new and assorted miracles; the derivation of man's moral code not from the animal world and society but from revelation and church sources; and—worst of all—the principle that it is a virtue to accept statements without adequate evidence. True enough, Protestantism, an offshoot of Catholicism, has shared and now shares in the propagation of all these items, while Judaism and Islam have shared in some. The main point of this particular remark is that the earlier and older teaching of the Church of Rome—even though that misinstruction had been entirely unsupported by any similar teaching during the past one hundred years—was able to establish antibiological thought and sentiment that continues alive in many minds and in some thought-strangling institutions of today.

But has or has not the Church of Rome materially modified its antibiological attitude during this last century of startling enlightenment? And has it left peoples freer than before to develop and operate their own educational systems in the light of new scientific truth? Or has that Church perhaps modified its old instruments and developed new ones for controlling education? These are the questions of concern in the following pages.

Fortunately, the attitude of the Catholic Church at this moment toward a few biological items is now fully available. The Pope may speak with the fullest authority on such matters of doctrine, and in 1950 Pope Pius XII issued an illuminating encyclical letter. From the official English translation of that document, dated August 21, 1950, the reader is offered the following pertinent paragraphs:

If anyone examines the state of affairs outside the Christian fold, he will easily discover the principal trends that not a few learned men are following. Some imprudently and indiscreetly hold that evolution, which has not been fully proved even in the domain of natural sciences, explains the origin of all things, and audaciously support the monistic and pantheistic opinion that the world is in continual evolution. . . .

It remains for us now to speak about these questions which, although they pertain to the positive sciences, are nevertheless more or less connected with the truths of Christian faith. In fact, not a few insistently demand that the Catholic religion take these sciences into account as much as possible.

This certainly would be praiseworthy in case of clearly proved facts; but caution must be used when there is rather a question of hypotheses having some sort of scientific foundation in which doctrine contained in sacred scripture or in tradition is involved. If such conjectural opinions are directly or indirectly opposed to the doctrine revealed by God, then the demand that they then be recognized can in no way be admitted.

For these reasons, the teaching authority of the church does not forbid that, in conformity with the present state of human sciences and sacred theology, research and discussions on the part of men experienced in both fields take place with regard to the doctrine of evolution in as far as it inquires into the origin of the human body as coming from pre-existent and living matter—for Catholic faith obliges us to hold that souls are immediately created by God. However, this must be done in such a way that reasons for both opinions, that is those favorable and those unfavorable to evolution, be weighed and judged with the necessary seriousness, moderation and measure, and provided that all are prepared to submit to the judgment of the Church, to whom Christ has given the mission of interpreting authentically the sacred scripture and of defending dogmas of faith. Some, however, highly transgress this liberty of discussion when they act as if the origin of the human body from pre-existing and living matter were already completely certain and proved by facts which have been discovered up to now and by reasoning on those facts and as if there were nothing in the sources of divine revelation which demands the greatest moderation and caution in this question.

When, however, there is a question of another conjectural opinion, namely polygenism, children of the church by no means enjoy such liberty. For the faithful cannot embrace that opinion which maintains either that after Adam there existed on this earth true men who did not take their origin through natural generation from him as from the first parent of all, or that Adam represents a certain number of first parents. Now it is in no way apparent how such an opinion can be reconciled with that which the sources of revealed truth and the documents of teaching authority of the church propose with regard to original sin, which proceeds from sin actually committed by an individual Adam and which through generation is passed on to all and is in everyone as his own.

Just as in the biological and anthropological sciences, so also in the historical sciences there are those who boldly transgress limits and safeguards established by the church. In a particular way must be deplored a certain too-free interpretation of the historical books of the Old Testament. . . .

For this reason, after mature reflection and consideration before God, that we may not be wanting in our sacred duty, we charge Bishops and superiors general of religious orders, binding them most seriously in conscience, to take most diligent care that such opinions be not advanced in schools, in conferences or in writings of any kind, and that they be not taught in any manner whatsoever to the clergy or faithful.

Let teachers in ecclesiastical institutions be aware that they cannot with tranquil conscience exercise the office of teaching entrusted to them unless in the instruction of their students they religiously accept and exactly observe the norms which we have ordained. That due reverence and submission, which in their unceasing labor they must profess toward teaching the authority of the church, let them instill also into the minds and hearts of their students. . . .

Nor must it be thought that what is expounded in encyclical letters does not itself demand consent, on the pretext that in writing such letters the Popes do not exercise the supreme power of their teaching authority. For these matters are taught with the ordinary teaching authority, of which it is true to say: "He who heareth you, heareth me"; and generally what is expounded and inculcated in encyclical letters already for other reasons appertains to Catholic doctrine. But if the supreme Pontiffs in their official documents purposely pass judgment on a matter up to that time under dispute, it is obvious that the matter, according to the mind and will of the same Pontiffs, cannot be any longer considered a question open to discussion among theologians.

From the above quotation one learns that "outside the Christian fold" not a few learned men "audaciously support the opinion that the world is in continual evolution." Also that those within the church who are experienced in the *two* fields of "human science and sacred theology" *may,* under proper safeguards, research or discuss "the doctrine of evolution" as this relates to the human *body* "provided that all are prepared to submit to the judgment of the church." And "the children of the church by no means enjoy such liberty" [*sic*] on the question of an individual Adam "as the first parent of all." Apparently we must await that later century in which some future Pope may discover, and perhaps tell us how to deal with, the fact that the "first man" was not of *our* species, but at least two or three species behind us. Any fragment of evolutionary biology capable of wrestling successfully with that impasse may thereafter fully exhaust itself in searching for a way in which the odd sin of Adam "is passed on to all and is in everyone as his own."

In still another way this encyclical letter has bearing on the question of the present attitude of the church to evolutionary science. The entire message is a warning to Catholics, particularly to Catholic administrative and educational leadership, to shun deviations and compromises that would undermine Roman Catholic doctrine and dogma. It is thus announced that some such deviations already exist. These deviations have two aspects of quite dissimilar significance. First, this movement within the church, known as modernism, was earlier condemned by Pope Pius X, in 1907. The warning letter of 1950 therefore indicates that within the present century the hierarchy is finding it difficult to force its frame of thought upon all of its own clergy and teachers. It should be acknowledged that a fair, or better than fair, proportion of these Catholic modernists are found in the United States. These men are charged with deviating from authentic Catholic teaching not alone in regard to science or to biology, but on items of history, philosophy and theology as well. Second, the letter frankly states and condemns the reasons commonly given by at least some of these deviationists for their departures from Catholic doctrine. But partly veiled is their suggested trick, or their intent, of assuming outwardly a congenial modernity, which is later to end in pressing for accepted Catholic dogma and doctrine. Parts of these disclosures will be found in the following brief quotation:

> There are many who, deploring disagreement among men and intellectual confusion, through an imprudent zeal for souls, are urged by a great and ardent desire to do away with the barrier that divides good and honest men . . . so today some go as far as to question seriously whether theology and theological methods, such as with the approval of ecclesiastical authority are found in our schools, should not only be perfected, but also completely reformed, in order to promote the more efficacious propagation of the Kingdom of Christ everywhere throughout the world among men of every culture and religious opinion. . . .
> Unfortunately these advocates of novelty easily pass from despising scholastic theology to the neglect of and even contempt for the teaching authority of the church itself, which gives such authoritative approval to scholastic theology. This teaching authority is represented by them as a hindrance to progress and an obstacle in the way of science.

Incidentally, we thus have the Pope's word for it that some within his own organization regard the "teaching authority" of the Roman Church as "an obstacle in the way of science." And his own words unite some of his venerable brethren with the present writer "in despising scholastic theology."

But our momentary supply of impossible Roman biology is not all that we are soon to have. As this is written, in the latter days of October, 1950, disbelief in the concept that the body of the Virgin Mary was taken up to heaven and there united to her soul is, in the words of Pope Benedict XIV, two centuries ago, merely "blasphemous, scandalous, senseless and unreasonable." Within another week, on November 1, 1950, Pius XII will have made this a Catholic dogma, and thereafter any disbelief in it will be a much more perilous matter. On the day indicated, and from the altar of the cathedral in St. Peter's Basilica, the Pope will declare the Catholic dogma of the corporeal assumption of the Blessed Virgin Mary into heaven. Vatican circles seem to have indicated (*The New York Times,* August 15, 1950) that 113 Cardinals, 2,523 Patriarchs, Archbishops and Bishops, 82,000 priests, and 8,000,000 faithful have declared themselves in favor of the new dogma.

Has the Roman Church since 1900 modified its old instruments and developed new ones for *controlling* education in wide areas of the world? The slightest inquiry shows that this has occurred, and with results so gratifying to the Church that the non-Catholic world has much reason for a concern that as yet it little shows. Still other instruments fashioned to promote Catholic influence in the social and political spheres—and thus too in that of education—are new to this period. The present account can properly do little more than refer the reader to fuller discussions[1] of this momentous matter and list the names of the principal newer organizations.

Except from behind the Iron Curtain, the Jesuits, or Company of Jesus, now operate in several countries from which this "theocracy within a theocracy" formerly had been expelled. And the powerful influence of this long-established group has now been immensely reinforced. In the last century, this reinforcement came first through the Salesians, a company of "lay priests" engaged in operating colleges and in caring for the "spiritual and physical welfare" of students. They serve in North America, in several European countries, and especially in South America. To them were added the Company of St. Paul, founded in 1920, which has the specific task of combatting socialism and communism. Its members wear only "lay" clothing. In many countries this group already operates missions, schools, technical training centers and printing presses, and publishes newspapers. In Great Britain, another organization, the Sword of the Spirit, operates on rather similar lines, under the direction of the local Cardinal Archbishop.

Of somewhat different character is the extremely potent and up-to-date instrument of Pope Pius XI called Catholic Action. The Catholic periodical *Commonweal* said the goal of Catholic Action is "to produce change and adjust all religious, moral and social and economic thought and procedure of modern life to Catholic standards of thought and action, in order to spread the Kingdom of God." Now spearheading this layman's movement in the United States is the revived Confraternity of Christian Doctrine, whose Ninth National Congress, held in Chicago,

---

[1] Avro Manhattan, *The Vatican in World Politics* (New York: Gaer Associates, 1949). Paul Blanshard, *American Freedom and Catholic Power* (Boston: The Beacon Press, 1949). Paul Blanshard, *Communism, Democracy and Catholic Power* (Boston: The Beacon Press, 1951).

in November, 1951, was reported by the press to have been attended by ten thousand delegates. Many well-informed persons privately state that in the United States it is common practice for a member of Catholic Action to use his own position to secure the preferential employment of other persons of Catholic faith. This is said to apply in *public* school systems, in business, and elsewhere.

With these many organizations operating under the definite control of the hierarchy, plus many others such as the Catholic Youth Organization, the Knights of Columbus Association in America, and the many Catholic political parties of European nations, a highly aggressive Catholic influence in educational, social and political spheres is unquestionable. To all this must be united the strength that was added, and is still developing, from the new political status and authority which the Vatican attained in 1929. Only one aspect of that new and developing strength is indicated in the fact that fifty-two ambassadors, ministers, and personal envoys are now accredited to the center of Catholicism in Rome.

Two conclusions of central importance in the present discussion can now be brought together. Catholic dogma and power today are, as in earlier centuries, the chief antagonists of evolutionary and related social thought in the Western world. That antagonism is exercised, first of all, in the schools. And, there is most convincing evidence that it is steadily becoming more powerful. If evolutionary biology, with its potentials of social release, is to do better than survive in a clan, it must now frankly do battle with an old and highly aggressive foe.

RESTRICTIONS THROUGH OTHER RELIGIONS

Protestantism, except, chiefly, its smaller sects, is a less consistent and less dangerous enemy of evolutionary biology and modern thought than is Catholicism. In part, this is a consequence of the disunity of Protestantism. At the highest level—that of the better universities of a very few countries—this Protestant antagonism usually is so largely an indirect, off-campus and precollege affair that the true extent of its restraint is little recognized or resented even by the biologists who teach in these universities. It would be much more resented if it were better understood. Not inclined to parade on this campus are the ghosts of the absent—those who should have been but were not encouraged to enroll here with zeal for further insight and service; nor is attention focused upon the resulting increased proportion of the aimless among the entrants to college, and the faculty's resigned acceptance of the necessity of dealing with a mounting proportion of a type of youth that is innocent of any tie to thought. On this superior campus no parades or memorials commemorate the casualties from thought strangled at all earlier levels of education. And these absentees—indistinct and overlooked—are scantly considered.

Moreover, the better universities often maintain chapels and theological schools that are not neutral within the local instructional pattern. Those local services to religion may be neutral enough toward mathematics, languages, music, art, engineering, law, physics and chemistry. But they unfailingly provide alternative and opposed answers to vital things in evolutionary biology, and they are often at variance with, or they strive to infiltrate, the departments of sociology, psychology, philosophy, history and education. The relative freedom displayed in these universities of the Protestant world is a priceless thing. Yet even their precincts are not free from religious opposition to the advance of unmitigated evolutionary thought. And just now many universities, large and small, are establishing or strengthening departments of religious studies. These great universities —these most modern and aberrant children of the monastery— have sprouted many wings, which they use freely and superbly. But to date not one of them in any land has been able to shed completely the monastic hood of the mother.

In most of the colleges and smaller universities that serve predominantly Protestant populations, the restrictive influence of religion is, however, much more effective. In the United States, a state-supported school is usually less restrictive than are private and church-supported schools. The forms under which restrictions are applied vary greatly. For the students, such items as compulsory chapel, required courses dominated by religious thought, and semireligious organizations sponsored by the college are utilized. For the faculties, particularly those directly concerned with evolutionary thought and its implications, there is usually no wholly invulnerable atmosphere. These sections of the faculty may be impressed merely with the need for much discretion, since here and there a teacher of evolutionary thought has been dismissed; since promotions and desirable posts more often go to frank conformists; and since still others may have been appointed to their posts because they were members of a particular church, or at least only after a "screening" had shown only a tolerable variety of view of diluted evolution.

An illustration of this last-named procedure can be cited from personal experience. While a young instructor at the University of Chicago, I was interviewed by the president of a small college in Iowa. The interview disclosed that he came to me thinking I was probably the person he wanted to head the department of biology in his college. But when he learned from me that the evolution I had been taught was not flat and innocuous, the entire matter was at once happily dropped.

At the all-important levels of education below the college level—those at which the formal education of most citizens ends—the diligent search of this writer has found no country in the world in which unmitigated evolutionary biology is or can be taught. The nearest approach to this is perhaps an occasional school in England, where a particular headmaster may elect to do or to permit quite free teaching. In general, the Protestant state and community permit the teaching of only a little more of evolutionary fact or thought than does the Catholic community. The greater difference lies in increased Catholic attention to the dogmas of the church that oppose or deny evolution. Further, the quantity as well as the quality of biology taught in the lower schools is usually influenced adversely by Protestantism. Of course, considerations wholly unrelated to religion may also influence the amount—though much less the kind—of biology taught in particular localities. Holland has long been a leader in the field of intensive agriculture both within its borders and in the tropics. In its secondary schools, most students receive five full years of training in zoology and botany. There, however, the implications of biological principles are as taboo as they are elsewhere.

Within the United States it is notable that only in the most Protestant and the most religious section of the country—the "Bible Belt" of the South—could state laws prohibiting the teaching of evolution be enacted. Two or three decades ago such laws began to operate in Florida, Tennessee, Arkansas and Mississippi. Florida did not enact a law but passed instead a "concurrent resolution" of the legislature. Similar sweeping legislation obtained fairly strong support in several other state legislatures, but failed of enactment into law. Less sweeping enactments, however, on both state-wide and community levels, were to follow. In Texas, for example, the word "evolution" had to be deleted from textbooks. And yielding to pressure, the authorities at the University of Texas passed the following resolution: "No infidel, Atheist or Agnostic shall be employed in any capacity in the University of Texas. . . . No person who does not believe in God as a Supreme Being and the Ruler of the Universe shall hereafter be employed." It should be added that, on the surface, these restrictions have become "dead letters" in Texas. There the governor no longer serves on the Textbook Commission, which nowadays (1950) approves four or five high-school texts from which local authorities may select the one they prefer. Thus many discreetly stated facts of evolution may now appear in Texas textbooks, and this is also true in Florida.

Both before and after the enactment of these antievolution laws—and in many or in all states of the Union—teachers of biology, geology and sociology were deprived of their teaching positions because they taught the principle of evolution. For this reason the writer's friend and mentor—to whom this book is dedicated—was ousted, in 1898, from the superintendency of the high school in Linton, Indiana. When a court in Tennessee confirmed the dismissal of teacher John T. Scopes from the Dayton high school in 1925, and also fined him one hundred dollars for teaching evolution, this outrageous practice got national and international attention. But since that date still others in lower schools and in colleges (some of them are the personal friends of the writer) have lost their positions in northern as well as in southern states because they taught the principle of organic evolution.

Catholicism is chiefly responsible for laws passed in recent years by two states—Massachusetts and Connecticut—prohibiting physicians from giving instruction to patients regarding contraception. This law is binding regardless of the importance such information may have for any type of patient. In 1943, a survey was made of the teaching of conception control in twenty-four American medical schools. It was found that some type of instruction was given as part of the curriculum in twenty-two of these schools and extracurricularly in the other two. In fourteen schools, students with religious objections, chiefly Catholics, were permitted to omit the course.

Early in 1950, the State of New York enacted a law providing that on the request of parents a child may be "excused from such study of health and hygiene as conflicts with the religion of his parents or guardian." Thus a fraction of the future citizens of New York may not be taught such things as the relation of bacteria to disease or the usefulness of vaccination and sanitation. Moreover, in order not to penalize these pupils for lacking this information, the Commissioner of Education has directed that no questions relating to these subjects shall be asked on the examinations of *any* pupil. Thus a possible way is paved for the ultimate elimination of these biological topics from the course of study and from the textbooks used in the schools of the state. This bill was strongly opposed by the New York Academy of Medicine and the State Charities Aid Association. But it was sponsored by the Christian Science Church and had strong Catholic support. Presumably Catholic support was given because the case would provide a precedent for later and hoped-for exemptions of Catholic children from lectures and films on sex education, and for such other items as Catholic convenience may dictate.

Particularly in Protestant England and America, the idea that "animals are God-created" and are uselessly employed in animal experimentation has been crystallized into national or state laws (antivivisection) that handicap biological and medical research and teaching. In England, the troublesome details and limitations of "licensing" are ever present and unavoidable (law of 1876, amended 1906), though the system does permit the use of animals for the more important experiments. In America, almost annually, busy medical scientists and teachers must go before several state legislatures to fight the perennial antivivisection bill presented there. In many states this fight is already lost. In 1946, biologists and physicians established a National Society for Medical Research whose job it is to combat continuously the misguided sentiment that presses for such laws.

It seems evident that, throughout its sphere of influence, Protestantism is somewhat better disposed toward popular education than is Catholicism. In general, too, it gives the private mind the right to interpret Scripture, which Catholicism forbids. But any marked difference in attitude between Protestant-sprouted

Fundamentalism and Catholicism toward teaching the animal origin of man would be difficult to establish. And Fundamentalism is very far from being localized in our own South. It is present as the thought of a majority or of an active minority of the religiously inclined in all communities of every state in the Union, and indeed in all other countries where Protestantism exists. It is true, of course, that here and there a Catholic scholar privately drops his dogma; and a Protestant pastor not infrequently finds a congregation that will give him what may seem like much rein—though it is, in fact, a stout tether. Some Unitarians and humanists may speak courageous thought. But past and present teachings of both Protestantism and Catholicism are everywhere crystallized in *community* sentiment. Throughout the Christian world that sentiment restricts or suppresses the teaching of the best of evolutionary biology in practically every classroom, from lowest to highest.

The existing "liberalism" of parts of Protestantism, particularly in England and some areas of America, is of importance on the religious scene. It is also of some importance to the present and future of science; it thus requires further notice and better definition. These liberals have shed, or they tend to shed, all belief in a devil, scriptural gods, a hell, and a Bible that may be read literally. They do not shed a belief in God, in the supernatural, and in one or another extra-earthly thing pertaining to the person, the career, or the teaching of Jesus; and usually an unpaved heaven is retained. They freely admit the evolution of religion itself; they admit that organic evolution proceeds, but only under the regulation and direction of their God.

Pulpit evolution is a most extraordinary thing. If ethics and moral law emerge from the animal world and from society—as these were described in Chapters 4 and 5—not man but God is nevertheless primarily involved. The earth still belongs to God, not to man. And a dictionary of items about the "soul" becomes the sourcebook of knowledge and reality. The pulpit evolution of the liberal clergy provides a subtle blessing of science that mutilates and rejects it. Liberalism, with its evolved God; like Fundamentalism with its crude God, is mainly Bible-born; in the United States it has drawn its recruits largely from those hitherto aloof to organized religion. The great advances of science—prominently Darwinism, old and new—have forced an appreciable shift of view of biblical things upon some millions of people, the liberals. There remain, however, those more numerous hosts who have refused to shift, the Fundamentalists. The unshifted bear most blame for the long and continuing fight about such things as antievolution laws and putting religion into schools and government. The shifted, the liberals, are the ones waiting at the college door to confuse with supernaturalism and a freshly crystallized dialectical theology the student who exits successfully with hard facts about the natural world that is his home.

The two extreme branches of Judaism, Orthodox and Reformed, are quite comparable with Fundamentalism and liberal Protestantism in their attitudes toward evolutionary biology. Though a minority group in many Western nations, Judaism usually rivals or exceeds the best of Protestantism in interest in general education. The extraordinary contributions of Hebrew scholars to scientific fact and thought, including the biological and psychological areas, suggest that within the Western world Judaism is not a stout resister of evolutionary thought. It is still too early, however, to know whether the new state of Israel will incline to modern thought or to a full-fledged theocracy. The following paragraph is from the January 12, 1951, issue of *The Jewish Newsletter:*

Another cabinet crisis in Israel was averted when the government yielded to the demand of the Religious Bloc, a group of clericalist parties in the Cabinet, that religious education be compulsory for children of Jewish immigrants arriving from Moslem countries. . . . The Religious Bloc have already succeeded in having numerous theocratic laws passed regulating diet, Sabbath observance, marriage, family relations and other manifold interests of the daily life of the average daily inhabitant, disregarding the fact that many Israel inhabitants do not hold Orthodox religious beliefs.

Islam directs education in many lands, but no one finds an oasis in which it fails to exclude and resist evolutionary thought.

In Hinduism, the religious and the social are so blended that in this brief word they may be treated as one. The sacredness of all animal life, including that of insects, is a major barrier to serious study and to all teaching of animal structure and function. University students of that faith have sometimes even refused to handle animal *fossils.* An eminent physiologist in

Calcutta told the writer that if he should undertake experimental work with monkeys—really needed in his studies and less expensive in India than elsewhere in the world—all the caretakers and helpers within his institute would quit their jobs. The established Brahmanic attitude, though tolerant of physics and of chemistry and of some botanical studies, bristles with resistance to much of zoology and to all penetrating biological thought. This point is touched upon later in this chapter.

Effects of Buddhism, Confucianism, Taoism and Shinto on the teaching of evolutionary biology in the Far East will be little treated in this survey. In the higher institutions of learning of prewar China and Japan, there was perhaps no more of such resistance than in the smaller colleges of the Western world. However, recently established conditions in China and Japan seem, temporarily at least, far more important.

The two condensed items that follow will perhaps suffice to recall the serious fact that, within a mere four or five years, the crown of evolutionary thought—the hope of building society upon trust in reality—has been either badgered or buried for the two most prominent peoples of the Far East.

A dispatch from Tokyo printed in the *Catholic Register* for May 3, 1948, stated:

> Like an earlier textbook on world history, a biology text for middle schools that caused considerable anxiety to [Catholic] missioners in Japan will be withdrawn. New books are being prepared by the Catholic university in Tokyo in conjunction with several Catholic middle schools, and will be submitted to the government.

We note elsewhere (Chapter 10) the present predominance of Catholicism in the Christianity that General MacArthur declared must accompany the democratization of Japan.

Soon afterward, the "missioners" of Russian communism began their work on the biology textbooks of China. According to *The New York Times* (December 19, 1949), the Communist-controlled Shanghai paper, *Ta Kung Pao,* described what was then being done to these "reactionary" texts in Communist China:

> In dealing with Darwin's theory of evolution care should be taken to differentiate between its progressive and reactionary features. . . . Reactionary theories on heredity propounded by Men-

del, Weismann and Morgan have already been deleted from biology textbooks for the senior and middle schools, but the progressive theory of Michurin should simultaneously be inserted to take the place of the discarded ones.

RELIGIOUS RESTRICTIONS IN HIGH SCHOOLS OF THE UNITED STATES

Much definite information concerning the restrictive influence of religion on the teaching of biology in the high schools of the United States has been obtained from 3,183 high-school biology teachers. It was derived from answers to a special group of questions included in a questionnaire mailed to these teachers early in 1940. The questionnaire was sent to approximately one-half of those actually teaching biology in the nation's high schools. But it went chiefly to teachers in schools presumed to be of the better grade, and to about 2,000 teachers who belonged to one or another organization of teachers of high-school biology. The better-than-average schools from which replies were received actually enroll rather more than one third (about 3,000,000) of the nation's total of high-school students, though fewer than one eighth of the teachers of high-school biology responded. The replies came from all forty-eight states, though the Southern states and the parochial schools throughout the country were inadequately represented. The questionnaire was sponsored by a committee of the Union of American Biological Societies, of which this writer was chairman. A report dealing with much of the information obtained in the extensive questionnaire has been published.[2]

Of this choice sample of teachers, only fifty-three per cent said that they had trained especially to teach biology. This is a meaningful fact, though the several items that account for it and enter into it cannot be appraised with confidence. In any case, it is probable that little more than one third of those now teaching biology in all high schools of the United States are trained to teach that particular subject.

The point of special concern here is to learn how two biologi-

2 O. Riddle, F. L. Fitzpatrick, H. B. Glass, B. C. Gruenberg, D. F. Miller and E. W. Sinnott, *The Teaching of Biology in Secondary Schools of the United States* (Lancaster, Pa.: The Science Press, 1942).

cal principles of first importance to an intellectual awakening fare in the high schools. First for notice is the principle of the genetic inequality of men. Twelve per cent of teachers in public schools, 31 per cent of those in parochial, and 10 per cent of those in private schools refused or failed to say whether they taught or did not teach this principle. Of teachers who supplied this information, the principle was taught in 2,551 public schools by 86 per cent, in 68 parochial schools by 65 per cent, and in 135 private schools by 87 per cent. But the Southern states and New England are inadequately represented in the sample, and in both of these areas acknowledged failure to teach this principle reached 23 per cent; and in parochial schools—also very poorly represented in the sample—the total failure may have been more than 60 per cent. When the failures or refusals to reply to the question are fully considered, it becomes probable that the principle was being taught by no more than 75 per cent of this choice sample of biology teachers; and of all teachers of biology in the high schools of the nation, probably less than 60 per cent taught it.

It was observed in Chapter 6 that an examination of the *implications* of this principle is what gives it significance in thought. And also noted there was the fact that these implications neither are nor can be examined at secondary and lower levels of education because of religious influence. The presence and effectiveness of this religious influence should become evident later, when this same group of teachers indicate the widespread religious restriction on their teaching of the evolution principle. In regard to the principle of genetic inequality, it is further notable that the restricted training of at least two thirds of those now teaching biology in the high schools of the United States is such that this large fraction of teachers is largely or wholly ignorant of the more important implications of the principle. Only an adequate training in genetics can bring full awareness of those implications; once aware of them, however, they can be taught as readily and easily as a lesson in history or arithmetic. This unawareness or ignorance in those who teach is ultimately partly born of religion, though it is immediately and greatly aided by educationists who insist that a "teacher" is a person who has been rocked insistently in a cradle of "methods." The implications of genetic inequality are everywhere taboo in the schools. And their frank presentation by radio would be forbidden as a "controversial subject"—religion again being the basis of the controversy.

The question of teaching or not teaching the principle of evolution "in my school" was made easy for all teachers to answer by supplying seven subdivisions of the question—for only one of which an answer was expected. This question was answered by 3,075 teachers, whose duplicate entries provided a total of 3,461 replies. Failing or refusing to answer the question were the following: teachers in public schools, 3.2 per cent; parochial schools, 9.9 per cent; private schools, 3.9 per cent. Of the 2,808 teachers in public schools, 124 indicated that the principle was either "entirely omitted" (109) or was "openly denied" (15). Of 90 teachers in parochial schools, 4 "omit" and 5 "deny." Of 177 teachers in private schools, 5 "omit" and 7 "deny." Various grades or types of teaching the principle were declared. It was either "taught as a fact" or "as a principle underlying plant, animal and human origin" by 1,651 teachers. This number represented 53.7 per cent of the teachers who returned questionnaires; and the number failing or refusing to reply to this question was, as noted above, 3.4 per cent. But 65 of these 1,651 teachers further indicated that their teaching was not open or direct but by "inference only"; and another group (approximately 60), who said that evolution was "taught as a fact," provided further comment, which showed that to them the principle did not include "human beings" but was "theistic" evolution, or that of St. Thomas Aquinas. The remaining large group of 1,541 replies indicated that the subject of evolution was dealt with in one or another attenuated form—"as applying to sub-humans only," "as a scientific hypothesis," or "by inference only."

Another item of the questionnaire asked the teachers who "avoided or lightly considered the principle of evolution" to indicate—by "checking" one of several printed headings, or by written comment—their reasons for doing this. Answers were given by 916 teachers, some of whom cited not one but two reasons, and thus a total of 1,237 of these entries were available for examination. Practically all of them pointed ultimately to a religion as the source of opposition to the teaching of this principle; they also served to locate the focal point of the opposition at some point on the line that extends from the teacher himself to his state legislature. Since no teacher was expected to sign his or her name to the answers supplied, there was no obvious incentive to give deceptive answers or to withhold true answers.

A total of 233 teachers, of whom 208 were in public schools, "checked" an appropriate column to indicate that the teaching of this subject was opposed by their own personal belief. From the written-in comments of several others it is clear that they, too, really belonged in that group. Only 73 indicated restraint by their "state legislature"; and, strangely enough, some of these replies came from states that had not passed antievolution laws. It is clear that some biology teachers believed or assumed that such state "laws" existed; whether this impression was confused with local law or city ordinance is unknown. Opposition to the teaching of evolution is assigned by 64 teachers to the "board of education"; by 90 to the "school administration"; by 287 to a "majority of the community," and by 109 to a "minority of the community."

The 381 written-in comments, which completed the total of 1,237 entries or answers, are so illuminating that they merit special consideration. From ten states—the four corners of the nation and six representative states in between—101 (of the 381) comments are offered below for the reader's information, and perhaps his entertainment. The public or private nature of the school and the size of the community (rural, town, or otherwise) from which each comment came are suitably indicated. It should be well understood that none of the comments reproduced here came from one of the 233 teachers who had elsewhere or *otherwise* indicated that evolution was opposed by his or her own personal belief. The following statements were replies to the question: "If the principle of evolution is avoided or lightly considered, the reason is:"

MAINE. *Public* (town): "Do not stress due to religious groups." "Very little time for discussion." (Rural): "No special reason except time."

NEW YORK (upstate only). *Public* (large city; New York City omitted): "Recent work in genetics shows discrepancies in Darwin's theory." "Call evolution 'development'; when evolution is mentioned we face facts." (Small city): "Only criticism has come from Catholics." "Not required in curriculum." (Town): "I've had fights but haven't lost yet." "No conflict between evolution and religion." "Present as much as I can get away with." (Rural): "Minister still objects but need not be taken seriously." "If openly taught as a fact would probably be strongly opposed by community." *Parochial:* "By scientific truth." "14- and 15-year-olds not capable of delving into philosophical reasoning so far beyond them." "Subject taught thoroughly according to St. Thomas Aquinas (13th century)."

NORTH CAROLINA. *Public* (large city): "Premature subject for age of 9th-graders." (Small city): "14-year-olds not sufficiently advanced to justify time required." (Town): "One could stir up trouble here if he cared to; take evolution for granted and make no issue of it and get along." (Rural): "Considered best to avoid." "If omitted would be due to pupil opposition." "Class time limited—text inadequate—taboo subject to most people." "Lack of opportunity for a clear, thorough treatment." "I can see no reason for introducing it; besides, it is a controversial subject."

FLORIDA. *Public* (large city): "Religious groups." (Rural): "Not essential for H.S. students to worry over." "Personal belief that hypothesis goes too far on facts known, certain cases of orthogenesis only do I accept." *Private:* "Evolution is not based on any fact or reason and Biblical account fits in with every natural law of science and nature."

WEST VIRGINIA. *Public* (small city): "Taught as it arises with no special emphasis." "Not much said about it." (Town): "Can teach what we like but do not think theory important enough to spend great deal of time on it." "Presented casually with definite proofs through plants and animals." (Rural): "Inability of students to grasp true significance." "Not entirely avoided but mentioned carefully because of possible misunderstanding."

WISCONSIN. *Public* (large city): "No point in bringing up as controversial issue; structural progress is obvious and treated frankly and scientifically." (Small city): "Who cares about evolution, my students don't; other topics are more important." "Taught as evolution, not as origin of species theory." "Teach evolution but never use word; community predominantly Catholic." "No definite opposition except occasionally from an individual." "Community highly religious." "Most H.S. sophomores cannot properly understand scope of theory; also many subjects of more interest and value to be taught." (Town): "Politics." (Rural): "90% Catholic community." "General educational and religious practice." "Not important in H.S.; time involved to explain can be used to better advantage." "Children trained in parochial schools immediately take issue if stressed and developed."

NEBRASKA. *Public* (town): "Conflicts with religious belief of some families." (Rural): "Content and nature of textbook." "If avoided, because teacher's place is not to break down what homes and churches have taught; besides, only a theory, not a fact." "I believe religious reasons." "Lack of time."

NORTH DAKOTA. *Public* (small city): "Conflicts with religious belief of some families." (Rural): "State requirements disregard it almost entirely; we follow the course of study." "We should take no stand in either direction." "By the bigoted ignorance of parents."

WASHINGTON. *Public* (large city): "Children have not yet good scientific background." "Introduced as theory; not made dogmatic." "Believe too deep for H.S. students." (Town): "Evolution meaning change—not from one-celled animal." (Rural): "Pupils asked to speak to their ministers; not necessarily avoided." "Text does not stress and I see no reason to when there are other things more important to H.S. students." "Just a touchy subject in a rural community." "Because lack of definite information that man came from lower forms; evolution within species definitely considered." "Stress from theory standpoint but tell them scientists do not have proof and they can believe or not."

CALIFORNIA. *Public* (large city): "Fundamentalist beliefs of majority of our students may not be attacked (Negro and Mexican)." "I do not think subject should be allowed to disturb any religious belief of adolescents." "Since we are not responsible for how we were created but are responsible for what we become I think it does not merit much time in H.S." "Title 'evolution' avoided." "Fear of public reaction and lack of support of administrators." "Don't use the word much; can't learn botany without learning evolution." "If taught as hypothesis and not fact would be little opposition anywhere." "Controversial subjects are dynamite to teachers." (Small city): "Seems relatively unimportant in helping individuals to live better." "To avoid argument and difficulty with a few parents." "Expected not to offend beliefs of students; do not refuse to discuss it." "Not much value to tenth-grade students." "Scientists' reasons given; own personal belief carefully guarded—no kick-back." "Mostly I believe in it but think time may be better spent on other phases at this grade level." "Little opposition in spite of Fundamentalists." (Town): "See no reason in our situation to be particularly concerned with this subject." "Unimportant until more scientific 'facts' are produced." "Not avoided but misinterpreted by parents; biology opposes teachings of their churches." (Rural): "Principal must kow-tow to all." "Enough biology can be taught without it." "No reason for omitting or stressing." "Objected to but not opposed." "Don't believe in giving too much." "General religious attitude of the group." "More valuable topics to use the time." *Parochial:* "Theistic evolution is taught in religion classes as well as in science." "Seventh Day Adventists oppose evolution as unbiblical and unscientific."

The above fragment of the comment of 381 teachers convincingly tells the same things the total of that comment tells. From even a rather choice sample of schools and teachers, with parochial schools and Southern states very inadequately represented, it is made clear and certain that religion-born sentiment opposes and chokes the teaching of the principle of evolution in most public high schools of every state in the Union.

The whole of the information obtained from the questionnaire shows that in the secondary schools of the United States biology is a cheated and battered subject. Its central principle, formally excluded by state law from use by textbooks and teachers in some whole states, is restricted and bludgeoned by more localized sentiment in practically all communities throughout the country. Because of these restrictions, and of related limitations on the amount of time granted to the teaching of biology, the all-important *implications* of the entire evolution principle —its extensions into the inorganic world and its mothering of morality in animal and human societies—are things practically untouched in *every* American high school. Finally, the principle itself is renounced by at least one tenth of those teaching biology in better public schools, and by practically everyone teaching the subject in parochial schools—these latter teachers presenting indeed a package labeled "evolution," but containing only a denervated ghost of its principle. Only one oblivious to fact and history could fail to assign each of the above-named elements of obstruction and suppression to an agency other than past and present teachings of organized religion.

Occasionally the public press has supplemented the information obtained from the questionnaire. Most educated people assume that the principle of evolution is rather satisfactorily taught in the public high schools of our *largest* cities. For them we take from the Philadelphia *Evening Bulletin,* of August 13, 1937, the following words of the principal of that city's most honored high school:

> The old theory of evolutionists as to whether man is descended from the monkey has been over these many years. Such teaching is discredited and is not representative of science and so will not be found in our textbooks.
>
> The public schools teach biology. In this study, the difference of species is indicated.
>
> The difficulty in teaching science often has been that it has been approached with an irreligious attitude. There is no such attitude among the public school teachers of Philadelphia.
>
> This old controversy is a recurrent thing among laymen but in science it has been generally discredited and is not representative of science.

If those bizarre, blind and circuitous phrases mean anything, they mean that the principle of evolution was not being taught in the high schools of Philadelphia; and that, in those schools, science in general was being used as an anesthetic, not as a stimulant. This is all the more regrettable and remarkable since in many other respects Philadelphia's schools are now recognized as among the nation's best. Probably in no high school in America can a parent or pupil now presume that uncurbed teaching of the evolution principle exists.

An additional reference should be made to a state of affairs in the public high schools of Greater New York, in 1953. In that city a considerable and increasing number of biology teachers are Catholics. Well-informed persons have told this writer that many of these teachers definitely refuse to teach "evolution" (as generally understood), and that nobody in authority ventures to consider or discuss this matter in any way. It is all a hush-hush affair. Of course, in sober fact, no teacher should be expected to teach a thing that he does not believe. The error resides in expecting a Catholic to teach biology in a high school. It is also stated that in history and some other subjects Catholic teachers likewise refuse to teach certain topics as they are taught by non-Catholics. The conditions described for New York are known to exist elsewhere. It must be assumed that they are destined to spread throughout the United States.

The foes of evolutionary thought have been only partially successful with state legislatures, but much more successful with their continuous and up-to-the-minute campaign against textbooks that deal in any forthright manner with the principle of evolution. The sale of the successful textbook for high schools is so profitable, and the opposition to suitable presentation of the evolution principle so widespread, that author and publisher are forced to compromises, silences and deadening dilutions. Shipley[3] and others have recorded many specific instances of this mangling of biological and sociological textbooks during the earlier years of this unholy crusade, and there is no intent to complete or extend that ugly record in these pages. Only one illuminating case will be cited here. An eminent biologist, a friend of this writer, is the author of a reputable high-school text in which the final five chapters rather well describe the evolution principle in a survey of the animal world. One day his publishers wrote him that from Mississippi had come an order for two thousand copies of his text—provided it could be furnished without those five chapters. The publishers wished his consent to the binding of this number of copies without the offensive chapters. When my colleague refused to agree to this emasculation of his work, the publishers simply scissored the five chapters from two thousand complete copies and sent them along to Mississippi.

Unquestionably it is a widespread religious sentiment that prevents the preparation, adoption and use of high-school texts that clearly present the vitally significant biological facts—even though their implications in thought are left untreated. Several of the most widely used texts are near marvels of clarity on all *other* topics considered—and these may include items such as gardening, contour plowing, antibiotic drugs, posture, budgrafting, and extensive identification and classification of insects. But clarity and elaboration will be—and must be—dropped when reference is made to the topics particularly vital to thought; and now and again one or several of these topics remain unmentioned. Some such are the concepts of earth formation under natural law; the extended evolution principle; the grow-

[3] Maynard Shipley, *The War on Modern Science* (New York: Alfred A. Knopf, Inc., 1927).

ing acceptance in science of the origin of the living substance from nonliving matter; the earlier existence on earth of apelike men and of still other species of man than *Homo sapiens;* and the method of inheritance that brings human inequality—talents and disease—by genes distributed on the rule of chance. These books can say something about natural selection, and indeed may use the word "evolution" by coupling it with some phrase about a new species coming from "pre-existent life." But they cannot clearly present or discuss any of the parts of biology that most awaken and transform thought.

Textbooks are used also by those being trained for high-school teaching in our normal schools and colleges. Are those texts for teachers exempt from the blight that has fallen upon the texts for high-school pupils? Most unfortunately many of the highly successful ones are not. As an item of evidence, excerpts will be cited from a text first published in 1934 and which, within a year, had been introduced into more than 131 normal schools and colleges in at least thirty-seven of our states and territories. A chapter of this book is entitled: "The Process of Evolution Cannot Yet Be Satisfactorily Explained." The final section of this chapter is dedicated solely to the proposition that what it calls "the doctrine of evolution"—but evidently only Darwinian work is included—"is quite compatible with a religious faith." To the untrained pupil or student-teacher, this, of course, usually means that it is quite compatible with whatever view of religion, or of the supernatural, he or she happens to have at the moment. From four paragraphs the following fraudulent lines are quoted: "The reader should remember that even Darwin himself did not believe acceptance of the evolutionary idea to be incompatible with a religious faith. . . . Why should the full-blown rose, the birds in the trees, the beasts in the field, and the stately oaks standing in the forest not be considered to be as much a part of God's world as the subjects of which the Bible treats?" There follow four lines from the Psalmist, and then this: "If this conception of the universe were kept in mind it would obviate much strife and confusion. The scientist can make no distinction between the natural and the so-called supernatural. What man can study, experience, and learn about through his senses is the natural; the supernatural is that part of the universe which he has not yet been able to understand [*sic*] or for which his powers of comprehension are too limited.

There is no difference between the two. The difference comes only in man himself. . . . So, then, since evolution neither denies the existence of God nor disclaims His directive influence over natural processes, it cannot be said," etc., etc. Then: "Finally, it must be remembered that the theory of evolution does not attempt to say when, why, or by whom life was first produced upon the earth. The honest scientist when pressed for an answer will say candidly that he does not know."

Why any textbook whose purpose is to outline and guide in the study of life-science should contain any phrase resembling those partly quoted from this book is beyond comprehension. Such full and far leaps into theology—with complete abandonment and betrayal of science—will never be made by the real biologist who loves and builds his science. And nothing is more absurd than to suppose that those broad leaps prepare one to teach others how to teach biology. Any biologist worthy of the name very well knows that his task is to facilitate an advantageous encounter between the student and the useful and meaningful phenomena and principles of biology. As a true scientist, he will not rob his teaching of the special and incomparable discipline that only the sciences can give—his student will have to wrestle with the facts and principles he finds. If and when astonishment at the inclusion of such material in a widely used textbook is sufficiently overcome, the biologist who well knows that his science today is very far beyond the point where Darwin left it will swear that he had not believed it possible—outside of theological discussions—to find words for extraneous paragraphs that would so defraud his science.

But how blithely may a student-teacher, trained from texts of this type, later patter this same deception to his or her own high-school pupils! It is both unquestionable and immensely disgraceful that many such teachers gaily mix this wine and brine and actually think they are teaching *science!* It is certain that very much of what passes as unrestrained teaching of the evolution principle in high schools consists in pouring this foul mixture into jugs with sieve bottoms thus carefully prepared. Brainwashing should have been patented by Christian peoples.

Finally, during the period of their prized growth and enrichment, the life-sciences have evidently lost part of their earlier and already inadequate place in the high-school curriculum. There is also much reason to suspect—though conclusive proofs cannot be presented—that a religious sentiment is partly responsible for this decline during the past forty or fifty years in the amount of biology taught, per high-school pupil, in the public high schools of the United States. Evidence as to the fact of a decline—except in Negro high schools established largely after 1910—is found in various reports of the Office of Education, in an earlier report by the writer,[4] and in the questionnaire that was cited above. From the latter source one finds that 11.6 per cent of 2,472 teachers indicated that in their schools a biological study had been displaced by a social study "during the previous five or ten years." It is true, of course, that in many high schools some pupils may and actually do obtain even more training in biology now than was offered forty years ago. The decline results largely from the fact that enormous numbers of pupils now receive no more than one third to one half of the amount of this study that was earlier required of all pupils. A report of the Office of Education (*Bulletin No. 17*) indicates the amount of science, all kinds together, studied by graduates of seven high schools located in Denver, Providence, and Long Beach (California) for the years 1900 and 1930. In 1900, science constituted 16.3 per cent of their total high-school studies; in 1930—following the antievolution legislation and a wider crusade—this percentage, in graduates of these same seven schools, was 6.9.

It can hardly be assumed that the religious pressure that in high schools deprives biology of its most exciting and illuminating principle—and its ability to interest and educate pupils—had no adverse effect on the ability of that subject to maintain or to extend its position in the high school. But if religion be held blameless, and the decline of this instruction be placed on the shoulders of educators—and if they, too, were in no way influenced by religious thought—there remains the fact that the best of biological science has lost ground in high schools of the United States during the past forty years. And certain it is that in this country the period of decline coincides (Chapters 9-10) with a growing movement to put religion into the schools, and with a religion-prompted period of antiscience, antimedical, antivivisection, and antievolution crusades.

4 Oscar Riddle, "The Confusion of Tongues," *Science,* January 17 and 24, 1936.

ON THE SIZE OF A CHALLENGE

Well-informed persons will not hesitate at drawing two conclusions: first, that unless the secondary school leads a youth to the beginning of an understanding of himself and of his place in nature, it has withheld or pilfered from him his modern birthright; and second, that a current national educational program that rejects or much neglects the ample spreading of key knowledge of the living stuff—in thought, sociology, agriculture, medicine and the rest—is inviting political failure and social inadequacy; this, because it does not come to grips with taproots of poverty, superstition, disease, fanaticism and communism. But people adequately informed on this or most other subjects are now, as always, only the thin edge of a very broad blade. Unlike Durkheim, most Americans rather hurriedly assume that it is precisely their educational program—restrained from the new and the true though it now is—that will provide their nation with a new and better society.

With the world's better universities and research institutes largely dedicated to truth, but with practically all of the gigantic rest of its educational institutions neglecting, twisting, curbing or suppressing basic evolutionary truth, there is little wonder that all peoples are still uninformed or misinformed on man's own nature and man's place in nature. Only formal and adequate education, beginning in the secondary school and earlier, can or ever could supply the missing information. Continued failure to face courageously this highly inconvenient fact is both easy and probable, although here and there such failure surely invites later social and political crises. The chief world-wide barrier to a solution is found to be organized religion. And no one lightly overlooks the stature of that barrier.

Biologists in nearly all countries, and particularly in America, have tried a compromise with religious creeds. That compromise has failed. Most youth of 1954, like those of 1859, leave our schools without having an opportunity to learn that the worthy facts concerning man's origin and destiny come not from religious traditions but from investigations made in biological and other sciences within the time of men now living. That compromise now robs most modern youth of opportunity to learn the outlines of what is known concerning his or her place in nature.

That compromise deprives society of its opportunity to attain a safer high level of thought and purpose. The present restrictive influence of organized religion on the teaching of the best of biology, psychology and sociology is intolerable. Such an influence, from whatever source, is too dangerous to the well-being of man—to modern beehive aggregations of men who live under ever-multiplied rules and laws that must wreck them if based on variegated tradition instead of a common knowledge. For moribund traditional beliefs to continue to exercise such governance over the educational and social programs of a country is a confession of either the apathy, the cowardice, the impotence or the intellectual bankruptcy of leadership in that country.

Some of the political and moral problems of all countries are now brought to the United Nations. The men who bring and must resolve these problems share no common view of either the origin and nature of man or the sources of values and morality. This is an outcome of the tragic tardiness and the deadly failure of education to press unmitigated evolutionary principles clean through religious diversity and across national boundaries. In the whole Western world, reason and duty have called for positive and aggressive action since, at least, the publication of Darwin's book in 1859. Some of the problems and some of the disagreements within the United Nations doubtless arise from that primary failure of education. Perhaps it can happen that persisting diversities born of this failure are capable of wrecking that present source of world hope.

The extensions and dimensions of biology's foes in the sphere of politics and government—all clearly born of religion—can be illustrated by four brief quotations. The first of these is not recent, but its atmosphere is neither local nor outdated. In 1879, a Select Committee of the (British) House of Commons included the following in its Report:

Physiology, besides being costly and useless, is an immodest subject. When the Author of the Universe hid the liver of man out of sight He did not want frail human creatures to see how He had done it.

William Jennings Bryan, who twice was almost elected President of the United States, and who had been its Secretary of State, said:

In one chapter Moses gives us three verses that more vitally concern men than all that can be found in the books that uninspired man has written.

Governor Miriam A. Ferguson, of Texas, after joining her efforts to those of the State Text-Book Commission in forcing deletions of all references to evolution in the many texts used in that state about 1926, said:

I'm a Christian mother who believes Jesus Christ died to save humanity, and I am not going to let that kind of rot go into Texas textbooks.

Rajrishi Purushottamdas Tandon, elected president of the dominant Congress Party of India, declared in 1950:

Cow protection is part of Indian culture and as such . . . the cow should be afforded full protection even if it leads to the collapse of the country's economy. . . . The cause for the deterioration of health [in India] is the smallpox vaccination.

It is a common and tragic mistake to underrate the extent and vitality of ignorance of simple biological truth in all lands today. The uninformed and misinformed, the credulous and infatuated, are everywhere. And always they are a menace to themselves and to others. The masses of men the world over seem more interested in myth than in mentally earned truth—perhaps because they have been so gloriously bathed in the one and only spoon-fed the other. Both a nation and a cult spoke through these recent words of a Gandhi disciple: "We have no right to take the lives of mosquitoes, flies, lice, rats, or fleas. They have as much right to live as we." Again, superstition—some call it religion—is now a more effective opponent of birth control in Connecticut and Massachusetts than in China and Japan. Every biologist must feel that India and China—now having the two largest assemblages of men—are much more restrained by ignorance of simple biological law and fact than by any deficiency in letters, law, arithmetic or the rules of trade.

Those who know what is involved in biology and in biological engineering will easily conclude that India's persistently basic problem is a staggering biological problem. A realistic program for health and agriculture alone in India would require the teaching of biology at all levels of education. And preliminary to that fruitful stage of the program, much and protracted national planning for both a supply of teachers trained in the life-sciences and biological research would be necessary. Nevertheless, one finds that, although India has made sacrifices and real strides in organizing chemical and physical institutes to deal nationally with these matters, she has taken no similar steps to wrestle with her immense and central biological problem. When the writer talked with Pandit Nehru on this subject in 1947, directly pointing to the relative lack of effort in the biological area, it became evident from his conversation that India's leaders were as yet unprepared to plan or venture into this field. A foreign observer of the Indian scene may surmise that advance on this front meets with almost insurmountable religious resistance, and that sovereign attention at present can be focused on little else besides pressing problems of survival of government and of national independence. The case of India is made the final item of this brief survey because it relates to about one sixth of the world's population, and because conditions there seem to represent the accumulated maximum of religious restrictions on the teaching of evolutionary thought and basic biology in a nation's schools.

Those who take the trouble to examine the goal of a public-school system agree well enough that this is not mainly to give the citizen more and more of any sort of skill or information, but to generate the outlook, capacity, discipline and habit that somehow accompany a comprehension of the social group, the world and ourselves. At least that was once a widely acknowledged aim. Within the community known as America, how well does practice conform to that end? Can it be denied that educational practice of the moment actually excludes this immense community from contact with and understanding of the potent newer knowledge of society, the world and ourselves?

To one who is familiar with the present range of thought-spreading evolutionary discovery, and who has also become partly aware of the over-all inhibitions in the world's classrooms, a condition of supreme concern becomes clear: in all lands practically every competent teacher of a cultural subject, from primary school through college, is now fully or partly restrained by religion-controlled public opinion from teaching the implications of an array of facts and principles of the highest importance to the mental stature, the welfare, and the survival of the race. Nothing in the spheres of intellect and social striving is more amazing than that this rather easily observed condition is so little recognized—and so feebly opposed—by most thoughtful people.

Dominant in a pervasive atmosphere of restraint upon the world's schoolrooms of today are the past and present teachings of organized religion. Whatever else it does, this one of society's oldest institutions partly changes school into a sepulcher for virile newer thought—virtually killing culture-building thought in the throes of birth.

9

What the Founding Fathers Meant: The Constitution on Church and State in Education

The culture of the United States has naturalistic and cultural foundations of strength which are connected. The naturalistic foundations of strength derived from two sources. The first source is the practically pristine nature, unobstructed by culture, which our founding fathers met when they came here. The second naturalistic source is the Newtonian mathematical natural science which came here by way of French and British culture and was woven into the legal institutions and the way of thinking and acting of our people by men like Jefferson, Franklin, and the Adamses. Whenever the arts and culture lose this essential, direct contact with nature, they become vacuous, merely precious, and wither.—*F. S. C. Northrop*

As to religion, I hold it to be the indispensable duty of all governments to protect all conscientious professors thereof, and I know of no other business which government hath to do therewith.—*Thomas Paine*

PAINE'S VIEW OF THE INDEPENDENCE OF RELIGION AND government, so clearly stated above, attended the birth of only one of the nations of the earth—the United States of America. Unquestionably his conviction was supported by still other founders of the Republic. Those Founding Fathers could not at once erase the many laws that had existed in the colonies and were in poor agreement with that conviction, but they struggled successfully to write the ideal into the charter of the nation. Largely due to the jealousies of the several religious sects, the following generation was able to extend *slightly* this nearly central theme of American democracy into the sphere of state and local law. Today, however, we are witnessing massive encroachments of religion into the areas of both education and government. Those encroachments fix the present lines of political battle between

200

secular method and supernaturalism. New decisions on this American ideal must now be reached by the nation's citizens.

The temper of those who had fought a war of liberation and later gathered to write a constitution is illuminated by the result of their discussion of a very early question. Several weeks after the Federal Convention of 1787 had started its work, and at a time when success in forming a Union seemed highly doubtful, the aged and ailing Franklin made a motion to open later sessions with prayer. "The Convention, except for three or four persons, thought prayers unnecessary" (note on a Franklin manuscript). But to this day, the great majority of the American people are unacquainted with the fact, well known to scholars, that many of those who most influenced the text of the Constitution were liberal dissenters in the religion then prevailing. Few equaled Paine in the thoroughness of their dissent, but the dissent was there. The Christian scheme of salvation was rejected, largely or wholly, by the dominant political leadership of the time. Though they all remained deists or theists, these men had obtained a new outlook on life from the then recent findings in Newtonian natural science and from the writings of Locke and Voltaire. "So great was the influence of this rational religion," wrote Morris Kline, "that no one of our first seven Presidents professed Christianity, though of course many made references to the Christian God in political addresses." It was not by accident or oversight that God was not mentioned in the charter of this new nation.

THE BIRTH OF TWO PROBLEMS

Popular and public education in America lay almost wholly in the future when Jefferson spoke a new thought and a prophetic truth in defining "the schoolhouse as the main buttress of any society of free men." Many of his countrymen and others elsewhere have fought and still fight valiantly to sustain intact that forward view of Jefferson. And a majority of Americans have joined in steadily extending at great expense a system that makes school attendance both free and easy, even compulsory. The end result to all, however, is a continual and dangerous sapping of the foundations of this bastion, the schoolhouse, together with mounting threats to the constitutional rights of everyone.

These two results flow from easily traceable sources, and this chapter is concerned with them. First of all, American Protestantism has refused to acknowledge or to observe, in the field of education, the principle of complete separation of church and state as written into federal law. Throughout the history of the American public-school system, children of Catholics, Jews and nonbelievers have been subjected quite generally to one or another amount—sometimes a small amount—of propaganda for the Protestant faith. And Protestantism is currently engaged in finding means for circumventing a recent decision of the Supreme Court and for further infiltrating the public schools. In another way, and very acutely in more recent decades, the foundations of the bastion that is the schoolhouse have been weakened and constitutional rights threatened by an increase of our Catholic population from fewer than forty thousand in 1790 to nearly one fifth of the total population today. But more particularly these results arise from the Catholic hierarchy's insistence upon its own education of the children of Catholic parents; and, at the moment, by coupling this with a demand that the American taxpayer provide aid of one kind or another to that Catholic education. In a nation committed to the utmost limits in the matter of individual and group freedom, this demand poses a problem insoluble within the intact ideal of a society that completely divorces civic from religious actions and is buttressed by the government schoolhouse. When fully examined, this Catholic demand for the conduct of education by the hierarchy proves to be not of local but of foreign origin— it is part of a Vatican mandate that operates as thoroughly as it can within the incompatible framework of our ideals and Constitution.

Most but not quite all of this chapter is concerned with conditions within the United States. Education is partially or largely a captive of the church in practically all lands. In the year 1950, there still exist two completely theocratic nations—Tibet and the Vatican. The Buddhist theocracy seeks complete isolation. The Catholic Vatican seeks a complete religious domination of the world that involves some essentials of political dominion—essentials already shared in Spain and a number of Latin nations. And it largely controls the educational systems of several additional nations. Where it dominates it commits the supreme outrage of taxing non-Catholics and nonbelievers for the support of Catholic education. Hierarchical aims in education are primarily concerned with extending the power of Catholicism; and no one can learn those aims from hierarchical demands already made in the United States, nor in any one decade.

The purposes of this book require a survey of the political and legal background of the current controversy over church-state relations in the United States. That political frame largely determines the extent to which our educational institutions may now serve an expanded human outlook and a rational society. Clearly, it is within the educational area that forceful evolutionary principles and social insights either become rooted or meet an opposition to their effective cultivation and use. Elsewhere the conclusion is reached that the survival of society itself already rests mainly on endless and unobstructed education. In part, too, this book appraises the service and disservice of organized religion in human affairs. This discussion is therefore an indispensable part of this book.

PROTESTANT PERVERSIONS OF THE PRINCIPLE OF CHURCH-STATE SEPARATION

One first remembers that in actual practice the separation of church and state in the United States was not accomplished at one stroke, nor indeed has it yet been attained. That idea or ideal had a slow and difficult birth, and actually an incompleted one. Following earlier notable experiences in colonial Rhode Island, Pennsylvania and Delaware, the idea of separation slowly matured throughout two decades. Leaders in that further development during those years, and for a generation later, were two Virginians, Madison and Jefferson. Their effective efforts began in 1776, with drafts proposed for the Declaration of Rights of the Virginia Constitution. Part of Madison's proposal read:

Religion . . . being under the direction of reason and conviction only, not of violence or compulsion, all men are equally entitled to the full and free exercise of it, according to the dictates of conscience; and therefore . . . no man or class of men ought, on account of religion, to be invested with peculiar emoluments or privileges, nor subjected to any penalties or disabilities. . . .

That of Jefferson read as follows:

All persons shall have full and free liberty of religious opinion; nor shall any be compelled to frequent or maintain any religious institution.

But even in the Virginia of 1776 it was possible to obtain only the disestablishment of the Anglican Church (with its enforced church attendance), exemption of all dissenters from payment of taxes to support that church, and a temporary suspension of such taxes on church members—this latter becoming permanent in 1779.

The natural-rights doctrines of the Declaration of Independence in 1776 also strengthened the cause of those who would later seek a complete separation of church and state in a new nation. Nevertheless, the rebelling colonists had been meeting together since 1774, and during this overstrained period they had accepted some old practices seriously opposed to the separation principle and which have never been corrected to this day. Thus, in that year, chaplains were appointed to conduct prayers for the Continental Congress, and the Revolutionary armies in the field were assigned chaplains of a variety of faiths. At the end of the Revolutionary War, in 1783, when asked by Franklin whether it would permit the establishment of a Catholic bishop in America, the Congress stated that it "had no authority to permit or refuse it, these powers being [then] reserved to the several states individually." In 1787, it was written into the Constitution that "no religious test shall ever be required as a qualification to any public office or public trust under the authority of the United States." In the Bill of Rights, the First Amendment, adopted in 1791, stated that "Congress shall make no law respecting an establishment of religion, or prohibiting the free exercise thereof."

Reference to the well-known facts just reviewed is needed here because they relate to two regrettable and generally overlooked truths that are essential to this survey. First, though it was from a predominantly Protestant culture that a hitherto unattained degree of separation between church and state emerged —by virtue of the hard fight of several rationalist statesmen and the rivalry of sects—it was that same Protestant culture that was responsible, even in 1791, for the failure to attain complete separation in either federal or state practice. Second, this supreme and well-begun effort was feebly supported and often shamelessly abused by the later Protestant generations to whom the unfinished effort was entrusted. The ideal of Paine, set forth at the head of this chapter, proved to be an unattainable ideal under American Protestantism. For more than a century afterward, the Protestant part of the population of America—little affected by Catholic power—gave nothing better than an occasional nod to 1791 and the Constitution; and it frequently and flagrantly transgressed that charter of rights.

Some significant items of that record of delay and of transgression belong here. Their total may be recorded as part of the high price Americans paid, and still pay, for the organized Protestantism of their past. In 1791, four of the thirteen states had not made the nominal break of separation. In Maryland and New Hampshire, "multiple establishments" were permitted at the discretion of the legislatures; in Massachusetts and Connecticut such establishments were compulsory. These states were still practicing on a broader scale the essentials of what many are now asking all states to do in the field of education—with tax money they were "co-operating" with the several churches and getting large lumps of religion into their schools. Their stiff and formal bows to the Constitution came later. Through eventually enacted state laws, these four states gradually dropped compulsory support of religion, but prohibitions against it did not appear in their constitutions until much later —1810 in Maryland, 1818 in Connecticut, 1819 (by statute) in New Hampshire, and 1833 in Massachusetts. Vermont's constitutions of 1777 and 1786 had eliminated compulsory support of religion, but a statute of 1783 permitted towns, by vote, to levy taxes for building meetinghouses and the support of a minister or ministers. That law was repealed in 1807. Puritan thought in New England proved itself a very firm crust not easily dissolved in the new oil of separatism.

Then as now states and state laws were laggards, reflecting Christian affiliations and impulses of majorities or organized minorities of the population. The Constitution, however, had declared and sought—not too clearly or specifically, it is true—to guarantee complete separation of church and state. In 1796, during the presidency of George Washington, this principle was reaffirmed indirectly or partly, in a treaty made with Tripoli.

It was stated therein: "As the government of the United States of America is not in any sense founded on the Christian religion —as it has in itself . . ." And Jefferson's terms were marked by a valiant fight by the chief executive to sustain fully the principle of separation. His letter of January 1, 1802, to the Danbury (Connecticut) Baptist Association was first submitted for criticism to his Attorney General with this comment:

It [the letter] furnishes an occasion, too, which I have long wished to find, of saying why I do not proclaim fastings and thanksgivings, as my predecessors did.

The address, to be sure, does not point at this, and its introduction is awkward. But I foresee no opportunity of doing it more pertinently. I know it will give great offense to the New England clergy; but the advocate of religious freedom is to expect no peace nor forgiveness from them.

Both in and out of office Jefferson stoutly maintained that the powers of government are entirely civil and wholly deprived of *any* authority to "co-operate" with some or with all religious groups. And, "To Jefferson," says Dumas Malone, "the freedom of religion meant the freedom of the mind."

President Madison fully shared these views. In 1811, he vetoed a bill to incorporate the Episcopal Church of Alexandria, then within the limits of the District of Columbia, saying, "The bill exceeds the rightful authority to which governments are limited by the essential distinction between civil and religious functions, and violates in particular . . ." His writings clearly show that he regarded as transgressions of the proviso against "an establishment of religion" all such things as tax exemption for churches, chaplains for Congress when paid by public funds, chaplains for the Army and Navy when paid by public funds, and religious proclamations by the President.

Long statements on this early history—though revealing survivals of old traditions and creeping transgressions of the new order—do not belong here. Recent and very full accounts of these church-state relations, including those involved in education, have been provided by Butts[1] and by Stokes.[2] For their bearing

---

[1] R. Freeman Butts, *The American Tradition in Religion and Education* (Boston: The Beacon Press, 1950).

[2] Anson Phelps Stokes, *Church and State in the United States* (3 vols.; New York: Harper and Bros., 1950).

on matters of history already reviewed, two wholly separate paragraphs are quoted from Butts:

And throughout his writings Jefferson used the phrase "freedom of religion" to include freedom from compulsory taxation for the support of religion as well as freedom of exercise of religious worship and belief.

Despite the clarity of the principle of separation of church and state as expressed in this authentic historical tradition, there have been many practices continued which are in effect holdovers from the pre-separation days of the seventeenth and eighteenth centuries. These practices include religious phraseology in several state constitutions, the appointment and payment of chaplains for Congress, for the armed forces, and for certain penal and charitable institutions, tax exemptions for religious institutions, religious exercises at official ceremonies, and certain requirements for religious oaths and tests for officeholders of a few state governments. The weight of evidence indicates that these practices are exceptions to the principle of separation of church and state rather than practices which prove the principle of "co-operation" between church and state. The principle is clearly "separation" and not "co-operation."

Canon Stokes regards the outcome of the bitter struggle for "separation"—in federal and state law—as practically nothing but a new "mutual" and "friendly" arrangement between two hitherto variously intertwined and everlasting agencies charged with directing basic human interests. Therefore, the persistence of church-state entanglements—and all later co-operation and immense transgression in Indian affairs, schools, chaplaincies, etc.— are viewed as *favorable* social developments whenever these seem to have aided the purposes and spread of organized religion. From his extensive studies (Volume I) we quote:

The colonial nonchurch-membership was a naturally liberal group and not interested in maintaining a state church. . . . Professor Sweet's studies have led him to the conclusion that [at the close of the colonial period] New England, the best-churched section of the thirteen colonies, did not have more than one church member to every eight persons in the total population.

It was only much later, in 1868, that certain of the original constitutional *restrictions* upon the federal government were made equally applicable to all of the several states. This was done through the Fourteenth Amendment, which said in part:

"No state shall make or enforce any law which shall abridge the privileges or immunities of citizens of the United States; nor shall any state deprive any person of life, liberty, or property, without due process of law; nor deny to any person within its jurisdiction the equal protection of the laws."

Meanwhile, during the first half of the nineteenth century—and particularly from 1820 to 1850—a system of public schools was being established throughout the United States. The Constitution had not made education a federal function, and indeed the states themselves placed little claim upon it since it was essentially in the hands of religious sects. But when sentiment for free and public schools developed, it appeared that public education would best be treated as a state responsibility. Here again the principle of separation was unequally expressed in the school laws of the different states. This was largely a result of the fact that Madison lost his fight in 1789 to make the principles of the First Amendment applicable to the states as well as to the central government. The private or sectarian predecessors of the new public schools were heavy with religion; and while the new schools were being established there was always present a group of problems relating to the use of public funds for religious schools, and for religious instruction in the public schools.

The new movement in education was propelled by a belief that a democratic citizenry could be best developed by providing *free* education and providing it for *all*—a belief unborn or unexpressed at the time the Constitution was written. There was now little local or community urge to change its religious character. Though the Fourteenth Amendment (making the First Amendment binding on the states) was not yet operative, it was early realized that separation statutes of the various states would prevent the use of tax money for *sectarian* education. And neglecting the aim and ideal of a complete divorce between civil and religious functions in the federal sphere, many state legislatures loosely concluded that "religious freedom" was sufficiently cared for by the elimination of *specific sectarian religious instruction*. This left a considerable place for Bible, prayer and other religious instruction since, for no moment were the nonbelievers and infidels—present in every generation of our history—accorded a thought or a right by those legislatures. Again, concerning the teaching of morals, many state legislatures reached the bigoted conclusion—unable then to know that morality arises naturally in animal and human societies—that this could be done by basing it on "those common elements of Christianity to which all Christians would either agree or take little exception." Thus it was decided that a nonsectarian religious public school of this type could be supported by taxation of all citizens.

Partial relief from this parody of separatism in education was eventually obtained. It came in some measure from the nonbelievers, who were most insulted by it, through the support they gave to the First and Fourteenth Amendments. It also came in part from the large number of Catholic immigrants who arrived from Ireland and Germany prior to 1860. The readings from the King James version of the Bible, which Protestants had put into the schools and treated as "nonsectarian," were called "sectarian" by the Catholics. This organized group wanted the readings eliminated, and sought a share of the public funds for schools of their own where Protestant teachings would not violate their rights of conscience. This initiated many new state laws, much litigation, and the partial solutions that project themselves into our own day. Protestants have persistently refused to follow the principles of Jefferson and Madison by protecting equal rights of conscience through the elimination of *all* religious instruction from schools operated by the state. Protestants have thus flouted the Constitution continuously, kept alive a most serious area of conflict and divisiveness, and still speak of "co-operation" in a country in which "separation" was accomplished in its Constitution.

The decade 1920-1930 was the center of a widely spread effort to write state laws forbidding the teaching of evolution in the schools. In March, 1921, the state of Utah passed a law whose First Section read:

It shall be unlawful to teach in any of the district schools of this state while in session, any atheistic, infidel, sectarian, religious or denominational doctrine and all such schools shall be free from sectarian control.

The words "atheistic" and "infidel," in addition to the usual use of the word "sectarian" in state laws and constitutions, were doubtless directed toward the teaching of evolution. No doubt whatever on this point was left by the joint resolution passed by the legislature of Florida in 1923, and by state laws enacted in Tennessee in 1925, in Mississippi in 1926, and in Arkansas in 1928. The trial of teacher Scopes at Dayton, Tennessee, took place in 1925, and that decade was one of widely exposed bigotry. Two separate editorials in the same issue of the New York *Evening World,* February 26, 1926, made reference to the two following occurrences: "It is a startling thing that a man is being tried in Brockton, Massachusetts, this week, under a law passed by the intolerance of three centuries ago, making it a crime to deny the existence of God." And again: "Witness a dispatch from Regensburg, Germany, wherein it is set forth that the study of biology has been interdicted in the Municipal Girls College as being incompatible with maidenly modesty."

Almost the whole of the United States became involved in the antievolution campaign spearheaded by William Jennings Bryan and several roving evangelists. The so-called Fundamentalists provided the aroused soldiery and the threatening votes. The history of five years of that movement, 1922-1927, was written by Shipley.[3] For nearly ten years many liberal-minded persons, in addition to scientists, gave much time and effort to checking this resurgence of medieval thought and act. Bitter fights in state legislatures, in some cases extending over several years, occurred in all Southern states except Virginia, and in eight other border or Northern states. In addition, state boards of education temporarily censored the textbooks used in California high schools, in North Carolina public schools, and in Texas, as earlier noted in Chapter 8. And several local school boards, in different sections of the country, dropped biology altogether from among the subjects taught in their high schools.

For a moment the atmosphere of this antievolution campaign may be recalled. First, here are four brief paragraphs from Shipley's book:

A mob at Morristown, New Jersey, the seat of Morris County, during the winter of 1924-25 made a bonfire of scientific books. Not one of the many newspapers in the county commented adversely on the affair. . . .

The Indiana State Superintendent of Instruction, Dr. Henry Noble Sherwood, did not hesitate to say, in an address before the

---

[3] Maynard Shipley, *The War on Modern Science* (New York: Alfred A. Knopf, Inc., 1927).

National Education Association, at its mid-winter session in Washington, D.C. (1925-26), that only those teachers should be employed who are active in church work. . . .

During the spring of 1925, patrons of School District 18, Jewell County, Kansas, voted 14 to 5 to order a set of "The Book of Knowledge" burned. The books had been purchased by the school board in 1923, before the rural pastors had been inoculated with Bryanism. . . .

On April 3, 1925, [in California] a large number of "resolutions" from Fundamentalist clergymen and churches were read before the State Board of Education; but a brief from the president of the Science League of America was neither mentioned nor read. The subservient board—not one of whom had any training in natural science—then and there, without discussion, passed a resolution that high school teachers might continue to teach evolution "as a theory but not as an established fact."

New York City showed both intolerance and wide areas of cowardice during this period. In May, 1926, that city's Superintendent of Schools, Dr. W. J. O'Shea, and Dr. Eugene Gibney, Director of Extension Activities, denied the Union of East and West and the League of Neighbors the use of the Morris High School for a Peace Week meeting, because of objections to three of the speakers, two of whom were ministers with independent outlook. The reported objection to lawyer Arthur Garfield Hays was that he had "defended Scopes in the evolution case, and we cannot allow people to speak in the schools who are against religion." Again, during this period of spotlighted strain (April, 1926), this writer's friend, Dr. Paul B. Mann, head of the department of biology of the Evander Childs High School, gave his approval to the inclusion of evolution in textbooks, but was constrained to advise that high-school instruction in the subject be given only by "believers in some religion." At a luncheon meeting in the Town Hall Club, New York, the press reported that "with a few honorable exceptions, the speakers resorted to compromises and evasions in the statement of their opinions."

Meeting with only a few victories on a state level, and finding that success was easier through local school boards and the intimidation of educators, this latter strategy was adopted by the churches almost automatically in subsequent years. Automatically also during the past decade this movement has joined forces with the "liberal" clergy—who largely opposed the antievolution legislation—in a common demand for religious instruction in the schools.

More recent events and current problems now bid for attention. Since the beginning of the century, and during recent years in particular, questions of religion in education and of church-state relations in education have appeared in new forms. In part, they have received the attention of the Supreme Court, and in part, they relate to legislation of a new type. Over all is the plain fact that the claims of religion on the schools increase year by year, and that some long-postponed solutions must soon be applied.

Two recent decisions of the Supreme Court are more appropriately discussed in the next chapter under the subheading, "The Catholic School Problem in America." At this point it is sufficient merely to mention the fact that in 1948 the Court's decision in the McCollum case ruled that religious instruction during school hours in a public-school building is a violation of the Constitution. The attitude of Protestantism to religion in education immediately before and after that decision now requires closer examination.

It should first be pointed out that wholly apart from pulpit pressures—both Protestant and Catholic—some prominent laymen have given their full support to the movement for "religion in the schools" and to the invasion of constitutional rights, which the Supreme Court decision in the McCollum case refused to uphold. According to *The New York Times* of January 24, 1926, an influential paleontologist and museum administrator, Henry Fairfield Osborn, with an equal disregard for the biology not found in bones and for the rights of free men written into our Federal Constitution, spoke in New York as follows:

I am a strong advocate of restoring the teaching of religion to our public schools, religion of the kind which has been abolished because of purely theological differences, not because of its inherent lack of force in education. As a man of science, I am not tongue-tied by adherence to any denomination, creed or dogma; I am free to speak from the scientific standpoint whatever may be my personal opinion and principles.

I would like to see all the religious men of this great city of 6,000,000 souls, of this great country of 100,000,000 souls get together and agree upon a simple, elemental and more or less primeval teaching of religion, in which all men, except those who persuade themselves that they are atheists, agree.

Assuming that the citation just made sufficiently indicates the extent to which some self-appointed spokesmen for science have betrayed it in the two fields of civics and religion, it is gratifying to note, as a further fact in this bit of history, that throughout the years some Catholic citizens have opposed their Church's drive for the educational segregation of Catholics.

The question now arises: How have religious groups and city or state educational authorities responded to the decision of the Supreme Court in the McCollum case? Parts of the letter and the whole of the spirit of that decision are a declaration that it is unconstitutional for any public-school authorities to "openly or secretly, participate in the affairs of any religious organization or group" and that "no tax in any amount, large or small, can be levied to support religious activities." How could these mandates be carried out conscientiously without permitting the public schools to wash their hands of all that pertains to religious education? Has organized Protestantism permitted the schools to do this? The continuing evasions and transgressions of this decision by religious groups provide at once a heavy indictment of current organized religion in America, and a sure proof of the foresight and wisdom of those who framed our Constitution but were thereafter unable to erase the existing intolerance of state laws. The nature or form of these evasions and transgressions by religious groups varies from state to state. Only an outline of the story, however, belongs here.

A favored method is to have complacent school boards dismiss classes, or a part of them, an hour early, one or more times weekly, with no reason specified as the purpose of the dismissal. Of course, the pupils know that they are expected to go somewhere for religious instruction—and it is the state-supported school that points out this expectation and indicates its approval of that instruction. Unquestionably, in classrooms here and there, the pupils are told also that if they do not use the hour for the intended purpose unpleasant consequences will result. Attorneys general in various states have endorsed this practice of "released time" and in some states lower courts have upheld it. Specific methods and events as these relate to two states only will be included in this account.

Three published items relating to the city and state of New York are informative. In New York, a state law forbids giving religious instruction in the schools. It provides instead that pupils may be dismissed for one hour a week, if their parents so request, to attend classes in churches or religious schools of their respective denominations. Local school boards determine what pupils who remain in school may do in the absence of those who have gone for religious instruction. A few days after the McCollum decision, letters of inquiry about proposed changes in these current school practices were addressed by editor Joseph Lewis, of New York, to the heads of the city and state educational systems.

The Freethinker, for May, 1948, published the two replies, from which the following excerpts are taken:

It is the opinion of the Board of Education that the program of released time adopted for use in the New York City schools violates no constitutional or statutory provisions and therefore the board has decided to continue the present practice.
(Signed) William Jansen, Superintendent of Schools.

For the state:

This office is not at all sure that under the United States Supreme Court the practice being followed in this State of releasing children for religious education under the provisions of section 3310 of the Education Law is not constitutional. . . . Until there is some new conclusion of the courts in this matter boards of education, upon my advice, are making no change in their programs.
(Signed) Charles A. Brind, Jr.
Counsel for the State Education Department.

On July 12, 1951, the State Supreme Court upheld this law with one sharply dissenting vote.

Something concerning the actual operation of the program for "released time" in New York City nearly two and one half years later is found in the following account, published in *The New York Times,* August 16, 1950:

The desirability of the "released time" system for religious instruction in school hours came in for frank discussion yesterday by representatives of leading religious groups and teachers. The question arose at a meeting of the sub-committee on youth activities of the New York State Citizens Committee of One Hundred for Children and Youth. The sub-committee's recommendations will be incorporated in the committee report to the White House Conference on Children and Youth next December.

The sub-committee considered whether it should recommend universal support of "released time." A schoolteacher told the meeting that she did not think the program was worth while "and most schoolteachers think that, too," in that it created many difficulties.

One of the difficulties she cited was that "time was wasted" for children who did not attend religious instruction, since teachers could not present any new subject matter to the class "while a fraction of the children were gone." This also represented a waste of taxpayers' money, she said.

She held that the program added to the truancy problem in that many children merely "wanted to get out of school," and instead of going to church or synagogue played hooky.

Msgr. Cornelius J. Drew, pastor of St. Charles Roman Catholic Church, said that "not all teachers opposed the program" and that difficulties should not be considered in endorsing the principle. W. Noel Hudson, executive vice president of the Federation of Protestant Welfare Agencies in New York, said that even if there were "particular and unique difficulties," nonetheless "schools share with churches and public and private agencies the responsibility to carry to youth an opportunity for religious education and experience."

Finally a tentative recommendation was drawn up by Msgr. Drew and the Rev. Joseph A. Belgum, director of the Lutheran Inner Mission Society, Brooklyn, which the group approved. It read:

"It is a mature national conviction that religion is a necessary aspect of a child's growth. In order to bring this to the child the committee endorses a close and resourceful relationship between the public school and the churches of the vicinity together with their related social agencies. This relationship should strive to improve their points of common contact in the following areas:

"1. The released-time program should be continued with a frank discussion of its difficulties in administrative cooperation and creatively seek to improve the liaison relationships, the possible content for the children remaining in the public school and the elimination of moving children from one institution to another.

"2. Parochial schools should be encouraged to participate as equals in the family of public and private schools in the community.

"3. Channels for clearing the schedules of both the church and the public schools in the community should be arranged by the local *superintendent of public schools"* [italics mine].

The conference also stressed that the subcommittee report include the statement that there is a need for a "total community approach" to all religious problems.

From the concluding words of the resolution adopted by this subcommittee—"that there is a need for a 'total community approach' to all religious problems"—it is clear that this group still does not recognize the American principle of separation of church and state. Elsewhere, in full defiance of the McCollum decision, the head of the Protestant Welfare Agencies in New York declared that the "schools share with the churches and public and private agencies the responsibility to carry to youth an opportunity for religious education and experience." Further, a Catholic priest and a Lutheran official, representing the two groups having most parochial schools, wrote the adopted recommendation, which asked that the "channels for clearing the schedules of both the church and the public schools in the community should be arranged by the local superintendent of schools." In asking or in receiving this service of a public official, a service paid from tax money, they flouted once more the Court's clearly expressed decision. Again, for this group, the unwelcome yet crucial testimony of a teacher—among other things that the released-time procedure was "a waste of public money"—was brushed aside and became no part of the recommendations that were to find their way to the White House Conference on Children and Youth in December. The importance of this fragment of action in New York is to be found in its clear evidence that organized religion in that state continues and intends to continue to break the basic law of our land.

Three and one half months after the McCollum decision, the state of Florida's legal adviser saw nothing in it that would change current educational practice, and religious teaching of a kind was continued there. A story in the Tampa *Morning Tribune,* June 24, 1948, sums it up:

Attorney General Watson today said Florida's non-sectarian Bible instruction plan was not made illegal by a recent U.S. Supreme Court decision banning religious teachings in the public schools.

He gave his opinion to State School Superintendent English,

who had asked for an interpretation of the Supreme Court decision's effect on four plans for religious education in Florida.

Watson declined to give a definite answer to English's questions on three of the plans. He said they were hypothetical and he was "unwilling to assume anything" with respect to the case in view of the importance of the subject. He asked for specific examples.

The plan, which he said is legal, even in view of the Supreme Court decision, now is provided for by regulations of the State Board of Education.

It permits teaching of a Bible course to those pupils who want to take it. English described it this way:

"The purposes of such an elective course would be to present necessary historical and literary background and to give general guidance in ethical values accepted by all our people.

"Emphasis would be given to general character development; no discussion of sectarian beliefs would be permitted. Such a course is administered most satisfactorily through the employment of a full time, regularly certificated teacher of Bible who is a member of the school faculty."

Watson said that it "does not offend and is not made illegal by anything laid down as law by the Supreme Court of the United States" in its recent decision.

Watson declined to answer questions by English on the legality of [three] plans [not quoted here] now in use in Florida schools.

In a letter to the writer dated January 10, 1951, Sam H. Moorer, Assistant Director of Instruction in Florida, stated:

In 1940, this department asked a group of ministers representing various faiths to prepare a bulletin, "Suggestions for Bible Readings in Florida Public Schools." This bulletin was prepared and distributed widely over the state in 1940. . . .

In order to have Bible placed on the face of a teacher's certificate in Florida, one must meet the general requirements for all public school teachers plus (1) twelve semester hours of Bible study or of courses in religious education giving literary and historical appreciation, (2) the applicant must meet the specialization for one of the following broad fields: English, mathematics, science, and social studies.

It was further ascertained that up to the end of 1953 the Attorney General of Florida had made no further rulings or interpretations of the law; and that earlier practices were being largely continued in the schools of Florida. In few states does organized religion—mainly Protestantism—exercise greater power over education, and over legislation, than in Florida. And in that state it is now hard to find public schools in which the McCollum decision has even slightly promoted the constitutional right to education free of religious intrusions.

Speaking on November 7, 1952, to 2,000 public-school teachers of the Tampa area, Dr. Doak S. Campbell, president of Florida State University, besought his audience, often in only lightly veiled terms, to indoctrinate their students "in religion." He declared it wrong "only to present the facts and let the students draw their own conclusions." (Tampa *Morning Tribune*, November 8, 1952)

This brief two-state review of measures taken or not taken by religious groups to abate their encroachment in the public schools could be extended, with like examples, to other states. It is true that many school systems have dropped the practice that was specifically complained of in the McCollum case. But an acceptance of the clear text and meaning of the McCollum decision is perhaps exceptional and not the rule. It might be supposed that by 1953 a majority of Americans would have learned that "the government of the United States is not in any sense founded on the Christian religion." Again, most Americans with or without a religious faith would presume that organized religion in our day would be among the first to obey the Constitution. That has proved to be far from true.

In a split decision, six to three, the Supreme Court, on April 28, 1952, affirmed the constitutionality of the present law of New York State concerning "released time." From the four opinions rendered in this case, we here quote some important statements. The first is from the majority opinion:

Views pro and con are expressed, based on practical experience with those programs and with their implications.

We do not stop to summarize these materials nor to burden the opinion with an analysis of them. For they involve considerations not germane to the narrow constitutional issue presented. They largely concern the wisdom of the system, its efficiency from an educational point of view, and the political considerations which have motivated its adoption or rejection in some communities.

Those matters are of no concern here, since our problem reduces itself to whether New York by this system has either prohibited the "free exercise" of religion or has made a law, "respecting an establishment of religion" within the meaning of the First Amendment.

From the dissent of Mr. Justice Black:

In this case . . . the sole question is whether New York can use its compulsory education laws to help religious sects get attendants presumably too unenthusiastic to go unless moved to do so by the use of this state machinery.

That this is the plan, purpose, design and consequence of the New York program cannot be denied. The state thus makes religious sects beneficiaries of its power to compel children to attend secular schools. Any such use of coercive power by the state to help or hinder some religious sects or to prefer all religious sects over non-believers or vice versa is just what I think the First Amendment forbids.

From the dissent of Mr. Justice Frankfurter:

Again, the court relies upon the absence from the record of evidence of coercion in the operation of the system, . . . but the court disregards the fact that . . . the petitioners were not allowed to make proof of it. . . .

When the constitutional issues turn on facts it is a strange procedure indeed not to permit the facts to be established.

From the dissent of Mr. Justice Jackson:

This released time program is founded upon the state's power of coercion which, for me, determines its unconstitutionality. . . .

We start down a rough road when we begin to mix compulsory public education with compulsory Godliness . . . the wall which the court was professing to erect between church and state has become even more warped and twisted than I expected.

Today's judgment will be more interesting to students of psychology and of the judicial processes than to students of constitutional law.

The associated question of "Bible reading" in the schools of today also requires consideration. Comment on the matter would be less necessary if present practice merely represented pockets of survival of an ancient wish or tendency. The fact, however, is far otherwise. We now deal with a relatively new movement, which has been consolidating its position during almost four decades by seeking and obtaining *compulsory* Bible reading in the public schools of various states. Five weeks after the Supreme

Court decided that New York's state law concerning "released time" is constitutional, an article by Associated Press writer Ronald Autry reported (June 1, 1952) that "a strengthened drive is under way throughout America to provide religious education for more children through public school systems."

In a comprehensive study of legislation in America on this and related questions, published in 1934, Professor Johnson said:[4]

In states this spirit is reasserting itself in a movement to make Bible reading in the public schools compulsory.

During the first fifty years of the nineteenth century only one state, namely, Massachusetts, enacted a statute requiring Bible reading in the public schools. During the last fifty years of the century no such law was passed, nor in the opening decade of the twentieth century; but during the next ten years, 1910-1920, the legislatures of several states enacted laws requiring Bible reading in the public schools: Pennsylvania in 1913, New Jersey in 1916, and Alabama in 1919. During the decade 1920-1930 six more states passed similar laws: Georgia in 1921, Maine in 1923, Kentucky in 1924, Florida and Idaho in 1925, and Arkansas in 1930; and in 1926 the Board of Education of the District of Columbia passed a ruling requiring the Bible to be read daily in the public schools of the District. . . . There has been a definite attempt to bolster religion by requiring Bible reading in the public schools. The incidental controversy and its significance is the chief concern of this book.

In 1931, the American Civil Liberties Union described the movement as follows:

Even more successful than the attempt to impose "Genesis as a state religion" has been the movement to compel by legislation the reading of the Bible in the public schools. In practical effect this is equivalent to attempting to impose the Protestant religion on the children of the schools, for the King James version is almost invariably the Bible selected. Any Bible reading in the schools was once generally held to be contrary to the provision for complete separation of church and state.

The movement did not spread until after the war [World War I] and developed under the influence of the Klan and the Fundamentalists. . . .

---

[4] Alvin W. Johnson, *The Legal Status of Church-State Relationships in the United States, with Special Reference to the Public Schools* (Minneapolis: University of Minnesota Press, 1934).

As Butts states:[5]

It is evident that Protestant groups and the Protestant outlook have been the great impetus behind the movement for Bible reading and that the fundamental outlook of judges who affirmed its constitutionality has conformed to the Protestant assumptions that the Bible is a common religious and moral heritage of all Americans.

The early and now accelerated movement for compulsory Bible reading and prayer, the effort to write antievolution laws in the states, the accent on religious festivals, the crusade for "released time," and the current drive for direct religious instruction in the schools are clearly all successful fronts of essentially the same movement. The fact that small segments of the "liberal" Protestant clergy opposed the passage of antievolution laws is absorbed in the larger group of facts. In recent decades extremely few in that "liberal" group have sought to check the several methods of infiltrating and crumbling the wall separating church and state. The multifaced movement is designed to advance the Christian religion through the use of the public schools. If *morals* were actually its center, there also would be calls upon books that deal with morality; compulsions would not be sought; the word and spirit of our federal Constitution would be accepted; and the rights of the nonreligious and non-Protestants would be respected.

In 1950, twelve states and the District of Columbia had laws requiring Bible reading in their schools. Most other states then had laws which specifically prohibited "sectarian instruction," though their courts or other local authorities have ruled (National Education Association, 1946) that this does not prohibit Bible reading. Thus, in 1950, thirty-seven states either required or permitted Bible reading in the public schools. Additions to this number since 1950 are not accurately known to this writer. However, during the early months of 1953, bills requiring Bible reading were pending in the legislatures of Maryland and Oregon, while permission was being sought in California. A pending resolution of the Ohio legislature calls "upon the authorities of the schools in Ohio to invoke prayer as a part of each day's program." And Pennsylvania, adding to its present outrageous requirements, was considering a bill that would require every

5 *The American Tradition in Religion and Education, op. cit.*

public, private and parochial school to place a plaque on each school building inscribed "In God We Trust."

Educator F. Ernest Johnson[6] records the following: "A New England school superintendent told me his school system had solved the religious problem in the teaching of history. 'How?' I asked. 'By omitting the Reformation,' said he."

In the same book the observance of religious festivals and holidays in the public schools of the United States is discussed from the Jewish point of view by Rabbi Simon Greenberg. Partly quoting from records of a conference held on this subject, in 1949, he writes:

Yet in one American community twenty or thirty per cent of whose population was Jewish, recently during the Christmas season assembly programs were conducted [in public schools] in which priests participated. They intoned the prayers, invoked the Holy Trinity, and Catholic students went into the aisle to kneel and cross themselves at the proper moments. Moreover, in a number of high schools in that city "speech classes would be assigned the task of spreading the story of the Nativity in sequence form over a period of three weeks, over the public address system, being channelled into every classroom, with priests coming every morning for invoking prayers during the three weeks preceding Christmas day, in which the story of the Nativity was going over the public address systems. . . ."

These practices (at Christmas and Easter), on one level of intensity or another, are becoming well-nigh universal. . . . In addition, these public school celebrations are becoming more rather than less religious in their content and presentation.

Though these celebrations, along with "released time" and the less offensive Bible readings, are inwardly opposed by the Jewish population, the rabbi further notes that this opposition is not expressed "because they [Jews] frankly fear that in touching upon such emotionally charged issues, deep anti-Jewish resentments will be stirred within the community."

Incidentally, some statistics are enlightening here. Surely the half of the American public that is (or lately was) unassociated with the churches has not pressed for these various invasions of religion in the schools. And Rabbi Greenberg's (unquoted) statements assure us that some millions of Jews, though asso-

6 F. Ernest Johnson, ed., *American Education and Religion* (New York: Harper and Bros., 1952).

ciated with churches, are overwhelmingly opposed to the invasions. So, too, are some of the clergy and laymen of "liberal" persuasion. Thus it cannot honestly be claimed that a "majority" of our people (see Chapter 12) pressed for these successful and multiform intrusions of the past decade. The real pressure comes from many militant clergymen and laymen, and a few school officials, all heedless of a citizen's constitutional right to total aloofness and uninvolvement in religion.

Again, some Americans have developed a slight degree of willingness to have all views on controversial political and social questions discussed in the schools; but they base this concession on the idea that study and discussion of this kind helps in *finding the truth*. The case of religion, however, is wholly different. The clergy and that part of the public that presses for the teaching of religion in the schools consider this in no wise a *search for truth*, but as arbitrary acceptance of highest truth previously established by divine sanction. For them religion most definitely is *not* a subject of discussion and debate.

A closer look at the statutes of two or three of the states that require Bible reading in their schools is warranted. In April, 1950, the legal advisor of the Superintendent of Schools of Pennsylvania quoted the existing law as follows:

At least ten verses from the Holy Bible shall be read without comment, at the opening of each public school on each school day, by the teacher in charge; Provided that where any teacher has other teachers under and subject to direction, then the teacher exercising such authority shall read the Holy Bible, or cause it to be read, as herein directed.

If any school teacher, whose duty it shall be to read the Holy Bible, or cause it to be read, shall fail or omit to so do, said school teacher shall, upon charges preferred for such failure or omission, and proof of same, before the board of school directors of the school district, be discharged.

In Florida, the state constitution itself declares: "That all schools in this state that are supported in whole or in part by public funds, be and the same are, hereby required to have once every school day reading in the presence of the pupils from the Holy Bible, without sectarian note or comment." And laws have provided that, in making their monthly reports, the teachers are required to show that they have complied with the above act. The county superintendent must make certain that the reading has been complied with before he can draw salary warrants on the public funds.

The Supreme Court of New Jersey recently upheld the constitutionality of a statute that requires the daily reading of five verses from the Old Testament, together with the recitation of the Lord's Prayer, in all public schools. On March 3, 1952, the United States Supreme Court refused to pass on the constitutionality of this law. Justices Douglas, Reed and Burton dissented. The opinion written by Justice Douglas held that "the issues are not feigned; the suit is not collusive, the mismanagement of the school system that is alleged is clear and plain. Where the clash of interest is as real and strong as it is here, it is odd indeed to hold there is no case or controversy within the meaning of the Constitution."

To many Americans it must be both surprising and nearly incomprehensible that a legion of great communities of the nation thus badger and humiliate that large host of teachers and principals to whom the nation looks for a modern and costly education. It would be difficult to devise a means more effective than these laws and practices for driving highly competent and self-respecting persons—especially the mentally emancipated or those trained in science—from the all-important field of public education. The continuance and increase of this stultifying practice provide much of the reason that a modern social outlook and all evolutionary thought must now fight for consideration.

In 1927, North Dakota passed a law making it the duty of the school authorities of every state-supported institution of higher education to display in a conspicuous place in every classroom a placard containing the "Ten Commandments of the Christian religion." A second thought on a similar proposal recently made in England is given in the following quotation taken from the Newark (New Jersey) *Star Ledger* of April 13, 1950:

An English clergyman admits that the Ten Commandments should not be taught to children because they might get a wrong idea about God. A local education committee in Ipswich, England, was considering a proposal to post the Commandments in nearby Wickham Market School to fight a "decline in moral standards" among pupils.

The Rev. W. G. Hargrave-Thomas, Church of England Vicar of Needham Market, objected that this would be "misleading and wrong." He quoted the Second Commandment about a "jealous God" who visits the father's sins upon their children. "No decent father would do such a thing," he said. The Committee took the Reverend's advice and voted down the proposal.

But such "second thoughts" now seem alien in these United States. In late 1953, local and state authorities gave their blessing to this practice in the fast-growing small city of Huntington, Long Island, thirty-five miles from New York. *The New York Times* (September 20, 1953) reported:

Adopting unanimously a proposal . . . that printed and framed copies of the Ten Commandments be placed in each school room in the nation, the Board of Education of Union Free School District 3 of this village has ordered such copies hung in each classroom of the two high schools and five elementary schools of the system.

Letters of approval have been received by the board from the clergymen of the Roman Catholic, Protestant and Jewish faiths, as well as from many patriotic organizations and from Dr. A. E. Getman, New York State Commissioner of Education.

When evolutionary thought was having an effect on the outlook of many intelligent American laymen two full generations ago, even the minister of Plymouth Church, Henry Ward Beecher, could say: "The God of the Bible is a moral monstrosity." In Beecher's country today, presidents, cabinets, congressmen, leaders of education and of industry regard that God as the one somehow infiltrated early into our government, and they regard that Bible as a requisite guide to morality in our public schools.

We proceed. To this moment the emancipated thought of teachers in *state* colleges continues to be a reason for depriving them of their positions. And to this moment state courts continue scandalously to grant ministers and religion powers and immunities never countenanced in our federal Constitution. Both of these shameful facts are well illustrated in the following excerpts from an Associated Press dispatch from Fairmount, West Virginia, dated December 21, 1951:

A minister refused to answer questions today in the $100,000 slander suit brought by a college art teacher who says she was fired because she had been called an atheist and a poor security risk.

The slander suit was brought by Dr. Luella R. Mundell, dismissed last summer as head of the art department at Fairmount State College. Thelma Brand Loudin, of Fairmount, a member of the State Board of Education, is the defendant.

The Rev. James Clair Jarvis, pastor of Fairmount's First Methodist Church, acknowledged he sat in on a conference with Mrs. Loudin about the Mundell case in September. When asked what Mrs. Loudin said, the minister said: ". . . The relations and conversation between a minister and those who come to him for counsel and guidance or confession are confidential."

Dr. Mundell's attorney asked him if what Mrs. Loudin said was in the nature of confession of sin. The minister said it was not. The attorney then contended the West Virginia law on confidential status of a minister's knowledge covers only the confession.

Judge J. Harper Meredith said he didn't know what the law said but he had great confidence in the Rev. Mr. Jarvis and would not require him to answer.

Religious student and author Espy[7] recently provided this item regarding a large state university:

A state university president who felt limited because of legal restrictions in his support of religious instruction or movements pointed with pride to sixty new appointments he had made in ten years—most of which he sincerely believed had strengthened the concerns of religion in that great university.

Still more recently, Dr. J. Hillis Miller, president of the University of Florida from 1948 to the autumn of 1953, frankly stated:[8]

It is extremely important that the administration of a state university exhibit a vital concern for religion, and that the staff members be selected on a basis of *what they are* as well as on the basis of *what they know*. At the University of Florida we do not care *what* religious faith staff members hold, but we are concerned that they have some sense of the importance of religion in personal and social life.

Let us continue for another moment this short visit to the campuses of our state universities. Shall we somewhere find a leadership earnestly and successfully striving to advance within

[7] R. H. E. Espy, *The Religion of College Teachers* (New York: The Association Press, 1951).

[8] *American Education and Religion, op. cit.*

the several states the hard-won separation principle written nearly two hundred years ago into federal law? The answer is best given in the words of one of them—words of information mailed by one of our great Midwestern state universities to its living graduates, in the autumn of 1951. Since the statement itself is quite typical of others of this powerful group of state institutions, it will here be veiled as from Blank University. It reads:

Blank University recognizes the importance of an active religious experience in the daily life of its students. All men and women entering the University are encouraged to participate in the programs of their chosen denomination. Although the University has no sectarian affiliation, it fosters activities of organizations and churches designed to hold the interest of students in their religious faiths. To accomplish this objective, the University works through a University Committee on Religion, which is composed of faculty and staff members. There is also a student Religious Cabinet which is an inter-faith group. Most churches in [name of city omitted] carry on special work with students and maintain meeting places near the campus. Students who are interested in courses on religion may select and arrange religious courses in a number of combinations. These may be applied toward fulfilling the requirements for both the A.B. and A.M. degrees.

The above statement makes it quite clear that no part of the unfinished effort of the Founding Fathers is exciting this fraction of the educational leadership of the nation into further positive action. On the contrary, the attitudes and practices here described as a state university policy were set down with satisfaction three years after the Supreme Court decision of 1948 had declared that "neither a state nor the Federal government can, openly or secretly, participate in the affairs of any religious organization or group" and that "no tax in any amount, large or small, can be levied to support any religious activities. . . ." How can a state university fail to *violate* these provisions when, supported by tax money, "it fosters activities of organizations and churches designed to hold the interests of students in their religious faiths" and when, "to accomplish this objective, the University works through a University Committee on Religion, which is composed of faculty and staff members"? With the spines of many and great state universities thus softened, who is it that stands within hailing distance of the Jefferson who said that "to compel a man to furnish contributions of money for the propagation of opinions which he disbelieves and abhors is sinful and tyrannical"? With respect to the education front, who is it that now fights to advance—or even to sustain—that American ideal of democracy? Indeed, which of our state universities are now complying with this mandate of our federal Constitution?

10

The Constitution on Church and State in Education (continued)

No religion teaching that love is the cure-all and that God, through Jesus, Mary, and the saints, is the unfailing source of help in time of need, can be seriously concerned with the progress of knowledge.—*James H. Leuba*

The greater the proportion of our youth who fail to attend our public schools and who receive their education elsewhere, the greater the threat to our democratic unity. To use taxpayer's money to assist private schools is to suggest that American society use its own hands to destroy itself.—*James Bryant Conant*

THE CATHOLIC SCHOOL PROBLEM

THE WORLD'S TIGHTEST AND MOST EFFECTIVE ORGANIZAtion is the Roman Catholic Church. And that superb organization, despite recent losses behind the Iron Curtain, is continually growing in *power*. Where, as in Spain, the Catholic Church is in full flower, it is a political, a business, and a religious institution. The unrivaled power of the Roman Church seems to have been derived over the centuries from three principal sources: first, by preaching a Christian creed built for or molded to its own type of organization, always with external attention invited to creed and ritual and with internal attention firmly centered on organization; second, by exercising continuously all attainable control over education and thought; and third, by developing, and partly by inventing, those abundant materials or methods that lend themselves best to supporting emotion and to restraining reason in the common man. This third achievement is a towering enterprise in which the resources of myth, miracle, music, pageantry, poetry and all the arts have been employed with a measure of success that all future societies may well remember.

Defensively, wherever Catholics are a minority group, this

tight organization regularly employs only a few simple tools—
though an overfull arsenal is readied for use by the Church.
The strategy is about as follows: for both defense and offense,
as in communism, be sure that several key words—such as freedom, morals, authority, religion—have one meaning for the hierarchy and another for opposing peoples; then develop in all
priests and churchmen the duty and arts of skillful evasion and
a confusing logic; permit no local ownership or management of
any of the property of the Church; press adherents continuously
for funds and lands that can be acquired by the Church—freely
using the purgatory club on rich and poor; assert Church solidarity in all ways, and elect or secure the appointment of
Catholics to key civil positions; maintain full secrecy regarding
the expenditures of the central Church authority in Rome; relentlessly declare the supremacy of the Church over the state in
the fields of faith and morals—and thereafter the indefinite range
of faith into ideas, and of morals into economics, medicine, education, society, law and politics can be adroitly exploited by the
trained and alert hierarchy as occasion requires; and, wherever
or whenever this is expedient, insist that the Church is wholly
responsible for education.

Such is the powerful organization that, on finding Protestants
had already unlocked the American door separating church and
state in the field of education, now insists upon a further opening of that door in its favor. It has become clear to many of our
people, both Catholic and non-Catholic, that here is an issue
that presses strongly for solution. It is a most unwelcome problem to most American citizens, and if it is not thoroughly clarified and resolved, it will persist as a festering threat to national
unity. To fail to understand the background of this conflict is
to forfeit all chance for an American solution of the problem.

It is beyond the scope of this book to provide the factual support for the several affirmations made in the preceding paragraphs. Most of those statements have been adequately documented in three recent and highly informative books by Paul
Blanshard[1] and by Avro Manhattan.[2] Statements not thus cov-

1 Paul Blanshard, *American Freedom and Catholic Power* (1949) and *Communism, Democracy and Catholic Power* (1951), both published by The
Beacon Press, Boston.
2 Avro Manhattan, *The Vatican in World Politics* (New York: Gaer Associates, 1949).

ered will be accorded further attention in these pages. Some topics discussed in this chapter, and some related topics wholly ignored here, are reviewed in Blanshard's reports.

The Catholic school problem has been forced to the front by two recent decisions of the Supreme Court—in 1947 and 1948— and by the fact that for more than six years Catholic power has blocked the will of a majority in Congress to give federal aid to public education in the states. Familiarity with the basis and meaning of this growing controversy is just now a topmost duty of citizenship. It is quite clear that American ideals of personal rights and the future status of evolutionary thought and of social development are all involved in this struggle.

In 1947, after a generation of uncertainty and some permissive state legislation on this matter, a suit was brought before the Supreme Court of the United States to restrain the state of New Jersey from using public funds to pay for the transportation of children to and from parochial schools. By a vote of five to four the Court decided that the state of New Jersey, under its general welfare powers, could charge the cost of bus transportation of Catholic parochial schools to the taxpayers. The majority opinion accepted this as a "welfare" service to children as children, but followed this specific concession with the statement:

The First Amendment has erected a wall between church and state. That wall must be kept high and impregnable. We could not approve the slightest breach. New Jersey has not breached it here.

The minority of four, however, wrote a more extensively documented opinion asserting that New Jersey *had* breached the Constitution. In this dissent, Justice Jackson quoted the Canon law of the Catholic Church that compels Catholic parents to send their children to Catholic schools as part of their religious discipline, and further stated:

Catholic education is the rock on which the whole structure rests, and to render tax aid to its Church school is indistinguishable to me from rendering the same aid to the Church itself.

And Justice Rutledge, after reviewing the debates and the history of the constitutional fight for separation of church and state, indicated that the purpose of the First Amendment was not to strike merely at the establishment of a single sect, creed or religion, outlawing a formal relationship such as had prevailed in England and some of the colonies. Necessarily it was to uproot all such relationships. But the object was broader than separating church and state in this narrow sense. It was to create a complete and permanent separation of the sphere of religious activity and civil authority by comprehensively forbidding every form of public aid or support for religion. In proof, the Amendment's wording and history unite with this Court's consistent utterances whenever attention has been fixed directly upon this question.

At the time of this decision approximately nineteen states[3] had provided bus transportation for parochial pupils at public expense, and five states had begun to provide some textbooks. The dissenting opinion of the four judges—overruled by one vote—further noted:

Two great drives are constantly in motion to abridge, in the name of education, the complete division of religion and the civil authority which our forefathers made. One is to introduce religious education into the public schools; the other, to obtain public funds for the aid and support of various private religious schools.

In this manner, in several of our states, it has now become true that part of the tax money of the nonbeliever can be used to forward the religious element in education to which he objects. And in this way the democratic ideal of those who framed the Constitution is now successfully attacked through Catholic power in the field of education.

A second decision of the Supreme Court, in 1948, related to the now famous McCollum case in which religious instruction within the public schools was the point at issue. Non-Catholics more than Catholics were thus primarily involved in this suit. Young McCollum, brought up as atheist or humanist, did not elect to take the sectarian religious instruction that priests, rabbis and ministers were then giving in the public schools of Illinois on "released time." As a result he was isolated in the hall of the school at the time of religious instruction, and not only missed the education he should have had for that period, but was also subjected to ostracism and ridicule. His mother

3 R. Freeman Butts, *The American Tradition in Religion and Education* (Boston: The Beacon Press, 1950).

brought suit against the public-school system of Champaign, Illinois, to prohibit the use of public-school time and school property for sectarian instruction. The case was carried to the Supreme Court. Ostensibly that Court's decision was upon the single narrow question: Can local American school boards use public classrooms for "released time" religious classes? A negative answer was given by a vote of eight to one. The only Catholic member of the Court, Justice Murphy, concurred in this interpretation of the Constitution. Soon afterward, however, at a solemn assembly in Washington, the Catholic bishops of the United States rebuked the Court for "this entirely novel and ominously extensive interpretation" and for giving "scant attention to logic, history, or accepted norms of legal interpretation." The bitter criticism of the Court's decision by the Catholic press clearly expressed its great importance to Catholic plans and hopes.

Other aspects of the current controversy over "released time" than that relating to the use of public classrooms for religious instruction are not covered by the Court's action. But its ruling included the following:

No tax in any amount, large or small, can be levied to support any religious activities or institution, whatever they may be called, or whatever form they may adopt to teach or practice religion. Neither a state nor the Federal government can, openly or secretly, participate in the affairs of any religious organization or groups, and *vice-versa*.

Two later decisions of state supreme courts, along with many lawyers, upheld the view that in these two cases the Supreme Court consistently defined the principle of separation, but contradicted itself when in practice it came to apply the principle to the two cases.

Two related matters require further attention: first, the way in which this abuse had become established and extended; and second, the extent to which this law of our land is now being observed or being flouted by the various religious groups—both Catholic and non-Catholic. After noting the original constitutional guarantees and their partial defense in earlier years, the following statement was written in December, 1949, by Algernon D. Black, Executive Leader of the New York Ethical Society:

Despite this clear trend, the traditional separation of church and state in education has recently been violated in many communities in the United States. In part it has been due to confused thinking, in part to those who, in their fervor for their religious faith, have become very aggressive in relation to school systems locally. . . .Thus, if a minister, a rabbi, or a priest, or the board of directors of a church, or a delegation of women from a church or temple come to the superintendent of schools, or the school board, or the principal of a school, or to some of the teachers, and press for the inclusion of religious instruction or worship, it is very difficult to oppose them, even if one is defending the traditional democratic principles. . . . In many places the school authorities will say that they did not want this to happen; they would have wished otherwise; but the pressures were too great and they could not afford personally to stand against them.

Wholly apart from this technique of conducting religious instruction in public schools, as practiced by many religious sects, other and more flagrant procedures have been used to one extent or another by Catholics in a large group of states and for several decades. In some school districts with large Catholic populations, priests succeeded in moving the parochial school into the public school. These hybrid or semipublic schools have been supported by public funds, though they have remained Catholic in control and spirit. Here, nodding to the First Amendment, the catechism may be taught immediately before or after the regular school hours. And whether nuns and brothers shall do the teaching, also whether they wear or do not wear religious garb, can be decided by the district's vote unless this is forbidden by state law. Religious garb is forbidden in twenty-three states and permitted in twenty-five others. But in a majority of our states—such is our decentralized and democratic policy in education—neither governor nor congress nor president can successfully challenge the school district's vote on this and most other educational matters. Also, through capture of school boards, the reverse of the above procedure has been carried out —the public school has been moved into the parochial school— rent being paid to the Catholic Archbishop and nuns retained as teachers. The case of the Cincinnati suburb, North College Hill, Ohio, brought to public attention in 1947, is one of numerous examples.

The root of the Catholic school problem in America stems

ultimately from Canon law in Rome, but more immediately from the decision imposed upon laymen in America by the Third Plenary Council of Baltimore, in 1884. Under that decision, every parish priest must have a Catholic school attached to his church. It was at that time made an obligation of every priest to raise money for the erection and maintenance of the school from the people of his parish. So close is the association of parish church and school that separate treasuries are usually no longer maintained. Against this pressure by the hierarchy, numerous Catholic laymen have long struggled, and many of them today risk the displeasure of the Church by sending their children to the public schools. Such Catholics well know the supreme hollowness and deceit of the Jesuitical argument, advanced to non-Catholic peoples, that the parochial school represents "the right of the *parent* to educate his child as he chooses." The parents who do assert this right do so under priestly persuasion or compulsion. The real sustainer of the parochial school is the long harsh whip of the hierarchy, not the spontaneous wish of the parent.

Though some Catholic officials in America regret that their Church has assumed an uncompromising stand for the separate education of Catholics, since it results in divisiveness and antagonisms, other members of the hierarchy now and then freely admit that the parochial school is a prime support and spreader of the Catholic religion. Two such admissions may be quoted. At the dedication of a Catholic school in East Elmhurst, Long Island, *The New York Times* of January 13, 1941, quoted Bishop Malloy as follows:

Children are the hope of the nation. We are dedicating not merely a new institution but a nursery of the future members of the church and citizenry of this great community.

In an article published in the Pueblo (Colorado) *Chieftain* of June 7, 1948, Bishop C. Willging said:

We must face this issue squarely. The Catholic Church has realized through the centuries the unqualified importance of her educational system to her vitality and growth, and that her progress is proportionate to the degree in which the laity are religiously trained.

It cannot be denied that funds given to Catholic schools support and extend the Catholic religion.

It was earlier stated that the Catholic school question in America has been forced to the front by the two recent decisions of the Supreme Court and by a prolonged blocking of federal aid to state educational systems through Catholic influence in Congress. The books of Blanshard, Butts, and Stokes, cited earlier, all deal extensively with the question of federal aid. While expressing no opinion on whether it is desirable for the federal government to provide funds to states for educational purposes, it is necessary here to consider the controversy arising from the stalemate of legislation by Congress—a stalemate based on Catholic desire to have parochial schools share in any funds thus distributed, and on Catholic opposition to legislation that would not accomplish this. The controversy now reaches far beyond the halls of Congress. States and individuals, associations and cardinals are necessarily participating in it. The entire situation has been well presented by Algernon D. Black, in the following statement:

The current controversy concerns the question whether the public funds from the federal government shall be used by states without stipulation that it be used only for public schools. When Mrs. Roosevelt, insisting on a clear and sharp separation of church and state and the importance of building up the public schools as a bulwark of this democracy, suggested that in this Federal Aid Bill we should not permit the use of public funds for private and parochial schools, Cardinal Spellman attacked her personally as a bigot. . . .

The conflict over education, whether it be sectarian instruction in the public schools or released time, or whether it be public funds for sectarian and religious schools, is particularly severe because the religious groups are so concerned about the outlook of the coming generations. The struggle over education is the struggle for the minds of the children and youth. Moreover, certain religious groups are concerned with indoctrination not only in regard to certain theological ideas, but in regard to conceptions of human relationships, attitudes toward freedom and authority, and the individual's right to make his own judgments. They would censor the individual's reading, and his contacts, and his thoughts. In New York City, for example, the Catholic church has an enormous influence over education policy—greatly incommensurate with its numerical strength in the population. From the point of view of many Americans, its authoritarian emphasis and control is anything but a democratic factor in education. The failure of the Board of Higher Education to put through the appointment of Bryn Hovde as president of Queens College, the destruction of the Youth Builders' program in the public schools, the inability of the schools to purchase certain films, the prohibition of *The Nation* in the libraries of the schools—in part, all these are the results of church policy in education. Do we want to strengthen this in the non-public schools of America with public funds?

Here, then, on educational issues, we find the most bitter area of church-state conflict in America—and in almost every country in Europe.

Some additional facts relating to the question of federal aid belong here. In 1951, against Catholic opposition, the National Education Association recommended that aid of this kind be limited to public schools. On March 1, 1950, the American Association of School Administrators voted, 7,000 to 2, for a resolution opposing the use of public funds for the support of parochial schools (*The New York Times,* March 2, 1950). In October, 1952, the United States Office of Education issued a report showing that 275 different agencies or divisions of the federal government are already participating in educational programs. For the year 1950-51 their total contribution was about two and one half billion dollars ($2,120,215,751 under the GI Bill). Among the other projects supported are specific educational programs vital to national health and welfare; education in communities markedly affected by federal activities, educational services essential to national defense, but which are not the responsibility of any local community or state; and promoting exchange with other countries of information and of students, teachers, professors, technicians and leaders.

Again, concerning one of the first bills presented in Congress, Senator Robert A. Taft said:

No state has come before us affirming its inability to deal with the educational problem. No legislature has passed any resolutions requesting assistance. The adoption of the present bill would undoubtedly embark the Federal Government in a gradually increasing expenditure from which it would never be relieved. It is unnecessary to expand on the tremendous danger of centralized control of education.

The later Taft bill, passed by the Senate in 1949 but never reported for vote by the House, would have given the states authority to use federal grants for private (parochial) schools when that use was authorized by the state. Most of the Catholic hierarchy and prominent Catholic laymen will accept the more extensive federal aid only if it provides funds for parochial schools. They argue that the parochial schools now relieve public funds of a heavy expenditure. And this is undeniable. Non-Catholics, however, regard costs of parochial schools as an unsatisfactory contribution to education in our democratic way of life. Moreover, to pay these costs from public tax money would destroy at once one of the most advanced principles on which our government is founded.

In April, 1952, Harvard's President Conant, discussing the broader problem of nonpublic schools, stated: "The greater the proportion of our youth who attend independent schools, the greater the threat to our democratic unity. Therefore, to use taxpayer's money to assist such a move is, for me, to suggest that American society use its own hands to destroy itself." His opinion was attacked extensively by Catholic educators and church authorities. Later, Dr. Conant wrote[4] the epigraph used at the head of this chapter.

Catholic parochial-school enrollment in the United States is impressive. The Official Catholic Directory for 1950 reported 316 seminaries with 25,622 students and 225 colleges with 252,727 students. Writing in *The New York Times* of April 12, 1953, Benjamin Fine placed the current enrollment in elementary and secondary schools at "more than 3,500,000 as compared with 2,500,000 in 1932-33. By 1960, the figures indicate, the Catholic enrollment will reach 4,500,000." Of this number 910,-000 are credited to high schools.

There can be some or much further postponement of proposed legislation for federal aid to lower schools. But a solution of the larger underlying problem involves either a surrender by the United States or by the Vatican. The "high and impregnable wall" that is already weakened or broken by both Protestant and Catholic intrusions, along with the Court's solemn declaration that "no tax in any amount, large or small, can be levied to support any religious activities or . . . to teach or practice religion," can be abandoned; or the native part of the Catholic hierarchy can be informed that a foreign power's claim to the right of directing the education of its adherents in the

---

[4] *Education and Liberty* (Cambridge: Harvard University Press, 1953).

United States cannot be given even a penny of support. Meanwhile, suitable measures might acquaint all Catholic citizens with that decision, and with the further fact—easily found in church documents—that the campaign for their parochial schools originates in Rome. If the people of the United States still retain enough power to make that decision, these two ways are open. Another way is not in sight.

PAROCHIAL SCHOOL MISCELLANEA

The parochial schools of various Protestant sects are becoming more important than they once were in the United States because of a recent revival of effort to establish them. On May 1, 1952, the National Council of Churches of Christ in the United States reported upon what it stated to be the first nation-wide census of elementary Protestant day schools in the nation. This report found 186,000 Protestant pupils thus enrolled in 3,000 religious schools, and stated that 3,035,000 Catholic children were attending parochial elementary schools. The study did not tabulate mission schools, boarding schools or secondary schools. Estimates of increased Protestant enrollment for the 13 years following 1938 were placed at 61 per cent and the Catholic growth at 35 per cent.

An article by Preston King Sheldon, in *The New York Times* (January 7, 1951), notes that such schools are now being established at an accelerated pace. He writes:

Among the new projects are ten Baptist schools with 875 pupils in Los Angeles, begun in 1947. A [Los Angeles] high school is being planned. . . . In the United States the Lutheran church now has 1,400 schools with 110,000 pupils . . . [and] the Seventh Day Adventists recently reported a total of 942 schools. . . . Evidence of new interest the last fifteen years among Protestant Episcopal communicants in parochial schools has been disclosed by growth under war conditions, which has continued. The Rev. David C. Colony, of Metairie, La., reports eleven new schools in 1948-50 in Louisiana.

The House of Bishops of the Episcopal Church declared on October 5, 1949, that "this house fully endorses the principle that sectarian schools be supported in full from private sources or from a church." In strong contrast to this Episcopalian statement, the locally powerful Lutherans, at their national convention in 1950, according to *The New York Times,* of June 27, 1950, decared as follows:

> The state may lawfully cooperate with and befriend religion on a non-preferential basis to the extent that it preserves the distinction between the two realms. Distinction, or differentiation, of the two realms allows for cooperation, whereas "separation," as many understand the term today, seems to forbid cooperation. . . . The children of religious parents may not receive religious education in connection with the daily public school program; the children of godless parents, however, are receiving at public expense the kind of education their parents want them to have.

This large religious group is thus aligned with the enemies of the secular public school. And it evidently has a better acquaintance with autocrat Martin Luther than with democrats Jefferson and Madison.

The above data and still other well-known facts clearly show that if the state were to support parochial schools several Protestant sects would have the state provide them with their own schools. In many regions the public school would disappear.

Concerning schools under Jewish auspices, operated in metropolitan New York, education editor Benjamin Fine wrote in *The New York Times,* of July 27, 1952, as follows:

> The more than six hundred schools of this city, Westchester, Nassau and Suffolk Counties under Jewish auspices will be studied. . . . There has been a tremendous growth during the past six years in the number of day schools under Jewish auspices. These schools are private religious institutions rather than parochial schools because all the Jewish schools, regardless of schedule or ideological persuasion, are independent units supported by tuition fees or by local neighborhood societies. Some of them are affiliated with national agencies reflecting their respective ideological differences, but they are neither directed nor controlled by them.

Referring to a new Catholic high school in White Plains, New York, the magazine *Time* (October 4, 1948) reported as follows:

> It was the new $4,200,000 Archbishop Stepinac High School, the nation's best equipped Roman Catholic diocesan school. . . . Last week the principal of Stepinac High, short amiable Father Joseph C. Krug, explained why Catholics have striven so hard

for this increase. Said he: "High schools are a graveyard for the Catholic religion. These are the years when adolescents take their first critical look at life, at themselves and at the world around them. The philosophy of this school, like every Catholic School, is to give a thorough academic training but always to emphasize the overall religious significance of life." . . .

At Stepinac every boy, whatever his course, has a 45-minute class in religion every day of his four-year high-school course. He attends regular services in the school chapel and auditorium. Just before Easter each year, the entire school will hold a three-day religious retreat. . . . In the ninth-grade religion class last week Father Joseph Sum reminded his young hearers: "At the end of the world our bodies will be reunited with our souls and either enjoy the beatific vision of Heaven or suffer the tortures of Hell." He led a careful discussion on the moral issues of the purposes of life.

The status of Catholic scholarship in America has been appraised by members of that faith. From the summary of a study begun a decade earlier by Professor J. A. Reyniers, of Notre Dame, as published in *America,* August 3, 1946, the following is quoted:

On the basis of scholarship we have no prominent universities. Among the schools which have reached the university status, we are at the bottom of the list of published research just as our medical schools are at the bottom of medical rating lists. The over-all picture is still blacker. . . . There is only one-fourth as much productive scholarship coming from Catholics as our numbers warrant. . . . Neither in its quantity nor its quality is there the slightest room for complacency about Catholic scholarship.

One decade earlier non-Catholics Lehman and Witty had reported that less than *one* per cent of the "starred" names in *American Men of Science* were Catholics.

Many hospitals are in part teaching institutions (nurses, interns), and an unusual proportion of hospitals of the United States and many other countries are Catholic institutions. According to *The Liberal,* of January, 1951, under the Hill-Barton Act passed by Congress, a total of $43,000,000 has been granted to hospitals of religious groups, and Catholic hospitals have received six sevenths of that sum. The opportunity of veterans of World War II to complete their education in schools of their own choosing at public expense has involved the payment of very large sums to all types of sectarian schools. Scholarship plans still in the making seem likely to invade still further and more deeply the "no state support of religion" principle. Some thoughtful people regard these "scholarship" proposals as the most serious of the several current threats to the principle of separation of church and state in education.

Thus the democratic practice of tolerance of sectarian education eventually becomes, for our nation, a plague to the separation principle through the very wide acceptance and development of such religious education by a sect or sects. The subsidized student must now either deny himself the school of his choice, if that is a sectarian school, or else the nation, established on the separation principle, must further strip itself of that ideal.

Nowhere in America has this basic American principle of complete freedom from involvement in religion been so long or so deeply outraged as at West Point and Annapolis. Those schools not only appoint chaplains,[5] they also continually affront every element of intelligent and self-respecting Americanism by making attendance at chapel compulsory. These facts are widely known. They are sometimes resented by youths who would otherwise seek admittance to these schools for military or naval careers in the service of their country. The writer personally knows of more than one such instance. The continuance of this practice, and its natural extension to many training camps, is a bigoted crime against citizenship and a flouting of the basic law on which our government is founded.

One example of the transfer of this infamous practice to Army training camps in and near New York will be cited. In the same year that young Scopes was tried, convicted and fined for teaching evolution in Tennessee, *The New York Times,* of June 1, 1925, reported as follows:

Major Gen. Charles P. Summerall, commander of the Second Corps Area, United States Army, declared last evening in an address at the third annual dinner of Trinity Choir Association, held in St. Paul's Parish House, 32 Vesey Street, that parents who did not like the order to have religious exercises in training camps this summer could keep their boys at home.

"There has been an attempt to attack the order to have religious exercises in the military training camps this summer," the Gen-

<hr>

[5] Anson Phelps Stokes, *Church and State in the United States* (3 vols.; New York: Harper and Bros., 1950).

eral said. "There are some who are opposing all forms of religious worship. However, the boys coming to the camps are sent by their parents and not by ministers, and if the parents do not approve the young men can stay away."

COMMUNITY CLASH ON SCHOOLS

Apart from the fact that several aspects of the educational problem confront the whole nation, involving divisiveness and bitterness on the broadest scale, some of its aspects continually exist in a more virulent form at local or community levels. There it engenders neighborhood strife, law suits, economic loss and civic disruption. All this, it should be noted, is part of the price we all pay for the failure of organized religion to accept the complete separation of church and state as declared in the Constitution; and by the refusal of organized religion to accept the American ideal—as expressed by Paine, Jefferson and Madison—of expecting only that government shall protect every individual's right to his own religious belief. That ideal too was to assure to every American citizen his right to total aloofness and uninvolvement in religion.

Only a few references to community clashes over the control or operation of the public school by religious groups shall be mentioned here. But those selected are only a small fraction of those reported in the press during the single year 1950. These instances will be recalled rather than described. Thereafter, the reaction of our educational and other leadership to parts of the underlying struggle will be reviewed.

In New Mexico, an earlier decision of a district court (Dixon case) was appealed, with trial not expected until 1951 or 1952. In the appeal it was alleged that though the earlier decision provided that certain "sisters, brothers and priests are forever barred from teaching in the public schools," the original purposes of the suit were defeated by the circumstance that "in many schools all that has been done is to hire different members of the same Order." It was stated that the "findings of fact and evidence [are] taken from over thirty widely separated schools and in different communities," and that Catholic Church buildings are still being leased for school purposes.

The Missouri Association for Free Schools has stated that at least $500,000 a year is being "leaked" from the state and federal funds in that state to the Catholic Church, and that more than one hundred garbed nuns are teaching in public schools there. The association brought suit to stop these practices.

In Minnesota, similar clashes were reported. And the Mississippi Supreme Court decided that state and federal aid may there be granted to hospitals run by religious organizations.

In Pennsylvania, the veterans' organizations took a hand in 1950, according to *The Liberal,* of October, 1950, in supporting a bill to be introduced in the 1951 session of the General Assembly to provide bus transportation for pupils of parochial schools. This action was bitterly resented by large numbers of Pennsylvania veterans. Nevertheless, in a letter to Bishop Gannon, the Secretary-Treasurer, J. Hugh McNeill, wrote:

Delegates to the Joint Veterans' Council, representing 800,000 organized veterans, on July 14th, 1950, approved a resolution to provide bus transportation for parochial school children in the commonwealth *en route* to and from their respective schools.

The *Jewish Newsletter,* of March 18, 1950, carried this item:

Mrs. William Zelemeyer, the head teacher of a nursery play school operated by the Jewish Community Center of Utica, New York, resigned in protest against a rule that three and four year old children in the nursery must wear skull-caps when they eat. "I am opposed to ritual in a nursery school," declared Mrs. Zelemeyer. "To my mind the wearing of skull-caps and the saying of a prayer before drinking a glass of juice has no place in a nursery school."

A situation in California was described by the *Christian Science Monitor,* of February 9, 1950, as follows:

Unknown to the principal of a San Francisco public school, two nuns visited the fourth grade class of the John Hancock School and interrogated the children as to whether they were attending church.

The nuns asked which of them attended catechism. Those who did not were threatened with not being promoted.

The Northern California Civil Liberties Union is protesting this proselytizing activity by a local church. The Union is not satisfied with the response by the principal of this school, who though not a Catholic himself, evidently fears to take strong action in this heavily populated Roman Catholic neighborhood. The most he would promise was that visitors would not be permitted to visit classes without passes. The Union insists on a strict rule forbidding religious representatives to propagandize in public schools.

This event would have gone unnoticed except that a frightened child mentioned it to her mother, who contacted the Civil Liberties Union. When the mother spoke to the principal, the latter admitted that the nuns had once before visited the school and given him a list of children who were not coming to church, after which the principal admonished those children to attend.

The small state of Rhode Island, viewed in relation to its history, gives accent to some significant items in this story. Some centuries ago Roger Williams and his associates asked and received from an English king a charter that distinctly said that the welfare of a state is best ensured when all men have liberty to believe as they please. Under that charter all those who obtained it, including a colony of Jews and a few others with and without religious beliefs, long enjoyed and helped to spread that freedom of opinion. Some months before the writing of this, it came about that the entire personnel of the school board of the city of Providence was Catholic, while previous to this a fair number of teachers had infiltrated into the school system who had a parochial-school background. There this march of Catholicism, foreshadowing restrictions on intellectual freedom, began its advance. Already, in 1930, published data (U.S. Office of Education, Bulletin 17) indicated that high-school graduates of Providence, compared with graduates of high schools in most other cities of the United States, had studied extremely little biological or any other science, and 7.5 per cent of them had studied no science whatsoever.

On January 20, 1952, Principal James P. McGeough, of East High School, Pawtucket, Rhode Island, suspended one of the student clubs that called itself the UNESCO Thinkers. His reason? A charge by a Catholic paper that UNESCO is "under atheistic control." He was congratulated by the local school committee, the Pawtucket D.A.R., and *The Visitor*, organ of the Roman Catholic diocese of Rhode Island.

WHERE ARE OUR GUIDES IN EDUCATION GOING?

From Maine to California many interested people are warrantably dissatisfied and clamorous concerning our educational outcomes. Many others suffer the pilfering of their unknown birthright in silence. But in no corner of this nation is the public prepared—emotionally, intellectually or legally—to permit professional educators to begin the serious and unobstructed cultivation of the intellect of modern youth. Unfortunately, others have documented the charge that, during recent decades, numerous professional educators have offered "proposals of the utmost vagueness and opened the schools to a most vicious and pervasive anti-intellectualism." (A. E. Bestor)

On the other hand, the educational air fills with the noise of organized groups whose knowledge of professional education is next to nil, and whose insight into modern thought is wholly unproved. Thus, the *American Legion Magazine,* of June, 1952, officially charged that "one of the strongest forces today in propagandizing for a socialistic America is the National Education Association." Some months earlier the Legion also took to the air to broadcast in favor of "putting religion into the schools" and asking the American public "to attend church every Sunday." In the summer of 1951 the Sons of the American Revolution made the extravagant charge that the N.E.A. is "the chief culprit" in a conspiracy to force socialism and communism on the United States. At their Sixtieth Congress (April-May, 1951) the Daughters of the American Revolution—fully failing in knowledge of Madison and Jefferson, and after myopic reading of our Constitution—passed a resolution favoring religion in the schools. They said that the removal of religious expression in public schools is "contrary to the spirit and principles under which this country was founded." Too, the D.A.R. has a National Defense Committee whose executive secretary, in the autumn of 1952, condemned financial support of UNESCO by the United States, and charged that the teaching of world citizenship to our school children is part of a subversive movement to undermine America. The Knights of Columbus, at its annual convention, in August 1952, passed a resolution warning all Americans against UNESCO. "[UNESCO], in some of its expressed views, advances theories which would support birth control as a truly scientific solution of problems of population and human betterment. . . ." Assorted labor unions also have placed various and quite effective pressures on several school systems.

The following paragraphs express this writer's criticisms of the leadership of the National Education Association. But this criticism is made from a point of view exactly opposite to that of the organizations mentioned above, and certainly it gives public education wholehearted support. Around the necks of most of our educational leadership, however, there hang two badges of personal failure: failure to learn that naturalism has supplanted supernaturalism in the world of thought; and failure to learn that, as regards the relation of education to religion in the United States, our Constitution says that there must be separation, not co-operation.

At this point a momentary digression is in order to reflect on the full validity of the statements earlier quoted from French sociologist Durkheim: "Education is only the image, the reflection of society. . . . Education cannot reform itself unless society is reformed." While conscious of the items treated immediately above, who will doubt that political action and modified religious views must precede or accompany the solution of the most pressing educational problems of our day? Students will learn or may be taught to think for themselves only if society permits it. American society seems sure only of its wish to put everybody in school. Most school officials and all statesmen of this century have cowered in sheltered corners in the face of increasing religious opinion—and meanwhile all may witness the slow erosion of society's higher hopes along with a creeping loss of the American citizen's right to total aloofness and uninvolvement in religion.

The materials thus far discussed in two chapters lead to a question of the highest importance: What does present leadership of public education in the United States now offer toward a solution of the problem of accomplishing within that field a complete separation of church and state? It can be said that this leadership is represented by a twenty-member commission that has just issued a one-hundred-page booklet that actually includes this subject. True enough, the real subject is of broader scope, and the book appears under a title that accords better with some of its evasions and cowardly discussions.[6]

Does this book refer to, consider or condemn any past transgression, or any present invasion, of religion in public schools? It does not. Its nearest approach is found in the statement that

6 Educational Policies Commission, *Moral and Spiritual Values in the Public Schools* (Washington, D.C.: National Education Association of the United States, 1951).

the teaching of moral and spiritual values in the public schools of the United States must be done without endangering religious freedom and without circumventing the policy of separation of church and state.

But one finds that to be "non-denominational" and "hospitable to all religious opinions"—irreligion or lack of religion being always omitted—is the recommended attitude for all teaching in the public schools. Again, not accomplishing, but avoiding, a complete separation of public education from religion is the desired goal. Thus:

> As public institutions, the public schools of this nation must be non-denominational. They can have no part in securing acceptance of any one of the numerous systems of belief regarding a supernatural power and the relation of mankind thereto. . . .
> The policy of the public schools is, in fact, hospitable to all religious opinions and partial to none of them. . . .
> Although the public schools are estopped from teaching any of the denominational creeds, they have their responsibilities toward religion, as is pointed out later. . . .
> Rejection of a state religion or of state religions is not the same thing as rejection of religion itself.
> The public schools faithfully reflect the religious diversity and tolerance which have helped to make our nation strong. In view of different religious faiths, a common education consistent with the American concept of freedom of religion must be based, not on the inculcation of any religious creed, but rather on a decent respect for all religious opinions. Such an education must be derived, not from some synthetic patchwork of many religious views, but rather from the moral and spiritual values which are shared by the members of all religious faiths. Such education has profound religious significance.

To the lines last quoted one would say—yes, most certainly so! In effect, it means teaching religion in the public schools, though the book also declares they should not teach religion. Statistics definitely show that fewer than half of the families represented in the *public* schools are associated with any church (see data given in Chapter 12). And when the commission says, "Such an education must be derived . . . from . . . values which are shared by the members of all religious faiths," it clearly means to disregard the views of those who are not members of a religious faith.

On one page the book quotes from the 1947 decision of the Supreme Court these words: "The people . . . reached the conviction that individual religious liberty could be achieved best under a government which was stripped of all power to tax, to support or otherwise to assist any or all religions." But on the following page it continues to point to ways in which the public school *should assist* and co-operate with religion. It states, too, that "the public school can teach objectively *about* religion without advocating or teaching any religious creed." Doubtless that is desirable. But which facts are to be used in teaching "about religion"? Are the facts sketched in the fifteen chapters of this book usable? Or does the commission have in mind the partisan views of the local churches?

Note the following, from *The New York Times,* of April 11, 1954:

The State Council of the California Teachers Association, representing 70,000 public school teachers, adopted here today a program to help "develop in pupils a greater recognition of God and religion as factors in our culture."

Dr. Hollis L. Caswell, Dean of Teachers College, Columbia University—writing in *The National Association of Secondary-School Principals Bulletin,* November 1953—thus remarked on the religious nature of the course of study now followed in the public schools of the United States: "In brief, just as our culture is permeated with elements of the Judaic-Christian tradition, so also is the school curriculum permeated with these same elements. The total impact cannot possibly be other than to encourage respect for religion and to foster the conclusion that religious beliefs contribute most significantly to the good life." A further unrelated but illuminating item from this same essay is quotable. When a check was made of the number of church members on the faculties of Teachers College and of the rest of Columbia University, "to the surprise of several in the group there were substantially more members of the church from our [Teachers College] faculty than from all the rest of the university."

Is it possible for leaders in education somehow to arrive at the fact that morals and religion are two separate areas with natural yet unlike origins? Can they somewhere meet with the evidence that "the Christian system is not primarily a system of ethics—it is a system of supernatural salvation"? Can these leaders in education accept the historical fact that a long and prominent list of the founders of this Republic repudiated Christianity's system of salvation? Can their progressive thought at all conceive that nearly two centuries and one intellectual revolution later our society might creditably seek to *complete,* or to *extend,* the trail of "separation" blazed by the Founding Fathers?

An entirely new, fast-growing, and dangerously perversive movement is now getting under way in the United States to finance private colleges and universities—including the smaller denominational colleges that directly teach and support religion —from the earnings of industry and business. Education editor Benjamin Fine wrote in *The New York Times,* October 12, 1952, that "the new movement, one that is expected to alter drastically the concept of college financing, is now sweeping the country." Note well that this is wholly unlike a gift from one's own estate. This development menaces the basic right of the citizen to withhold any personal support of religion. First, since such gifts are tax free, but taxable when used as dividends or capital gains, *every* taxpayer shares in the loss of tax money to government by this diversion of earnings. Second, the total amount of the gift is taken—through cost of living—from unbeliever and believer alike. This is a means by which a corporation, licensed by government, virtually taxes the unbeliever and presents the proceeds to a propagator of religion. In the industrially powerful United States this practice carries the threat of a great increase of religious education as against public and secular education. Industry and business will thus exercise a power that government itself may not exercise—they initiate a plan that results in forcing every unbeliever to support the teaching of religion. The stature of the foes of evolutionary insight, and of the separation principle, increases with each decade and year of the twentieth century. The failure of leaders of industry and business to comprehend and support the Founding Fathers has developed, during three years following 1950, a new and vicious menace to the separation principle.

Speaking on this topic at Yale University, in the autumn of 1951, Irving S. Olds, Chairman of the Board of United States Steel, said:

In my opinion, every American business has a direct obligation to support the few, independent, privately endowed colleges and universities. . . . And unless it recognizes and meets this obligation, I do not believe it is properly protecting the long-range interest of its stockholders, its employees, and its customers. . . . If it is necessary for us to spend millions of dollars to beneficiate the ore which goes into our blast furnaces . . . then why is it not equally our business to develop and improve the quality of the greatest natural resource of all—the human mind?

The following statement is taken from an editorial in the New York *Herald Tribune* of April 25, 1953:

Mr. Eugene B. Grace, chairman of the Bethlehem Steel Company, has formulated a plan [under which] Bethlehem will pay the sum of $3,000 to certain privately endowed educational institutions for each of their graduates who is recruited for the company's training program and who remains with the company for at least four months. The payments are made in recognition of the fact that four years of education cost a college more than it receives from a student in tuition and other fees. The institutions receiving the money are completely free to use it for scholarships or any other purpose which will best meet the needs of the particular institution.

Reporting from a meeting in Cincinnati of "presidents, deans and other administrators of 500 of the Nation's foremost colleges and universities," Benjamin Fine wrote, in *The New York Times,* January 17, 1954, as follows: "One of the greatest boosts for the colleges has been the backing higher education is now receiving from business and industry." Also, in a further note: "One hundred and twenty-three American corporations have contributed funds to the University of Notre Dame during the past year."

Some of these private colleges now propagandize religion to a relatively insignificant extent though originally founded largely for that purpose; but a much *larger* number of the private colleges, including all Catholic colleges, are governed, staffed and operated on a definitely religious bias.

A different though related failure, and now current transgression, by leaders of industry will be cited. The Texas and Pacific Railway (Dallas, Texas) makes use of a full-page advertisement (e.g., *Newsweek,* February 2, 1953) in which a sort of radiance projects from a "pulpit" on which rests an opened Biblelike book. Written on the left-hand page is "Have Faith," and the opposite page says "In God, In Ourselves, In Our Fellow Men, In Freedom." The few lines of script beneath the "pulpit" include these two phrases "IN THE FAITH WE PLACE IN GOD . . . in ourselves as individuals . . . in our fellow men and . . . in freedom—rests the future of the nation. . . . We need . . . Faith in God, who answers prayer. . . ." Certainly the full and heavy cost of this religious propaganda is paid for by the public—the devout and the atheist alike. Thus a corporation, licensed by government, is now flagrantly doing what government itself may not do. It supports religion, and it compels the nonbeliever also to support religion.

Midyear 1952 saw the beginning of a movement designed to remove the opposition of many Protestants to religious instruction in the public schools. The movement was started through resolutions passed by the Congregational Christian Churches' Council for Social Action, and published by the church magazine *Social Action.* The following fragments of the resolution are quoted from *Time* magazine (July 7, 1952):

> The first counter-heresy [of Protestantism is] the negative and sterile view of separation of church and state. . . . If we choose to ignore the public school system as an avenue of religious instruction so as to "keep the Catholics out," then in effect we allow secular theologies to become the source of meaning and motivation for the public school system.

The National Council of Churches of Christ in the U.S.A. speaks for some 35,000,000 American Protestants. A committee of this group released a letter—published by the press at the middle of December, 1952—from which the following is quoted:

> In our country, religion and government have not been like contiguous squares, but rather like circles that intersect at two points. These points have been the reverent awareness of God, on the one hand, and the recognition of absolute moral values, on the other.
>
> Inasmuch, therefore, as this nation was intended to be a religious nation, we should use all legitimate means to prevent it from becoming a secular state in the current sense of the term. . . . In some constitutional way provision should be made for the inoculation of the principles of religion . . . within the regular schedule of the pupil's working day.

On December 22, 1952, the Associated Press quoted President-elect Eisenhower, speaking to the Freedoms Foundation, as follows:

I was quite certain that it was hopeless on my part to talk with him [Russian Marshal Zhukov] about the fact that our form of government is founded in religion. Our form of government has no sense unless it is founded in a deeply felt religious faith.

General Eisenhower was then president of Columbia University, where his predecessor for many years was scholar Nicholas Murray Butler. In the statement just quoted, the General overlooked not only simple constitutional history but a worthy public statement of his eminent predecessor. Scholar Butler once said:

The government of the United States is in no wise founded upon Christianity. A barrier was erected by the fathers for a complete and what they thought would be an effective separation of church and state. Militant efforts are being made, as we have seen, to tear down that barrier. We must war against such efforts.

On January 16, 1953, speaking to the faculty of Columbia University on the current "cold war" with communism, the distinguished general further said:

We are engaged in a war of great ideology. . . . It is not just a casual argument . . . but freedom against slavery—Godliness against atheism.

The separation principle in the United States clearly has never had so much to fear from a Chief Executive as it has today.

Though it compromises our boasts of American freedom and democracy, it must be observed that no generation since the one that wrote the Constitution has equaled it in service to intellectual and religious liberty. And no president since Jefferson and Madison has equally observed the civil limits of his power and action. Meanwhile, in the sphere of education, new and nation-wide problems have arisen, the power of the foes of the secular public school increases daily, and the area of conflict widens greatly and continuously. The principle of separation, always insecure and only partially acknowledged by the several states of the Union, is now under heavy pressures never before encountered. The present generation of Americans can further fail, or it can return to a great ideal. If it avoids decisive action to secularize its public schools, or if it permits industry to subsidize denominational colleges, it will have further failed.

PEOPLES AND OUTLOOKS

It remains to sketch the bonds that bind education to religion in several countries of the present world; though for no nation can a few sentences disclose either the force of the underlying conflict or the depth of the intellectual anesthesia that everywhere attends this admixture. Nevertheless, such a review can provide contrasts and alternatives to the secular education that America has tried haltingly and vainly to obtain. It can supply evidence that the cultural impasse of our age is truly world-wide.

In England and in the Commonwealth all early education is mixed with religion to an extent not found in American public schools. When Britain's government schools were set up in 1870, the most that secularist opinion of that day would permit was optional classes in religion, and the matter was left entirely in the hands of local education boards. The practical result was that few teachers trained for it and few pupils took such instruction. Near the end of World War II, in 1944, the Archbishops of Canterbury, York and Wales scored a great victory for the church. God was put back definitely into education and, to all pupils whose parents have not definitely objected, the whole fabric of Christianity has been taught during the recent past.

Leadership in England looks often to Eton to prepare its sons for college. At Eton an unexcelled instruction and discipline are mixed with compulsory daily chapel and prayers. Under such educational practice and ideals, the wonder is that any English leaders ever become intellectually free. Yet, some do—perhaps because the colleges are usually good, and because the general population of England is less religious (Chapter 12) and more literate than many other populations. But even an occasional scientist there can make a very low score. The English physicist Sir Ambrose Fleming, president of the British Association for the Advancement of Science in 1935, stated that the entire Darwinian concept of evolution is baseless and quite incredible.

Catholic schools in England receive some tax support and apparently they are always trying for more. Catholics recently proposed a plan for reducing the cost to the church of reorganizing and building Catholic schools under the 1944 Education Act. Their plan was rejected by Mr. Tomlinson, British Minister of Education, who in a letter and memorandum to the bishops declared, in January, 1950, that the proposal would "wreck the Education Act. . . . The Roman Catholic hierarchy have always aimed at throwing the whole cost of their schools upon public funds, and have not ceased to do so."

Traces of Germany's past, and of Western Germany under the Occupying Powers in 1950, are indicated in the following personal communication from a friend and colleague in biology who recently participated in the program of exchange professorships with Western Germany:

The German universities tend to retain or regain their former character. They try especially to maintain the spirit of academic freedom. They succeed in this fairly well, in spite of the attitude of the Occupying Powers, which tends to favor church domination in education. We profess democracy but actually lend support to the most persistently authoritarian institutions in this world—the Catholic and Lutheran churches. Even though neither shows much love for American ideals, we trust them more than any other organization—supposedly because once upon a time they too had some trouble (though not outright war) with Hitler. Germany now has Protestant schools, Catholic schools and free schools. The latter are from most sides discriminated against, and are not doing well. All three are tax supported. But Lutheran school children are taught by Lutheran teachers who are trained in Lutheran teachers' colleges; and the Catholics have an equally tight and closed system. A Lutheran boy never meets a Catholic either in classroom or on playground. Only the teachers in the "gymnasium," the preparatory school for the university, are university trained. Freedom from the churches (relative freedom) thus exists only for the educated class, which is another separate segment of the German Volk—only 4 per cent of the university students come from the farm-labor class, i.e., from nonacademic families. At present, many church organizations gain increased influence because the Powers are using them for the administration of charity and the selection of fellowship students. The latter is indeed deplorable, because it fosters intellectual dishonesty.

Incidentally, the American occupation of Japan has likewise reflected un-American aid to both Protestants and Catholics. When the defeated Japanese asked for advice, General Douglas MacArthur told them: "You can't have democracy without Christianity." He also issued a call for 1,000 Christian workers to hurry the job. By the end of 1947, more than 1,100 Catholic and 220 Protestant missionaries had responded.

Two other religion-pointed MacArthur acts relating to the spreading of information on birth control will be cited here. The first item is from *The New York Times,* of Feb. 13, 1950:

Margaret Sanger, world-renowned birth control advocate, has been barred from giving a series of lectures in Japan, a spokesman for the Tokyo newspaper *Yomiuri* said today.

The spokesman said his newspaper wanted to sponsor the "planned parenthood" lectures to help find a solution for Japan's overpopulation problem, but officials in Gen. Douglas MacArthur's headquarters vetoed the idea.

A military government source familiar with the *Yomiuri* request said: "In view of pressure from Catholic Church groups it was believed impossible for General MacArthur to allow her to lecture to Japanese audiences without appearing to subscribe to her views." . . .

This pressure last month forced General MacArthur to halt distribution of a comprehensive book on Japan's natural resources problems. General MacArthur ordered subordinates to delete all birth control references in the book's conclusions.

Six weeks later, March 26, 1950, the same newspaper carried the following item on the same topic:

An immediate solution for Japan's rapidly worsening population problem must be found if disaster for Japan and possibly her Far East neighbors is to be averted, Juitsu Kitaoka, Japanese population expert and pre-war delegate to the International Labor Organization at Geneva, warned last week.

"Ninety million Japanese," he said in an interview, "will not starve to death quietly. More than 83,000,000 people [in Japan] are being rescued from starvation by the taxpayers of the United States of America today. This cannot continue." . . .

He added that emigration to less densely populated regions of the Far East "is only a dream which pleases the expansionists and encourages the opponents of birth control in Japan."

The work done by Paine, Jefferson and Madison in separating civil from religious functions seems quite unknown to those charged with the prolonged American administration of Germany and Japan. Indeed, that work seems unfamiliar to practically all military leaders trained in the United States. Equally unfortunate is the further fact that the courses of study followed at West Point and Annapolis provide far less opportunity than do those of our better universities for students to learn those insights into evolutionary science that end in the rejection of the supernatural. With each passing month these two blind-spots in the education of American military chiefs assume an increased political importance in the international scene.

The educational systems of most countries of Northern Europe are as much and sometimes more entangled with religion than is that of Germany. Rather exceptional is Sweden, which until 1860 had religious laws equaling in intolerance those of Spain today. Until that time, any attempt to get a Lutheran to change his confession was a penal offense, and apostasy from the state church made a Swede liable to banishment for life. Though some of those laws have since been somewhat relaxed, and though active agitation for still further change exists, non-Lutherans must now pay one half the usual state tax for the support of the established Lutheran Church. Only in 1951 did the Swedish Parliament pass a law permitting Swedes to leave the church without a declaration of faith in an alternative religion. Non-Lutherans are now barred from teaching in elementary schools. Nonmembers of the state church are currently barred from all cabinet posts, and even the proposed new law would bar them from heading the posts of education and religion.

Holland, long a Protestant stronghold, is rapidly becoming a Catholic nation. In 1952, Catholics operated 43 per cent of its primary schools; Protestants, 28 per cent; the state, 29 per cent. The government pays all the expenses of primary religious schools, and it subsidizes secondary schools without regard to creed. According to a Hague dispatch to the Chicago *Tribune,* August 10, 1952, the proportion of Protestants declined from 60 per cent in 1900 to 42 per cent in 1952. At that time, the proportion of Catholics reached 40 per cent. "Non-religious Dutch increased from 3 per cent in 1920 to 16 per cent."

In theory, France has a fair measure of secular education. In practice, as in the United States, this is limited and in imminent peril. Under the Vichy regime, in December, 1940, Jacques Chevalier, of the Ministry of Education, declared that "the Godless school has passed out of the picture for good." Since the end of World War II, perhaps the most meaningful church-state fight in all the world has raged in France over the question of state aid to Catholic schools. The significance of this ominous struggle rests on the fact that the issue cuts across party lines and becomes a part of the reason why stable governments cannot be formed in France since World War II. This in turn weakens the ability of France to exert her full power in the world against communism. The welfare and possibly the liberties of every person in the Western world is thus involved in that internal religious conflict. There is no reason to expect an early or definite end to that contest.

The Catholic schools of Belgium are already larger than the public schools and receive nearly equal funds from the state. There at present the hierarchy centers its fight against the Socialist minority for a final concession of government funds for Catholic higher schools.

Canada has never sought a separation of church and state to the extent that this was attempted in the Constitution of the United States. In its Constitution Act of 1791, large grants of land were made to the clergy; these Clergy Reserves were sold in 1853. Today its public schools are not secular, and in the most populous provinces of Ontario and Quebec one hour daily (Ontario) or two hours weekly (Quebec) of religious instruction is compulsory. In most other provinces, on order of the school board, religious instruction is given by the teacher or a clergyman during the last half-hour of the school day, or on "released time" after school hours, but within the school. A child may usually be excused from this instruction upon written request by parent or guardian. Normal-school students in Ontario must take "religious instruction" as part of their training to become teachers. Bible reading and prayer at the beginning of each school day are either obligatory or permitted in every province. Both Catholic and Protestant public schools are supported by public taxation.

For Spain, Italy, prewar Austria, Hungary and Poland, and for much of Latin America (in nine of the twenty republics Catholicism is the state religion and receives state aid; and except in Uruguay, Chile, and "periodic" Mexico, it is highly favored everywhere) the extent of the threat of educational bondage is best given in the form of a quotation. The quotation simply deals with the demands of the Roman Church, and it is from the Jesuits' publication in Rome, *La Civiltá Cattolica*, quoted by *Time* magazine, in its issue of June 28, 1948:

The Roman Catholic Church, convinced, through its divine prerogatives, of being the only true church, must demand the right to freedom for herself alone, because such a right can only be possessed by truth, never by error. As to other religion, the church will certainly never draw the sword, but she will require that by legitimate means they shall not be allowed to propagate false doctrine. Consequently, in a state where the majority of the people are Catholic, the church will require that legal existence be denied to error, and that if religious minorities actually exist, they shall have only a *de facto* existence without opportunity to spread their beliefs. If, however, actual circumstances . . . make the complete application of this principle impossible, then the church will require for herself all possible concessions.

In some countries, Catholics will be obliged to ask full religious freedom for all, resigned at being forced to cohabitate where they alone should rightfully be allowed to live. But in doing this the church does not renounce her thesis . . . but merely adapts herself. . . . Hence arises the great scandal among Protestants. . . . We ask Protestants to understand that the Catholic church would betray her trust if she were to proclaim . . . that error can have the same rights as truth. . . . The church cannot blush for her own want of tolerance, as she asserts it in principle and applies it in practice.

Thus, in the very broad empire of Catholicism, there is and there will long continue to be a forceful struggle to put Catholic doctrine into, and to eliminate the whole of evolutionary outlook from, the educational systems, social policies and laws of nations. Wherever Catholics are in a majority or in a strong minority that steady drive is on. And it is earth's most elaborate and efficient organization that contests church-state relations, and relations of both to education, wherever it has followers. That organization must indeed fight communism. And some now say that the fight must be endless and strength-consuming, since it appears that—through little or inadequate education, wherever that is acceptable, and through an associated spread of poverty—the Church creates Communists with one hand while fighting them with the other. Wealthier nations, for self-preservation against the poverty and communism beyond their borders, now begin to pay and plan to reduce the poverty that lies in the wake of Islam, Catholicism and Hinduism. Certain it is that the consequences of past and present battles of organized religions against free minds and secular education can neither be denied nor wholly escaped by any living person.

With the stakes so large, will citizen and secular teacher firmly meet their proud and common tasks? May the teacher, too, taste freedom? Can an unattained ideal of liberty, endlessly attacked, now get the support of forward-looking citizens? The American people might be considered truly democratic if only they had learned that freedom's house was not fully completed by an ancestor. They might be regarded as politically mature if they knew (Sulzberger) "that freedom is never a heritage, but always a new conquest for each generation." Whoever, anywhere, conscientiously strives to serve a liberty-preserving state and not a church is engaged in a very modern and nearly unsung adventure.

Democracies do not seem to have a really effective means of dealing with Catholic demands where that Church has as much strength as it now possesses in countries like the United States and Canada. Non-Catholic voters divide firmly on still other issues than those relating to faith; but Catholic voters in most countries are more nearly under the control of a single authority. Where this condition exists, and where one must obtain a majority vote to hold office, many statesmen feel—and logic seems to be with them—that they cannot afford to lose the large and nearly solid Catholic vote. Unless prepared to step from office, this writer does not see how they can do otherwise than trot along the path assigned them by the Church despite their misgivings about following such a course. Does the reader find a different or better answer? Are citizens of democracies deceived in presuming that they can deal successfully with the Catholic Church?

Quite generally we Americans are uninformed; worse, we are complacently drifting on or within the borders of anti-intellectualism. We are postgraduates only in gadgetry and in the hoop-la and skills of production, sports and marketplace. Our daily existence, though the gamut of diversity is as that of other peoples, is now largely bedded down in comfort, when not in luxury. Our portentous apathy to new or earned truth could flourish only in people born to a land overblessed by nature, and in an isolated nation initially provided with much liberty by the battle and foresight of its Founding Fathers.

For those who have found nothing sinister in the current and long-continued political transgressions and pretensions of religion in the sphere of public education, the preceding outlines of transgression and intent are submitted. If citizens—particularly those of the United States—remain unable to recognize and effectively defend the precious personal rights partially wrested from theocracy by the Founding Fathers, those liberties will soon belong only to the past.

For those who currently view with unconcern the problem of place for modern evolutionary thought and outlook in society, the materials of this chapter provide parts of the evidence that the bases for that unconcern are unsound. The eyes of this writer have witnessed the full eclipse of one or several principles of biological evolution in at least three of the leading nations of our time—Italy, Germany and Russia. When the Vatican Concordat of 1929 placed all basic education in Italy under Catholic direction, several biological principles lost even their chance for effective survival in Italy. Hitler's Germany perverted a central element of genetic science, threw at once a mendacious substitute into general education, and later threw millions of Jews into gas chambers. Russia, since 1936, has gradually liquidated the whole of genetic science—the key to all organic evolution. With personal friends and colleagues in all those countries, and having visited or studied in all those lands both before and after these tragic events, this writer has been a witness to a variety of successful killers of his science. Any scientist or citizen who now views the scuttling of evolutionary insight as no threat to sane and durable societies may well look further into the moving impulses of men. On the broadest of horizons those insights now fight a stygian brood for a place in private thought and for recognition in education, in social policy, and in law.

Contemporaneous encroachments on the principle of separation of church and state—those recently completed by the Executive Branch of the federal government, and acts of transgression pending in Congress at the end of May, 1954—are unequaled at any point in the earlier history of the United States. Within slightly more than a year, God has been brought by prayer into every meeting of the cabinet. The first United States postage stamp of regular issue to declare "In God We Trust" has just been issued. On February 7, 1954, with President Eisenhower in attendance, the sermon of his pastor in Washington urged that the pledge of allegiance to the United States be so changed as to acknowledge God. Three days later, the chairman of the Senate Republican Policy Committee, Senator Homer Ferguson of Michigan, introduced such a bill in the Senate; in the House fifteen resolutions of similar intent were introduced during the following weeks. In the amendment that seems to be most favored for adoption, the pledge is made to "one nation *under God,* indivisible, with liberty and justice for all." Pending adoption was Senator Capehart's bill—sponsored by Secretary of Commerce Weeks—to let clergymen ride free or at reduced rates on airlines. Also then pending in Congress was Senator Ralph E. Flanders' proposed amendment of the Constitution which reads: "This nation devoutly recognizes the authority and law of Jesus Christ, Savior and Ruler of Nations, through whom are bestowed the blessings of Almighty God." These entries—the latest that can be included in this book—do not fully measure present prostration of liberal thought in a nation that once could take a long step forward. Perhaps they merely record concluding ceremonies at the interment of the core of a freedom once won by men willing to battle for freedom.

11

The Reach of Religion in the News

There is not a single New York editor who does not live in mortal terror of the church groups. When I started in journalism I learned the first lesson, namely, that one must never write on controversial subjects, the first of which was religion, and that one must never report even the truth in any case in which the Catholic hierarchy might be offended. Every newspaper in the world is scared to death when any religious sect is mentioned critically.—*George Seldes*

Whoever would overthrow the liberty of a nation must begin by subduing the freeness of speech.—*Benjamin Franklin*

THE DAILY PRESS, THE RADIO AND THE MOTION PICTURE are everywhere wide open to a spread of religious thought and performance. But they all constitute, in effect, an Iron Curtain that shuts out effective criticism of widely accepted religious beliefs and discussion of desirable alternatives to those beliefs. This has become ingrained by custom. In this mid-century, it is taken for granted and rather readily tolerated by most peoples of the world—except those that have recently fallen to communism. Too, in much or most of the Western world, educated people and most students nowadays definitely restrict even private discussions in the realm of religion. All of this is an especially crippling blow to *democratic* societies, since in them action, advance, adjustment and welfare are presumably based on informed and intelligent votes. However, the extent and effectiveness of this Iron Curtain, in terms of suppression and censorship, is quite unknown to nearly all Westerners. Were it not so, many astonished citizens of twoscore or more nations would decide at once that no society can pay so high a price for a single by-product of organized religion. Since this blackout protects the front and flanks of well-entrenched medievalism, it rivals religion-regulated and emasculated formal education for the title of Enemy Number One in a society that would become

263

genuinely modern. It therefore must receive extended attention in these pages.

A first word, however, should warn that a thoroughgoing treatment of this subject is far beyond the competence of this writer. Indeed, no such account could probably be written at all. The more decisive items—the fears, cautions and prejudices of many thousands of editors, of some hundreds in the national and international press services, of numerous columnists, and the multitude of executives and owners in the theater, movie, radio and television fields, along with the plethora of boards and city agents that censor them—are nowhere set down in the form of documents and quantities. No one has obtained or is likely to obtain from these groups, within any large nation, the kind of information supplied for Chapter 8 by thirty-two hundred teachers of biology in high schools of the United States. The restraints on personal conversation in a huge population is also both notable and incalculable. In these pages, therefore, the task must be that of providing instances, quotations, events, and documents of current practice—all in amount sufficient to show that suppression and censorship of news and of criticism adverse to religion prevail widely, persistently and dangerously. The other matter—the habitual use of all these several agencies of communication for purposes of religious propaganda—is familiar to everyone and requires little documentation.

RESPONSIBLE SOURCES IN THE SUPPRESSION
OF CRITICISM OF RELIGION

The role of the spoken word, of face-to-face conversation, is far from negligible as a medium of news and criticism. Let us then first recall one way in which we are all lax and negligent—and the writer is as guilty as anyone. Our neighbors, associates and casual acquaintances include at least a few persons whose religious attitudes and thoughts are scarcely above magic and the Dark Ages. In our random or other contacts with them do we undertake to enlighten or criticize their attitudes and views? In general, we do no such thing. We all withhold information or criticism, when it applies to this dark field of feeling and confused thought. We can rightly plead that this personal reaction—and our definite personal failure—is based broadly on manners, on comradeship, and on an inner urge to respect the attitudes of others. But it is also in the minds of many of us that we have, and at considerable expense maintain, impersonal or less personal agencies whose job it is to clear up these caricatures of thought. In this connection we think of the school, the newspaper, the movie, the radio, the college, the various platforms and forums where men meet for discussion—and some untethered optimists would here include the church. We look to them to deal with this delicate class of blind spots in our neighbors. But what is the result when those agencies, formed and sustained by large groups of us, also fail?

In still another serious and quite local situation, large numbers of us are involved. The maturing youth of nearly all nations are currently advised by elder and trusted friends that their assertion of agnosticism or atheism will or may prove harmful or ruinous to their careers. The tragedy of this advice rests on the fact that it is often or usually true. Where influential organized religion exists, no one is born free. In such counsel to our youthful friends we must give the true answer; yet we sustain and perpetuate this dilemma of awakened youth when we do nothing to change it. Here is censorship in an especially bitter form. The enquiring mind between the ages of ten and twenty —when minds are most likely to awaken and welcome truth— must often stultify, hide or inhibit itself if its possessor looks forward to any one of several careers, or even if he is to avail himself temporarily of some of the usual forms of employment that will enable him to obtain an education. For this latter purpose, a hitch or two at teaching in common schools is used frequently in many parts of the world. Nearly everywhere, however, this avenue is closed to the known agnostic.

Bearing on this particular point is an early experience of Robert G. Ingersoll. When a young man of nineteen or twenty, he became a teacher in a little country school at Metropolis, in Massac County, Illinois. The school was maintained by private subscriptions of neighboring farmers. Much of the time there was no salary for the teacher, but he was able to stick to his task by "boarding 'round" with the farmers. All went well until a few days after the arrival of a small group of Baptist ministers, who had come to conduct a revival. They, too, "boarded 'round" and soon were placed at the same table with the young teacher. Though the ministers endlessly talked religion, the discreet young agnostic took no part in their discussions. Finally, one of their number pointedly asked young Ingersoll what he thought about baptism. Having already formed many of the opinions that later made him famous, and knowing that a candid reply would be dangerous, he sought a softer answer. "Well, I'll give you my opinion: with soap, baptism is a good thing." But the shocked brethren gossiped with the farmers, and as a result of what they learned the young man was obliged to give up teaching the school. Being without funds, he made his way on foot to his distant Illinois home.

Again, in more than one way, an individual's open rejection of religious creeds, even more than his acceptance of a creed, tends to restrict his field of choice in marriage. And, as is well known, those myriads in nearly all lands who look toward elective public office are almost necessarily conformists, trained in publicizing their conformity. These are compulsions and censorships of news and views—of the spoken word—operating at the local or community level. These are parts of the exacting price we pay for the presence of organized religious groups in our midst.

If some doubt that these situations prevail widely, the writer can give a list of the cities, states, and countries in which he has resided and personally observed their prevalence. He could provide a list of his own students and personal friends who have been deprived of teaching positions, in high school and college, because of their known acceptance of the evolution principle. He is aware, too, of the serious clashes that occurred in Mexico when, during a brief period, the Mexican government actually sought "secular" teachers to replace Catholic teachers in its public schools, about fifteen years ago.[1] Less adequate personal observation, strengthened by some definite information, indicates that these local censorships by religion exist in every state of the United States and in nearly every country of the world. Present-day Russia and her satellites might seem to be marked exceptions, since a mangled form of atheism represents one of the columns in the huge structure of communism. Nevertheless, nowhere more than in Russia is conformity with the prevailing political-religion desirable or necessary in order to obtain public office or a preferred position in any walk of life. One may conclude that practically everywhere, at the local or community

---

[1] Reports on this brief experiment in Mexico included numerous instances of attacks upon, and the occasional killing or crippling of, these "secular" teachers by excited religious populations.

level, organized religion exercises these restraints on every generation of maturing youth and on many adults as well.

One surprising attitude that relates both to teaching and to censorship should not pass unnoted here. Just as manners lead us to overlook the myth in our neighbor's religious thought, some atheistic professors purposely omit pressing the supernatural-destroying implications of scientific principles upon their college students. One English and one American friend and colleague in biological science—both frankly atheist—told this writer that such evasion was their own definite practice, and they believed it to be not uncommon among their associates. They viewed the early development of a mystical outlook in these young men and women as a calamitous but completed event; and for a few of them a shift to rationality would mean months and perhaps years of too serious a struggle with self. Such a fact again underscores the extent to which religious thought is sheltered, and also the need for more thorough comprehension—even by scientists—of prevalent positive dangers from organized religion.

The editor of newspaper and magazine disseminates his product in many communities. His financial success and his prestige depend directly upon maintaining a widespread demand for that product, and this in turn is sensitively linked to the profitable advertisements that can be quickly withdrawn if those who pay for them encounter opinions adverse to their brand of religion. In areas where religious belief is most openly active, the chances of offense are greater, and consequently blind censorship and suppression more likely. Some organized religions are prepared to act in groups, by boycott; and on the topic of religion editors can be only fearful, not free. Some of an editor's acts of suppression and coloring are doubtless based upon his personal religious prejudices rather than on fear or caution; but in either case religious teaching is the true source of the censorship and suppression.

The radio has a so-called self-imposed censorship that, in effect, goes far toward suppressing all criticism of any part of the religious area, while providing fully for religious propaganda and attacks on atheism. In the political sphere, this radio censorship, in actual practice, has freely permitted such un-American doctrines as those of Father Coughlin, when they seemed to have some backing from a powerful religious organization. Ultimate restraints on radio and television are clearly similar to those earlier imposed upon newspapers and motion pictures.

In the case of motion pictures, the suppression and censorship of all that may displease the credulous Christian or Jew is—despite recent adverse rulings of the Supreme Court—a finished and flourishing art. Since audience appeal did not earlier exclude some such thought-provoking materials, both Catholics and Protestants long ago placed their censors over Hollywood and provided strong barricades for the local scene. This power usually is exercised through ever-present threat—and through knives used on parts of an integrated and expensive product—but always producers know that lethal bludgeons await any film that attacks religion in the way many films freely attack atheism. Though this is an incomplete and preliminary statement, the religious net and blackout have become complete enough. Films that the churches do not wish you to see, you will probably not be permitted to see.

It is true that these censorship boards also deal with the wholly unrelated question of decency in films, and indeed that remains the only function they admit or declare. Discussion of their service to decency does not belong here, except to note that it may have been indispensable, though often noisome.

Fixing the responsible source of all this widespread blackout is not a difficult matter. Whatever difficulty attaches to this arrogant practice relates entirely to the unreadiness of citizens to acknowledge its religious sources, to their failure to comprehend the threat to democracy and the general welfare hidden in every form of such censorship, and to their delinquency in doing so little about it. In this introductory glance, it seems that at least some small blame falls upon the citizen who fails to enlighten his neighbor; upon the editor who suppresses or colors news; and upon executives of screen and radio who do their share of suppression. A heavier load of responsibility rests upon those particular members of public or private boards who vote to suppress and censor within the field of religion without recognizing that the Constitution attempts a complete separation of church and state in this country. But beyond and behind each and all of these offenses, and basically responsible for the blackout with its belly-thrust at democracy, is militant organized religion.

SUPPRESSION AND CENSORSHIP IN ACTION

"No more shocking fact exists than this: self-appointed groups have decided that you no longer have the right to choose your own movies, books, radio programs—and ideas." Thus wrote Collie Small in a magazine article[2] on censorship. He further stated: "Censorship by pressure from religious and racial groups has increased during recent years, as these groups have become better organized and more articulate. Somewhat insultingly, various of these groups express through their censoring activities the unflattering view that the rest of us can't be trusted to form our own opinions."

Motion Pictures and the Theater

Producers in Hollywood adopted a Production Code, in 1930, under the pressure of local censors and private groups that found Hollywood films increasingly offensive. The Code itself is mainly a Catholic product. It was framed largely by the Reverend Daniel A. Lord, a Jesuit editor, and Martin J. Quigley, trade publisher and prominent Catholic layman. With few and slight amendments, the Production Code still stands as originally written, and one branch (the Johnston office) of the Motion Picture Association of America administers it. If a motion picture company's own censorship departments slip up on stories or scenes that Catholics may object to, Joseph I. Breen, of the Production Code, will remind them. That Code restricts the treatment of religious subjects in the three following ways:

1. No film or episode may throw ridicule on any religious faith.
2. *Ministers of religion* in their character as ministers of religion should not be used as comic characters or as villains.
3. *Ceremonies* of any definite religion should be carefully and respectfully handled.

Released films are examined by the National Board of Review of Motion Pictures, an unofficial body, which is opposed to censorship but reviews all pictures for their artistic, educational and entertainment value. Meanwhile, other volunteer and unofficial

2 *Redbook,* July, 1951.

groups conduct previews of the film. Chief of these is the Catholic Legion of Decency, "whose moral standards are closely followed by all responsible movie producers," according to writer Lester Velie.[3] Another is the Joint Estimates group, which represents thirteen national organizations, among them the Protestant Motion Picture Council, the American Jewish Committee, and the United Church Women. These groups regularly report "back home" on the various qualities of new pictures. In addition, there are official boards of censorship established by seven states, by nearly one hundred boards in American cities, and also the many town ordinances that are so enforced as to constitute virtual censorship by local authorities.

The circumstances attending suppressed showings of two foreign films have been described as a threat to American unity by editor Henry Hart.[4] Since his views represent those of the National Board of Review, the following excerpts are quoted from his comment on the handling of two pictures made abroad, *Oliver Twist* and *The Miracle:*

Oliver Twist has been kept from the American screen because a large number of Jews—but by no means all Jews—believe that the character of Fagin, as played by Alec Guinness, will create or increase anti-Semitism. The means employed to keep this picture out of American theatres were covert, and consisted for the most part in the threat of a boycott of the picture by exhibitors, usually tacit, but not exclusively so. . . .

In this connection it should be pointed out that a picture of the quality of *Oliver Twist* should gross in the United States about two million dollars. A boycott of it would therefore have a financial, as well as an ideological, purpose. . . .

The Miracle opened at the Paris Theatre in New York City last December and was shown with two other non-feature-length pictures under the over-all and decidedly misleading title of *Ways of Love.* It had only been on a week or so when the License Commissioner of New York City, Edward T. McCaffrey, declared that he found *The Miracle* blasphemous and "personally offensive," and that if it was not taken off he would take the Paris Theatre's license away.

Such a tyrannous abuse of authority elicited offers of legal assistance for a fight in the courts, which the distributor of *The Miracle,* Joseph Burstyn, undertook—in the end, successfully.

Cardinal Spellman then denounced the picture as sacrilegious

---

[3] *Collier's Weekly,* May 6, 1950.
[4] *Films in Review,* May, 1951.

and warned all Catholics against seeing it. Thereupon picket lines were thrown around the Paris Theatre by Catholic war veterans and other Catholic organizations.

Next, another official of New York City, the Fire Commissioner, began to levy fines upon the Paris Theatre because people were found to be standing in the aisle. When it came out that inspectors from the Fire Department received gratuities from many theatres for allowing this practice, the management of the Paris Theatre was fined for having given such gratuities in the past.

Meanwhile, *The Miracle* continued to play at the Paris Theatre to audiences that were much larger than they would have been had the attempt to scuttle *The Miracle* not been made.

But those who had undertaken the censorship of *The Miracle* were not through. . . . After conducting a hearing, and seeing the picture, the Regents declared that *The Miracle* is sacrilegious and cannot be shown in New York.

As in the case of the Jews and *Oliver Twist,* not all Catholics approve of Cardinal Spellman's attempt to dictate what the American people shall see. Not a few Catholics believe that such action is wrong per se and also creates anti-Catholic feeling and does the Church harm, both from within and from without. . . .

Whether it is a case of Jews using our economic apparatus, or of Catholics using our state apparatus, or of any group using any means, censorship results in a repudiation of one of the fundamental American tenets—freedom of expression. Further, it is divisive, and a threat to the unity and strength of our country.

Not until May 26, 1952, was it declared by unanimous decision of the United States Supreme Court that this censoring of *The Miracle* on grounds of sacrilege by New York law was unconstitutional. Did this decision of the highest Court assure Americans their right to see *The Miracle?* Not at all. Note the later outcome in Chicago, as reported July 22, 1952, by *The New York Times:*

The Chicago police censor board, by a vote of 4 to 1, last Friday banned showings here of "The Miracle." . . . The Chicago board's decision, Police Commissioner Timothy O'Conner explained, was based, not on "sacrilegious" grounds, but on the basis of violation of a city ordinance prohibiting the exposure of adherents of a religion to contempt.

On January 18, 1954, the Supreme Court again ruled unanimously on two other censored movies. Without written opinion on these cases, it cited its earlier ruling on sacrilege *(The Miracle)* and declared that state boards cannot ban a picture on grounds that it is immoral or tends to promote crime; films enjoy the constitutional "free speech" guarantee just as do newspapers and other communication media. Whether this latest ban on religious reprisals will prove more effective than that on sacrilege the future only can tell.

Present problems relating to censorship of the *theater* in the United States are described by critic Monroe Lippman in *The New York Times,* July 8, 1951. From his account, the following statements are taken:

As is frequently true in times of crisis, there is abroad in the land today a large group of professional patriots and moralists who would like to inflict on the theatre an arbitrary and irresponsible censorship, based not on taste and reason, but on ignorance and bigotry. . . .

Whatever complaints may be lodged against the Broadway theatre, one thing is certain: compared with the theatre in many other parts of our country, Broadway has been and is relatively free from bigoted censorship. . . . In one university, censorship was attempted of a production of "The Male Animal" because of a scene in which a professor and one of his students unite in a drinking bout. Another recent stab at censorship was aimed at a production of "Rain" because in that play a clergyman is presented in an unfavorable light. "Family Portrait," a highly reverent and Christian play about the life of Jesus, has been ruled off the stage of one of our best regional theatres by the protests of a self-righteous censor who insisted that the play was sacrilegious simply because it was not completely in accord with his own faith. . . .

If we value our heritage of freedom of expression and our tradition of a free theatre, we must exert every effort to see that the theatre remains not only a medium for entertainment, but also for the free, honest and fearless expression of ideas.

RADIO AND TELEVISION

Supervisory power over these agencies is held by the Federal Communications Commission, in Washington, through its power to issue and periodically to renew or withhold licenses to broadcasting stations. A letter to the writer from the secretary of the Commission, dated July 5, 1951, states that "the Commission has consistently considered that religious programs make up an integral and important part in this well-rounded program service. For this reason, when a station comes up for renewal of its license, it is required to indicate in its renewal application the amount of time which it has devoted to religious programs and the amount which it proposes to so devote for future operation."

First of all, a citizen who lives under a federal Constitution that erected "an impregnable wall between church and state" will perhaps ask for a better understanding of the reason for a federal agency's solicitude concerning these "religious" programs. But a somewhat clearer view of the Commission's attitude on that question is given later in this account. The National Association of Radio and Television Broadcasters, like the motion-picture industry, several years ago adopted a code. Its Standards of Practice, under which it has operated since July 1, 1948, does not mention specifically atheistic or irreligious programs.

Two dissimilar episodes of radio history in the United States are of special significance to this account. The first is concerned with rulings and practices relating to the rights of atheists to the broadcasting facilities of stations that broadcast religious propaganda. Atheist Robert G. Scott, of Palo Alto, spoke once from a San Francisco station but thereafter was refused the facilities of three stations in California. Mr. Scott asked the Federal Communications Commission to revoke the licenses of those stations. In 1946, the Commission ruled on that request. Concerning punishment, the ruling noted that "the issue here involved is one of broad scope and is not restricted to the three stations which are the subject of Mr. Scott's complaint. We therefore do not feel that we would be warranted on the basis of this single complaint in selecting these three stations as the subject of a hearing looking toward terminating their licenses, when there is no urgent ground for selecting them rather than many other stations." The ruling, however, demolished the reasons given by the three stations for refusing broadcasting time to atheists. It maintained that, under rights of free speech, atheists have a right to be heard over the radio, and that "immunity from criticism is dangerous to the freedom of the people generally." Further, "It therefore appears . . . that the question here presented does not involve blasphemous attacks upon any religious belief or organization, but only such criticisms as would necessarily be implied in the logical development of arguments supporting atheism."

This liberal ruling of the Commission was attacked two years later by a Congressional committee. At the end of the two-day hearing, the United Press included the following statement in a dispatch dated Washington, September 1, 1948: "The Federal Communications Commission assured all radio stations that they will not be punished if they refuse broadcast time to atheists." The flavor of the Congressional committee's action is best given in the words of Marquis Childs, in his column of September 30, 1948:

Public opinion has come down like a ton of bricks on an extraordinary document recently released by a special House committee investigating the Federal Communications Commission. . . . The committee, of which Rep. Harness of Indiana is chairman, uses as a basis for attacking the commission two recent rulings.

One, known as the Port Huron decision, reaffirms what is clearly stated in the communications act—that radio broadcasting stations do not have the right to censor political talks.

The second—the Scott decision—reaffirms another basic principle, which is that where fundamental controversies are involved radio stations must accord time to the side that is attacked to reply. The decision, rendered in response to the demand of a professed atheist for revocation of the licenses of stations that denied him time, makes it perfectly clear that his argument is not to be arbitrarily interpreted and that the controversy must be one of genuine public interest.

In short, there was no attempt to say that atheists should be given equal time to answer all religious broadcasts. Yet, by distorting the language of the decision and calling in prejudiced witnesses, that is just what the Harness committee tried to make out. With the most remarkable effrontery under the circumstances, the committee report accuses the Federal Communications Commission of "thought policing" when that is exactly what the committee is proposing.

The Report of the House Committee, discussed above, would have it that "apostles of unbelief [are] numerically infinitesimal" in the United States. The identity and the handling of witnesses at this "investigation" are a matter of high interest, but it can receive but brief space here. It is notable, however, that the liberal and tolerant views of Dr. Robert Calhoun, professor of historical theology at Yale University, were given slight attention in the Report, while the testimony of Father Edmund A. Walsh, S.J., vice-president of Georgetown University, was given much prominence. Incidentally, at the end of Father Walsh's testimony, chairman Forest A. Harness expressed his appreciation "for this very fine presentation. It has been very helpful. I might say that it makes me a little prouder that Georgetown is my Alma Mater."

It would seem that this low-pitched investigation of the Federal Communications Commission by a Congressional committee has weakened the Commission's earlier policy concerning an atheist's access to radio time. This writer's direct question to the Commission on that point has been left unanswered—an understandable caution. Probably radio-station resistance is so well supported by intolerant public opinion that atheists in America may expect to exercise little of their clear rights on the radio.

In July, 1948, six Philadelphia stations either refused or ignored a request by editor Whitten [5] asking opportunity for the Friendship Liberal League of that city to broadcast its views on religious subjects. One such station wrote in reply that "it is our policy to exclude religion from the realm of controversial subjects." More recently, two New York stations permitted an irreverent broadcast. In Chicago, the Czech Rationalist Society now broadcasts in that language every Sunday.

The second episode of radio history that must be mentioned here concerns certain lessons from the radio career in the United States of Canadian-born Father Charles D. Coughlin. During the nineteen-thirties, Father Coughlin's weekly broadcasts were carried to all parts of the nation, and were widely considered pro-Nazi, anti-Semitic and un-American. His propaganda was little hindered, and he could boast many millions of followers. He had the approval of his bishop. An International News Service dispatch from Rome, July 29, 1936, said: " 'Father Charles Coughlin's social ideas outclass those of President Roosevelt,' Bishop Michael J. Gallagher of Detroit, the radio priest's superior, told International News Service today. Staunchly defending Father Coughlin, Bishop Gallagher said: 'Although he is not an official exponent, his political and social program is rigidly Catholic in an orthodox sense of the word.' "

Not until December 12, 1938, did Cardinal Mundelein, Archbishop of Chicago, "formally dissociate the Catholic Church from Father Coughlin's radio utterances." And the same column that carried that bit of news in the New York *World-Telegram* also contained the following item:

[5] *The Liberal,* October, 1948.

The Brooklyn Diocesan Union of the Holy Name Society, claiming to represent the sentiments of 200,000 members, threatened to boycott the advertisers of Station WMCA which had barred Father Coughlin from the air. The New Jersey Council of the Knights of Columbus also took Father Coughlin's side and said that its members had given Father Coughlin a "vote of confidence and approval" for his broadcasts of November 20 and 27.

During this period, and without using Father Coughlin's name, Dr. James Rowland Angell, president of Yale, was quoted as follows:

Consider the utterly ridiculous condition which compels President Conant of Harvard, under the Massachusetts law as it now stands, to take a teacher's loyalty oath, while at the same time it allows a recently naturalized foreign priest to escape such an oath and pour out weekly over the radio, under the blessed name of social justice, the most poisonous and inflammatory economic and social nonsense.

The reorganized Voice of America announced that henceforth it would emphasize a belief "in religion" along with belief in the worth of the individual. Business-educator Robert L. Johnson, President Eisenhower's director of the International Information Administration, is quoted (April 12, 1953) as follows:

We must stress the importance to our national life of religion. In general religion is foreign to the Communist. The Communist people want to believe in God, and we hope to reach them by showing that we, as a nation, are a religious people.

To the slander and falsehood of this concept (see Chapters 9 and 10) these broadcasts themselves are to add breaches of our own Constitution, and betrayal of the right of the liberal American to freedom from compulsory taxation for the support of religion.

Four days before Christmas, 1952, while televising in New York (CBS-TV) a program entitled "This Is Show Business," playwright George S. Kaufman remarked: "Let's make this one program on which nobody sings 'Silent Night.' " Telephone protests—perhaps two hundred, perhaps five hundred—began coming in immediately. The Columbia network fired the playwright. Mr. Kaufman explained: "This was not wittingly an antireligious remark. I was merely speaking out against the use and overuse of this Christmas carol in connection with the sale of commercial products." He was permitted to resume his work on January 24, 1953.

The few incidents just reviewed indicate that radio and television are overflowing sources of religious propaganda but contracted outlets for curbing prevalent ideas of the supernatural. They illustrate the paralyzing influence upon government, radio and television of our militant religious groups. They demonstrate that the boycott and similar resources of a single religious group are greater than the combined strength that constitutional guarantee and unmitigated evolutionary thought have as yet been able to muster in America.

Books and Magazines

The Post Office Department of the United States has long exercised a bureaucratic censorship. Following a recent decision of the Supreme Court that department may no longer bar any publication without a hearing, nor may it refuse or revoke second-class mailing privileges because of the character of a publication. It nevertheless still exercises important powers of censorship. At present, however, state and city governments, along with private committees representing religious groups, are the really effective agencies of censorship and suppression in the United States.

From a brief commentary in *The New York Times Book Review* (July 1, 1951) the following excerpt is taken:

A growing and potentially virulent form of book censorship has infected several states in recent months, according to the anti-censorship committee of the American Book Publishers' Council. It involves the pressuring of wholesale distributors by local District Attorneys acting upon the recommendations of committees set up to screen pocket-type books. The latest victim is Signet's "The Naked and the Dead," which has been ruled off the stands in Massachusetts.

In New Jersey, Governor Driscoll has appointed a committee which, in effect, proscribes books that it considers "unsuitable." The findings are "enforced" by local District Attorneys who employ "warnings" and occasional raids to keep wholesalers in line.

Methods used against a not-yet-published scientific book by a religious congressman and by a Catholic organization are partly indicated in two accounts published by *The New York Times* on August 30, 1953. The first item stated:

Representative Louis B. Heller, of Brooklyn, asked the Post Office Department today to bar from the mails the forthcoming book "Sexual Behavior in the Human Female" by Alfred C. Kinsey, pending a Congressional investigation of its text. . . . He indicated that he planned to put a resolution before the session [of the U.S. Congress] starting in January to install an investigating committee that could be put to work at any time.

The other report, from West Springfield, Massachusetts, noted:

A Roman Catholic organization announced today that it would conduct a panel discussion of "sex pure and proper" as an answer to "Kinseyism." . . . Bishop Christopher J. Weldon of Springfield cited the words of a national commentator on Kinseyism— "The only really important factor in determining sexual behavior is religion."

City school authorities in the New York area banned *The Nation* magazine from the school libraries. A part of *Time* magazine's (October 25, 1948) comment on this matter is here quoted:

. . . A little digging uncovered what the board of school superintendents [of New York City] had not announced. The board had voted not to renew its 18 *Nation* subscriptions, on the ground that the weekly (circulation 42,000) had printed articles by Paul Blanshard, one time New York City commissioner of accounts, criticizing the Catholic Church's stand on fascism, science and censorship of books and movies. The offending copies were yanked out of the school libraries.

Neighboring Newark has gone farther. Last winter, after an earlier series of Blanshard articles, the *Nation* had been removed from the libraries of Newark's four high schools by the school superintendent. When *Nation* Editor Freda Kirchway protested, the Newark board of education (five Catholics, three Protestants, one Jew) unanimously backed the superintendent. In Trenton, N.J., school officials clipped the articles from the magazines before they were put in the libraries.

By last week the controversy over the *Nation* had boiled up into a first-rate argument over freedom of the press. In the current issue of the *Nation,* 107 educators, lawyers, clergymen and writers . . . signed "An appeal to Reason and Conscience, demanding that the New York City board change its mind." That demand was made in vain.

The important part played in magazine and book censorship by a committee of the National Federation of Catholic College students has been fully told by the weekly *In Fact*. From its account (September 19, 1949) the following excerpts are quoted:

The National Press Commission of the National Federation of Catholic College Students (NFCCS) announces a nationwide campaign to censor, control, boycott and otherwise try to dictate what magazine literature the American people are to read.

The new magazine campaign was announced on the University of Dayton campus a fortnight ago by John W. Lynch, chairman of NFCCS. The drive, set to begin this month, is under the auspices of Most Rev. John F. Noll, bishop of Ft. Wayne and chairman of the National Organization for Decent Literature. . . .

So far as this weekly knows, the nation's newspapers have not reported the foregoing facts—credit, however, must be given the Dayton (Ohio) *Daily News* for a column on this subject. . . .

The first campaign of Bishop Noll's National Organization for Decent Literature (NODL) began in 1942 and (against it) the American Civil Liberties Union issued a protest. . . .

To the annual meeting of the NODL, Nov. 24, 1944, Bishop Noll stated: "All but 71 of 300 editors have brought their magazines into conformity with the NODL code. . . ."

In 1947 Bishop Noll admitted that his group censoring the U.S. was going much further than in deciding what was obscene. . . . The 1943 campaign against magazines was termed highly successful, but the Bishop deplored the fact that "respectable family magazines" had now begun to publish fiction and articles which were "the most violent attack on the very foundations of the Christian home."

The fact was that at this time the big circulation and most respectable family magazines had been discussing divorce, birth control, the giving of sex information in the public schools, how to conquer venereal disease and other subjects the Bishop and his followers want suppressed. . . .

Lynch [chairman of the new campaign of 1949] explained that the group would use the "direct approach" to the problem. He said student workers will visit bookstands in their communities, and inform proprietors which books on the stands are condemned. Students [250,000] will visit the stands once a week. Lists of condemned books, Lynch said, will be mailed to magazine dealers.

It remains to note that some of the very best and most important newspapers in the United States refused to accept paid advertisements for Blanshard's recent books in which Catholic power and practice are vigorously examined. Again, when the first of these books was put on sale at the large Macy store in New York, Catholic threats of picket lines or other similar pressure forced that great emporium to discontinue its usual public sale of the book.

In Canada, "an anonymous member" of the Revenue Department decides whether a particular foreign publication may enter that country. According to *The Liberal* (April, 1950), the following statement was published in the magazine *Protestant Action* of Toronto:

During the war years there was a constant complaint that Roman Catholic influence effected a censorship on Protestant and anti-Romanist books and publications from entering Canada. . . . Readers will recall the denial of the use of the mail to The Protestant Book House, located in Toronto, because certain books —which frankly discussed and exposed Romanism—were advertised through that agency.

The Herald of the Epiphany, a U.S. monthly, was stopped, although the Ottawa censors were never able to explain why. . . . Several issues were mailed to these puppets of the Roman Hierarchy and when they were asked what violated the regulations, no answer was received. The two so-called censors were then bluntly told that Canadians were not going off to a war to fight for liberty which permitted two persons in an office in Ottawa to tell the citizens of Canada what they should see, hear or read— without an opportunity, as British subjects, of a right of defense, in court. The ban on the Protestant Book House continued, without explanation or review, until the war ended. . . .

Now it appears that the censorship is in operation again, or has it been working all these years since without notice?

Senator Rupert Davies (Liberal, Ontario) has charged in the Senate (Dec. 5th) that "an anonymous member" of the Revenue Department decides what books shall be banned in Canada. He said that in 1948, 45 books and 23 newspapers were refused entry, and the number was 81 books and 22 publications banned in 1949. Let us know who is doing the censoring; let us know what books are banned, and why?

The Catholic hierarchy's Canon 1399 forbids Catholics to "read, borrow, buy or sell . . . any book which attacks Catholic discipline and dogma." This is not a local ordinance but a mandate for practically all Catholics everywhere. Only getting a nation to prevent the publication and distribution of such books—as is done in some countries—could improve upon this device for the preservation and expansion of Catholic error, since the hierarchy has and uses its own means of punishing both the Catholic reader and all daring writers.

THE NEWSPAPER, DEMOCRACY, AND THOUGHT CONTROL

SUBGOVERNMENT BY RELIGION

A democracy tries to give a full measure of liberty to men. Meanwhile, the various sects and religions within its domain clamp down their catalogue of special restrictions and punishments on as great a number of citizens as their influence can reach. And these sects have the effrontery to ask the favor of the state in making more of its citizens the captives of the church. These organized religions usually have, and actively exercise, ways of punishing the small or large groups of citizens included in their membership. In this way a prime aim and function of the tolerant democratic state is openly and continuously sabotaged by the churches it permits and protects. Thus heaviness falls on him who sips tea or coffee (Mormons), installs a telephone or buys an automobile or tractor (Amish), eats pork without a special dispensation on Fridays (Catholics) or on any day whatever (Orthodox Hebrews and Hindus), consults a physician (Christian Scientists), neglects church and fishes on Sunday (many Fundamentalists), is married by the state only or fails to attend mass (Catholics), and paints a picture or takes wine with his dinner (Mohammedans). Again, within the Western world, prominent Protestant groups are especially guilty of having forced "blue" and "Sunday" laws upon cities, states and nations, so that a wide range of sports and other activities are forbidden to every citizen and to his foreign guests; and, to crown it all, transgressions of these laws are made punishable by the state itself.

The preservation and possible extension of individual liberty, in democracies depend upon more than one thing. But they depend in part upon the *habit* of freedom—upon being accustomed to feel and to exercise one's elemental rights—and upon *not* having a habit of nonreasoning acquiescence in rights taken away. Every deprivation of a natural right by a church, or by the state acting under the urge or influence of a church, is an underhanded blow at the habit of freedom and at readiness to protect it. Individual freedom and democracy are best preserved and extended through citizens kept conscious of attempts to invade their personal rights. The task of encouraging this necessary vigilance in the citizen has long fallen partly or largely to the newspaper. Who will suggest that newspapers now effectively inform and defend the American citizen against these encroachments of religion?

Very early in this chapter it could be concluded that relief from suppression and censorship of news in a democracy is prevented chiefly by the unreadiness of citizens to resent these actions and deal with them. No one is likely ever to measure and know how much of that unreadiness is a result of the politically vicious habit—early ingrained and daily practiced—of meek acquiescence in unethical, religion-sponsored suppressions, but it would seem to be appreciable. The habit also makes organized religions more harmful to democracies than to totalitarian governments. No democracy, and indeed no government of any other type, has yet been born free of these subgoverning religious agencies. Democracy, like all organisms, is a born carrier of its own diseases. Democracies have not hesitated to turn long-established economic, political, social and professional agencies upside down and inside out. But, by overhopeful definition, democracies have not had the wish, and by circumstance they have not had the power, to deal similarly with the subgoverning agency called religion. Democracies exist and advance through political parties; but they continue to be parasitized by sects. In a limited but peculiarly dangerous way, therefore, religions compete with governments for men's loyalties, and for the power to impose penalties and restrictions upon citizens.

Newspapers may not be considered as of a single kind. The printed matter prepared by a sect or faith, and intended solely or chiefly for members of that faith, may be expected to provide a limited coverage of the news and to avoid or suppress whatever it dislikes. No matter how harmful this parasitic invasion may be, a democracy has no defense against it so long as such faiths and such newspapers exist. In the United States, several million citizens still read only or chiefly this type of newspaper. Many leading newspapers disclaim any religious bias. As already noted, however, the fear of church groups, dreaded reactions from the all-important advertisers, and the financial success of the enterprise, all lead to one or another degree of suppression

and coloring of the news. And this is an extremely far cry from
a defense of the citizen. Still other important newspapers, in several large American and European cities, are definitely prejudiced in favor of one or another faith, and without verbal acknowledgment of this fact they propagandize that faith and suppress or color news unfavorable to it. Bearing on parts of this
statement is the following excerpt from an article in *Time* magazine (April 10, 1950) on Boston's newspapers:

But when news breaks that might offend an advertiser, such
as fire or robbery at a department store or a suicide at a leading hotel, either the story is not covered or the location is
thoughtfully omitted.

Because of Boston's predominantly Catholic population, the
papers are equally careful not to print any news which might
offend the church, even though top newsmen know of no instances where it has tried to exert pressure on the newspapers.
Nevertheless, such stories as the debate between Paul Blanshard
and Father George H. Dunne at Harvard in February over the
political power of the church are virtually ignored (only the
Globe printed a story on the debate). Such sacred cows, real or
fancied, tend to blunt the nose-for-news of even the best re-
porters.

Still other specific instances of punishment or of restrictions
placed upon citizens by religious authority may now be cited. In
America and in many another country a Catholic mother who
sends her child to a public school over the objection of a priest
can be denied absolution in the confessional. No civil authority
of the community or nation can protect such a citizen and mother
from this theocratic coercion. In 1935, following her divorce and
remarriage, a Vienna news dispatch announced that Catholic-
born Maria Jeritza would never again be engaged to sing at the
Vienna State Opera, where she began her career. In 1949, Ivan
Obolensky, a grandson of the late John Jacob Astor, but brought
up in the Russian Orthodox faith, was married to Catholic Claire
McGinnis at St. Patrick's Cathedral, in New York. Soon thereafter, following an additional ceremony in Manhattan's Russian
Orthodox cathedral, the chancery office of the Catholic archdio-
cese of New York announced that "both parties solemnly prom-
ised in writing that there would be only the Catholic ceremony.
. . . . Therefore the Catholic party automatically incurred excommunication." Though excommunication might fall as lightly

as a leaf upon a non-Catholic, it is the gravest moral censure at the command of the Catholic Church. To the believing lay Catholic it is hellfire that may overtake him or her while forbidden to attend mass or other church services, receive sacraments, and enjoy indulgences.

A Jesuit institution, St. Louis University, in St. Louis, Missouri, penalized a citizen mainly for a political act. *The New York Times* (December 11, 1939) reported the incident in part as follows:

The dismissal, after almost twenty-five years of service, of Dr. Moyer S. Fleisher, Professor of Bacteriology at St. Louis University, is condemned as unjust in a report of the Committee on Academic Freedom and Tenure of the American Association of University Professors. Dr. Fleisher's dismissal resulted principally from a controversy over his sponsorship of a meeting in favor of the Loyalist faction in the Spanish civil war, the committee found.

Similarly, Catholic hospitals, which here and there receive state and federal aid as well as exemption from taxes, commonly punish physicians for their personal views. On this point, the April 21, 1947, issue of *Time* magazine reported as follows:

In Connecticut, it is still a misdemeanor for a doctor to advise, or a citizen to practice, birth control. This 68-year-old blue law, a relic of Anthony Comstock's crusades, is widely disregarded, seldom enforced. But Connecticut's doctors, who object to being lawbreakers, even technically, have tried eleven times in the past 24 years to get the law repealed. Last week, as they tried once again, Connecticut medicine was shaken by one of its biggest rows in years.

It all began when the medicos formed a Committee of 100 to sponsor a mild bill which would permit a doctor to give birth-control information to a patient whose health or life might be endangered by pregnancy. Roman Catholic spokesmen promptly opposed it. But the doctors had some unprecedented support: The Hartford *Courant,* first major Connecticut newspaper ever to come out on their side, 500 ministers, an Elmo Roper poll which showed that 85% of the state's citizens (including 75% of its Roman Catholics) favored the bill in principle.

The battle was raging in speeches and in letters to the newspapers—until last week. Then professional blood began to flow. Six angry doctors, members of the Committee of 100, announced that they had been kicked off the staff by Roman Catholic hospitals in Waterbury, Stamford and Bridgeport. Explained Father Lawrence E. Skelly: "The [hospital's] action was self-defensive. . . . You gave your name publicly to the support of a movement which is directly opposed to the code under which the hospital operates."

Another of these cases was reported from Poughkeepsie, New York, by the Associated Press under date of January 31, 1952:

St. Francis Hospital, a Roman Catholic institution, demanded today that seven physicians resign from its staff or sever their connections with the Dutchess County League for Planned Parenthood.

In further comment on this case, three days later, *The New York Times* added the following information:

St. Francis has 200 beds. . . . This hospital completed two wings last fall with the aid of Government funds and grants from two Protestant philanthropists. It withdrew several years ago from the Community Chest.

Enforcement by the state of church-inspired restrictions upon citizens is a matter that affects the daily acts, if not the consciousness, of everyone. Two examples only will be cited here. An Associated Press dispatch, dated May 26, 1951, from Albany, New York, stated: "Roller skating derbies cannot be legally held in state armories—or any place else—on Sundays, Attorney General Nathaniel L. Goldstein said today." Next, an item from just over the border. Since 1906, Canada has had a law, sponsored by the Lord's Day Alliance with headquarters in Toronto, which makes illegal any unnecessary toil or business on Sunday. The result of a recent vote in Toronto on a part of this restriction was given in *Time* magazine (January 16, 1950) as follows:

In Toronto's municipal election last week 88,108 people, a majority of 6,315, voted in favor of allowing commercial sports on Sunday. . . . Backed by a group of citizens called the Sunday Sports Committee he [Allan A. Lamport] pulled a surprise victory against the solid opposition of the three leading newspapers and all the churches.

Of different nature but notable here is a report upon a survey of seventy-two major campuses in the United States, published (May 11 and 12, 1951) in *The New York Times*. The report relates especially to political and social thinking, and states that "a subtle creeping paralysis of freedom of thought and speech is attacking college campuses in many parts of the country, limiting both students and faculty in the area traditionally reserved for the free exploration of knowledge and truth." Two themes run throughout the report: that restrictive measures taken by college administrators and legislatures are a very real danger to academic freedom, and that a still greater danger is the self-censorship by students in the face of "McCarthyism." Students feel that it has become "unwise" to think and talk about the most pressing problems of the world community.

Especially galling is the effective enforcement upon ourselves of church-inspired restrictions by the officials of a *foreign* power. The hierarchy and the Pope continue to restrain the peoples of many entire nations from courses of action that most non-Catholics and some Catholics regard as useful or necessary. To an audience of 1,800 members of St. Luke's Association of Medical Practitioners, as recently as 1944, the Pope is reliably reported to have declared that no power on earth has authority to take human life except in carrying out a death sentence against a guilty person. And "any act tending to destroy an innocent human being's life, whether as an end or as a means toward that end, whether that life is embryonic or fullgrown, or nearing its end, is forbidden."

Under the doctrine thus stated by the Pope—and long ago written by Jesuits into the Catholic medical code—a Catholic physician is not clearly permitted to remove a fetus, even though it be a monster, to save a mother's life. Despite this prohibition, it is greatly to the credit of many Catholic physicians that they do perform such operations. A tiny loophole has been developed by the Jesuit casuists, who now find that an *indirect* killing of the fetus, by an operation done to save the mother, is permissible; such, for example, is the removal of the appendix, which results in abortion. A resourceful and partly emancipated Catholic physician expands this permission.

In the courts of the province of Quebec, a witness is not allowed to take an oath or give evidence if he does not believe in God and in the existence of recompense and punishment after death. Further, from Montreal, on the eve of Ascension Day, 1952, the Associated Press sent this information:

A special 100-man police squad will visit stores throughout Montreal tomorrow—the Feast of the Ascension—to check violations of the new bylaw ordering stores closed on Roman Catholic holidays. The penalty for being open is a 40-dollar fine or two months in jail.

In full defiance of the federal Constitution, the state constitution of Tennessee makes ineligible for public office any individual who denies "a system of rewards and punishments."

Subgovernment by three of the religions of Asia is illustrated by news dispatches reporting three recent occurrences. From the new state of Israel, *The New York Times* (March 18, 1951) printed this item:

Three men and a 14-year-old boy, all members of an extremist religious organization, were arrested on suspicion of having burned seven private automobiles during a recent week-end in Tel Aviv, Israel.

Warnings to "observe the Sabbath" were found smeared on the damaged vehicles. Members of this extremist group are pledged to enforce the Torah (Holy Law) by violence, but they are small in numbers. However, *many others* [italics mine] are in favor of an ordinance to prohibit automobiles from being used on the Sabbath.

Concerning Mohammedan Saudi Arabia, the same newspaper (March 5, 1950) published this story:

The Saudi Arabian Government has issued a communique stating that any non-Moslem found inside the sacred areas of Mecca and Medina is subject to a prison sentence of up to five years and a fine of 5,000 Saudi Arabian rials (about $1,300). After serving the sentence, he is to be deported from the country. The communique adds that the Saudi Arabian Government will not be responsible for the safety of any non-Moslem found within the areas of Mecca or Medina.

An Associated Press dispatch (January 25, 1951) from India stated:

The government of India, faced with an annual food grains deficit of 2,000,000 tons, is perplexed by the passionate opposition of an influential section of Hindus in western India to destruction of locusts (grasshoppers).

The annual southward migration of locust hordes from the

Sind-Rajputana region has been a widespread menace this year, with swarms winging their way right into the heart of Bombay city where they had never been seen before.

Non-violent Hindus in Gujerat, Mahatma Gandhi's home province, took to violence to prevent the killing of the pests by official agencies. This enabled locusts to devastate some 3,000 square miles of fertile tracts in the Mehsana and Banaskantha districts. A mob attacked a government depot where modern insecticides were stored. Locust-fighting personnel were assaulted so severely that police had to arrest 30 persons in Mehsana. The populace actually threw a temporary bridge of shrubbery across the Banas River near Deesa town to facilitate the movement of locusts from barren regions to green fields.

The prevailing ideas of the powerful Dutch Reformed Church in South Africa were sketched in *Time* magazine (May 14, 1951) as follows:

In Pretoria, South Africa's Dutch Reformed Church (1,400,-000 members) held a synod, solemnly condemned: (1) cremation ("a heathen custom"), (2) commercial radio programs on Sundays, (3) American comics ("doing untold harm"), (4) Freemasonry, (5) The U. N. Declaration of Human Rights. The churchmen rejected racial and sex equality ("God spoke to Adam, not to Eve"), as well as freedom of speech and opinion: "Heresy and untruth may not be spoken freely. . . . The devilish tendencies in man place very definite limits on these freedoms."

Dean Carl W. Ackerman, of the Columbia University Graduate School of Journalism, with politics and McCarthyism doubtless largely but not wholly in mind, was quoted in *Time* magazine (April 13, 1953) as follows:

Today the vast majority of teachers in all fields of instruction have learned that promotion and security depend upon conformity to the prevailing community or national concept of devotion "to the public welfare." . . . There are not many classrooms in the country today where students are advised to be "drastically independent." . . .

After twenty-two years as dean I am now discontinuing my practice of cooperating with the federal, state and police agencies, except on written request and advice of counsel. . . . Silence on controversial subjects during private conversations, as well as in classrooms, is becoming so prevalent that it is dangerous to our liberties.

SUPPRESSION OF THE NEWS

The suppression of news can be widely prevalent without being apparent, and particularly so when it is practiced by virtually all newspapers, and by television and radio. The chief inducements to suppression—fear of losing advertising or sponsors, and the active resentment of racial and religious pressure groups among readers, listeners, and advertisers—are evident enough.

The weight and pull of advertising can be measured in terms of dollars. The trade publication *Printer's Ink* (August, 1952) estimated total 1951 expenditures for advertising in the United States at $6,496,500,000. The approximate total for newspapers was $2,200,000,000; for magazines, $573,700,000; for radio, $712,-300,000; and for (newly born) television, $390,000,000.

During the past several years George Seldes' weekly newsletter *In Fact* has maintained that there are honest newspapers in the United States but has challenged the publishers to prove that their number is greater than one per cent of 1,750 dailies. An item from that publication (July 3, 1950) referring both to a habit of handling news of religious import and to an instance of the suppression of news is here quoted:

Restrictions on freedom of religion usually make frontpage news when they concern Eastern Europe and the Vatican. The stories are buried or suppressed when they concern fascist countries or nations which tolerate only Catholicism. Latest evidence was suppression June 20 of a story sent from Madrid by Associated Press. Except for a 12-line item carried by the New York *Herald Tribune,* which tucked it away among the stock market ads, no New York daily mentioned it. Headed "Spain Extends Curbs on Protestant Activities," the story listed new edicts by Franco Spain against Protestant and other non-Catholic worshipers. "The Ministry of Government issued a new statement of its policy Saturday in reply to a Protestant appeal to Generalissimo Francisco Franco for protection," the story said. The new statement banned collections for the support of non-Catholic churches, prohibited public demonstrations and forbade the establishment of schools or recreation centers for Protestants.

In Fact (April 10, 1950) had earlier credited *The New York Times* with publication of a religion-pointed item, sent from

Europe under the date of March 28 by the Associated Press, but suppressed or buried by all other New York newspapers. Also credited was this instance:

A most significant item from Italy following news of peasant uprisings and police violence, resulting in injuries and death following attempts to seize land, is a 3-inch item which no American newspaper correspondent in Rome, nor the Associated Press, International News Service and United Press services reported. It was the British agency, Reuters, which sent it out—the New York *Times* printed it March 22. "Vatican Land Excluded," read the head. "The biggest landed proprietor in Italy . . . the church, will not be subjected to the new government bill to lop 4,500,000 acres off the estates of big landowners."

In Hungary it was the opposition of Cardinal Mindszenty to giving up church lands and his general opposition to governmental laws—treason in that country—which was the chief reason for the conflict of church and state, his arrest, and imprisonment.

Confession of newspaper suppression of news on an indefensible scale, and by a person then perhaps in as good a position as any American to know its extent, is contained in an Associated Press dispatch dated June 13, 1953:

The president [N)rman E. Isaacs] of the Associated Press Management Editors Association asserted today, "We in newspapering are continuing to lose prestige. And we can trace our losses in influence in large part to the one-sided, biased and inefficient coverage we provide in our local communities. . . . We have only one function—one basic function. It supersedes everything. That function is to convey information. We are (privileged) common carriers. . . . But freedom of the press cannot possibly mean the license to keep the people from knowing. And we keep them from knowing what the real score is every single day of the year by our backward and arrogant methods of operating newspapers."

When news of violent religious intolerance in a bordering nation is suppressed by newspapers and radio stations of the United States, it becomes worthy of examination. And whoever presumes that at least a start toward religious tolerance has been made in all northern parts of North America should inform himself —if that were possible—concerning recent performances in the province of Quebec. A brief reference to the misadventures of Jehovah's Witnesses and of Baptists in that province during the past five years will illustrate a state of mind and a depressed phase of Western culture that is little known or discussed outside of Canada.

First, the treatment accorded the Witnesses. Readers within the United States shall be left to decide whether their various news agencies have acquainted them with these occurrences just beyond our northeastern border. Insofar as this is known to the writer, the only organ of news seriously trying fully to acquaint citizens of the United States with this ocean of intolerance at their doorstep is the little-known semimonthly journal *Awake.*[6] That journal, apparently, is the official organ of Jehovah's Witnesses; it accepts no advertisements and obtains its own uncensored news reports. Its issue of April 8, 1950, stated:

These sincere ministers [Jehovah's Witnesses in Quebec] suffered from hatred, bitterness, and hundreds of arrests on trumped-up charges of peddling, distributing circulars and handing out printed Bible sermons to interested persons. Children were expelled from school or dragged into court as juvenile delinquents because of their faith. Police invaded places of worship and made arrests of Jehovah's Witnesses for celebrating the Lord's Supper. Respectable Christian girls were arrested, stripped and held in filthy jails. . . . Mob assaults were made on peaceable gatherings. At the peak of the frenzy of police and priest opposition to Bible teaching and freedom to preach the gospel of God's kingdom, in the year 1946 cases against Jehovah's ministers reached the staggering total of over eight hundred charges.

The same issue of this publication carried detailed accounts of persistent violence and persecution directed toward two women Witnesses in the city of Joliette, forty-five miles from Montreal, in December, 1949.

A recent and rather similar experience of a group of Baptists, in La Sarre, Quebec, was given little publicity in the daily press of the United States. An adequate account was nevertheless provided by *Time* magazine (August 7, 1950), and a part of that statement is quoted:

In the early evening, people began to gather around the rickety Victoria Hotel on the main street of the farming village of La Sarre (population 3,100). Soon the Baptists drove in from their little church outside of town. Six men and four women, they

6 Published by the Watchtower Bible and Tract Society, Inc., 117 Adams Street, Brooklyn, N.Y.

gathered on the corner, opened hymn books, and began to sing in French: *What Will Wash Away My Sins?*

The onlookers jeered and whistled. A 19-year-old in a white sweat shirt drew loud laughter by pretending to direct the hymn. Then, some in the crowd picked up dust and debris from the road and tossed it over the singers, but they paid no attention. A loudspeaker truck blaring jazz music raced up and pointed its horn directly at the group. Chief of Police Edward Carpentier arrived and ordered the singers to move on. When the Baptists refused, a dozen burly men hit them in a flying wedge. For 15 minutes the mauling went on.

It ended at last with the arrest of the Baptists. Five men were held in jail three days, charged with illegal assembly, until bail of $1,900 was raised. Dark-eyed, quiet-voiced Leslie G. Barnhart, leader of the band, was neither surprised nor dismayed. "We fully expected this to happen," he said. The group had been attacked twice before in La Sarre—once by a crowd, once by a pressure hose of the town fire department. The pattern was familiar in many towns of French Canada during the past decade since Baptists and other Protestant groups have stepped up their efforts to organize churches in solidly Catholic towns and villages.

A year later *Time* (August 27, 1951) referred again to the Reverend Mr. Barnhart. When he "investigated the disappearance of the sermons he mailed to the village of Ste. Germaine, he made a startling discovery. They were being turned over to the priest at Ste. Germaine, the Rev. Alfred Roy, who burned them. Such letters, said Father Roy, 'would give people wrong ideas. They can't take me to court for that, can they?'"

It should be stated that during the early nineteen-forties several communities in the United States came to regard Jehovah's Witnesses as "pests," passed ordinances against some of their noise-making activities, and enforced them. But neither the Quebec incentives and techniques nor the churches were involved in these contests, and newspapers apparently gave a reasonable and adequate account of such minor items of community strain. But when the entire principle of religious freedom is persistently flouted by the unrepressed mob activities of the Catholic people and the civil authorities of Quebec, newspapers and radio alike in the United States seem to lose their voices.

One may consider Jehovah's Witnesses socially irritating— even contemptible—but he may not forget that they are rooted and grown in the Christian Bible. Reasons for the considerable attention given these items from Quebec rest on the fact that the events described do much more than illustrate the place and power of religion in the suppression of news. They illuminate an area of virulent religious intolerance that is immensely disruptive in North American society. They indicate the present inability of a foremost organized religion to liberalize its thought or to democratize its social and political acts when or where it is the thoroughly dominant force. Those events also notably supplement evidence considered in other chapters of this book: organized religion's divisive power; religion today as a foe of inquiry, knowledge and change; and the Catholic faith as an enemy of the American type of democracy.

Americans may note the threat of an over-all censorship contained in a recent decision (May, 1952) of the United States Supreme Court. In a five-to-four decision, the Court upheld the conviction of a man for printing scurrilous material about Negroes, under a 1917 Illinois law that makes it a crime to hold up any "race, color, creed or religion to contempt, derision or obloquy." Only Illinois, Indiana and Massachusetts have enacted such laws on "group libel." For the Court's minority Justice Black stated that "sugar-coating" the law by calling it a "group libel law . . . does not make the censorship less deadly." Editorials in many newspapers deplored or feared the Court's action. The Chicago *Tribune* stated that "the Illinois statute . . . could be interpreted to outlaw books and plays about Okies. To call something 'a dirty Irish trick' could be actionable. Legislation intended to prevent this kind of thing . . . would not be worth what it can cost the people of this country in restricting freedom of expression."

A few—very few—books and magazine articles have documented the extent and fatal consequences of the suppression of foreign news by owners of American newspapers—and of the role of religion in that suppression. Involved in the story is Germany both before and after two world wars, Russia from the Bolshevik revolution onward, the entire course and nature of the Spanish Civil War, and some important phases of conflicts in most foreign countries during and following World War II. In other words, it relates to American unawareness and misinformation concerning the long train of developments that now threaten —and force action and sacrifice upon—practically all mankind. That default, during four decades, has increased greatly the constantly present peril to the peace and security of the world. The most recent of those few well-documented books[7] deals with some of these subjects. Certain items from it relating to the Spanish Civil War are quoted here.

When, in 1937, the Philadelphia *Record* published accounts of this war, as witnessed by reporter Seldes in Madrid, the Cardinal and hierarchy in Philadelphia threatened and boycotted that newspaper. There

a pastoral letter was read by every priest at every mass over a certain period of time. The faithful were urged to stop reading the *Record* unless it changed its policy from pro-Loyalist to anti-Loyalist.

Unable to persuade its followers to stop reading a liberal newspaper, the Philadelphia hierarchy then quietly went to work among the big advertisers, notably department store owners, Protestant and Jew as well as Catholic, and these men, all frightened by the Church, in turn frightened Mr. Stern [the publisher]. On August 10, 1937, the publisher sent his humble apologies to Cardinal Dougherty and called his attention to a new editorial he was running "denouncing the Spanish Government's action against the Catholic Church." . . . Finally, Mr. Stern said to the Cardinal: "I would very much appreciate your advice as to what I should or should not do in the matter."

Altogether, through American politics and newspaper boycott, the hierarchy was to prove itself perhaps the decisive factor in the overthrow of the Spanish Republic—the curtain raiser to World War II.

In 1936 the papal nuncio, Cardinal Pacelli, the most political prince of the Church in modern history (later known as Pope Pius XII), had visited the President (at Hyde Park) for purely political purposes . . . but Sinclair (in *Presidential Agent*) said he learned from other sources that "the head of the hierarchy . . . talked cold turkey" about Spain and the Catholic vote in 1936, saying, "either you keep arms from the Spanish Reds or else we defeat your party." . . . Louis Adamic wrote (in *The Eagle and the Roots*, 1952) . . . that a mutual friend of his was present in the White House when Jim Farley told President Roosevelt bluntly that unless he embargoed arms to the Spanish Republic he would lose all the Catholic vote (and therefore the election of 1936) because the hierarchy, etc. . . . The Gallup and other polls showed not only a good majority of the American people on the

7 George Seldes, *Tell the Truth and Run* (New York: Greenberg, Inc., 1953).

Loyalist side, but that majority increasing. Protestants voted 83 per cent against Franco. . . . In the course of time the Roosevelt error was realized—by F.D.R. himself, and the State Department. Mr. Sumner Welles, after his retirement as Secretary of State wrote (*Time for Decision*) that "of all our blind isolationist policies the most disastrous was our attitude on the Spanish Civil War."

Yet Americans of 1953 wonder why great political parties in countries like Italy and France thoroughly distrust the professed liberal position of the Government of the United States. Americans now wonder why highly influential blocks of fairly well informed Europeans remember that, when the chips were down, Russia could spare some technicians for the Spanish Civil War though the United States refused to permit the flow of arms, which it, as a neutral, could properly have done. The swarm of Europeans who perhaps rightly recognize the hierarchy and fascism as their durable and intolerable enemies are not sold on our nation's recent record. The blame rests far less on the ill-informed American people than on their newspapers and the hierarchy. And on the hierarchy rests most of the blame for the fateful subservience of the American press to Catholic power. One specific way in which organized religion now threatens both the present and the entire future of society and the race could not be better indicated.

MISCELLANEOUS RELIGIOUS PROPAGANDA

In the United States, flagrant religious propaganda is carried on by many officers and agencies of the federal government. As set forth in Chapter 9, federal officers swear to uphold a Constitution that "erects an impregnable wall between church and state"; and, presumably, a Constitution that embraced Jefferson's view that "to compel a man to furnish contributions of money for the propagation of opinions which he disbelieves and abhors, is sinful and tyrannical." Nevertheless, unlike Jefferson and Madison, the presidents of the Republic in our day reach for opportunities to put God into addresses relating to all our people, proclaim Thanksgiving Days, attend official ceremonies opened with prayer, and sign all bills that use public money for paying chaplains, building chapels, and other related purposes. The military and many other agencies of government, seemingly entirely oblivious of the constitutional "wall," repeatedly and shamelessly join in further breaches of it. Thus the many agencies of government endlessly and outrageously offend those millions of agnostics and atheists whose undivided loyalty is to government—not to a religion—and whose topmost reason for loyalty to the United States is the brave effort of its Constitution to guarantee to its citizens total uninvolvement in religion.

Congressional and executive acts—timed exactly with this mid-century—have done all they conceivably can do to change Memorial Day from an occasion for honoring those who gave their lives to their country in war to a religious day of prayer. Early in 1950, a joint resolution of Congress provided that Memorial Day "should be set aside as a day for nation-wide prayer for permanent peace," and requested the President to issue a proclamation calling upon the people to observe each such day in that way. Presidents Truman and Eisenhower have unfailingly shared in this retreat from unstinted gratitude to our heroic dead, which is also a further assault upon the constitutional separation of civil and religious authority under our government.

An Associated Press dispatch of December 10, 1953, reported:

> The Senate Committee said today the Commerce Department has endorsed a bill to let clergymen ride free or at reduced rates on the airlines. A Committee spokesman said Secretary of Commerce Weeks wrote the group, saying the bill ought to be passed to give ministers the same privileges on the airlines they have had on railroads for many years.

Other governmental controls over both these types of transportation *compel* users to pay costs plus a profit on the total service the companies render. Thus, at this late moment, both the executive and legislative branches of our government plan to force others, including atheists, to pay *additional* transportation costs of ministers, priests and rabbis.

Late in June, 1951, the scientifically important and half-finished forty-million-dollar hospital of the National Institute of Health was dedicated at Bethesda, Maryland, by President Truman. That public building, a skyscraper for which the President's signature had provided the public funds, also contains a *chapel*. Earlier in 1951, the Armed Forces had established—with public funds—a school for chaplains at Fort Slocum, New York. The chaplains also procure and distribute religious propaganda to members of the Armed Forces. Bearing on this point are excerpts from a news report published in *The New York Times* (May 6, 1951):

A series of pamphlets explaining the reasons for current American military operations has been prepared for distribution by chaplains among Army, Navy and Air Force personnel under Protestant, Catholic and Jewish auspices. Preparation of the pamphlets was directed by leaders of the three faiths operating through the Commission on Religious Organizations of the National Conference of Christians and Jews. . . . The first pamphlet is entitled "I Thank God." . . . Accepting the copies, Admiral Salisbury said: "The pamphlets clearly indicate that a belief in God is the real basis and motive in our struggle against aggression. . . ."

Not only is religion propagandized by the presence and present authority of the chaplains, but often the rights of the nonbeliever have been meagerly protected. Some of the printed "blanks" currently used for the registration of personal data of soldiers lack any such notation as "no religion," "agnostic" or "atheist," following a listing of specific religions. In these particular cases one may wonder whether the Armed Forces hope to embarrass the nonbeliever, or perhaps persuade his acceptance of one or another of the religious labels duly printed on these propagandist sheets.

On April 8, 1954, President Eisenhower personally launched by television the first United States postage stamp of a regular issue to incorporate a religious tone. He and Postmaster Summerfield inserted "In God We Trust" on a three-color stamp intended—at the rate of 200,000,000 yearly—chiefly for postage to foreign lands.

The Post Office Department regularly grants discounts on third-class mailings and second-class publications of religious organizations, and annually collects the deficits thus incurred from public tax money. An Associated Press dispatch dated May 10, 1952, from Nashville, Tennessee, stated that since 1950 "brief scripture readings followed by a prayer now form a regular prelude to the day's work in Nashville's postoffice and sub-stations." No criticism or comment accompanied this item of news.

The judiciary of many states has a long history of neglect of, or of ways of breaching, the "impregnable wall between church and state." And this history stems from the centuries-old parasitization and poisoning of the law by religion. From Swancara's [8] arousing book on this subject the following single item is quoted:

> Ben Wells was not permitted to testify in a case in Charlotte, North Carolina, in 1929, concerning the identity of the persons who had flogged him because of the following question and his answer thereto:
> Q. Do you believe that if you would tell a lie God would punish you either in this world or in the hereafter?
> A. No, I don't; but I won't tell a lie because of my own conviction!

According to an Associated Press dispatch dated March 2, 1949, reporting a divorce case in Chicago, Superior Court Judge Rudolph Desort enjoined citizen Duane Free, a twenty-six-year-old printer, from teaching atheism to his daughter and stepdaughter.

In the matter of elections, the Catholic Church has found that spreading news by hand is quite practicable. An instance is provided by excerpts from an account (October 24, 1949) in *Time* magazine:

> In defense of bingo, the Roman Catholic Church plunged deep into New Jersey politics. . . .
> Last week, officials of the Archdiocese of Newark summoned 400 nuns from parochial schools, handed out copies of a four-page circular urging the election of Wene [candidate for Governor]. Explained Auxiliary Bishop James A. McNulty: "The interests of the church would be better served by Wene and other Democratic candidates."
> The nuns were instructed to explain the pamphlet to their pupils. Other copies were distributed through the Catholic War Veterans, Knights of Columbus, Holy Name and Rosary Societies. Parish priests were briefed. In all, the church expected to distribute close to 250,000 copies.

In some other elections the Catholic Church uses the heavier weapons discussed earlier in this chapter, according to the following account of physician Upham in *Freethinkers of America* (March-April, 1944):

> Its full force as a powerful minority pressure group [the Catholic Church] is being used to influence legislation, to bar medical advice on birth control from public hospitals, to deny meeting

[8] Frank Swancara, *The Obstruction of Justice by Religion* (Denver: Courtright Publishing Company, 1936).

halls to planned parenthood speakers, to frighten radio stations, newspapers and magazines into silence on the subject.

The people of two of our states, Massachusetts and Connecticut, are being denied a type of medical service available in the other forty-six states largely through Catholic political threats and pressures. Dr. Karl Sax of Harvard University, analyzing the Massachusetts election on the birth control amendment in 1940, declared that "the only important opposition came from the Roman Catholic hierarchy" which carried on "a campaign of lies, scare propaganda, suppression of press and radio and intimidation of voters." He found that "three of the major metropolitan Boston newspapers, and the only newspaper in Fall River, and the three major Boston radio stations refused to accept advertising in favor of the amendment" for fear of Catholic reprisals.

One might presume that a planetarium, operated not in Tibet but in New York City, would be more concerned with science than with mythology and religious propaganda. But read these excerpts from a letter written by editor Joseph Lewis, of *Common Sense* (April, 1951), to the chairman of the Hayden Planetarium in New York:

A few weeks ago (Sunday, March 4th), I visited the Planetarium to see unfolded the marvelous array of stars in the firmament before the vernal equinox, or the coming of Spring. Shocking beyond words was my experience when, instead, I saw the prostitution of this display. I saw it turned into a cheap piece of propaganda for religious observance.

Where in the Bible will you find the basis for the science of astronomy? To cap the climax, the speaker sought out some stars in the firmament as indicating the Cross of Christianity. Only a deluded mentality can find in the firmament stars to represent the myth of the birth of Jesus. To add further insult to injury, the Cross of Christianity was flashed upon the ceiling as if this symbol covered the dome of the universe.

And what an insult was the showing of a choir singing (as in Heaven) of the resurrection. This is beyond forgiveness! What a prostitution of the science of astronomy!

Rumor from Catholic sources teams up periodically with the Catholic and secular press to spread widely weird accounts of alleged happenings in the world of today. The effect of these allegations is not merely the negative one that goes with the suppression of a fact in the news; their effect is to carry to millions a false and medieval view of the nature of the world in which we live. In 1948, a Brooklyn publication (*The Tablet,* Sept. 11) provided an account of a crown of tea roses and ferns that had been placed by an eight-year-old girl on the usual statue in the Church of Our Lady, in Stockport, England. Though untreated and unwatered, this wired bouquet was said to have retained its freshness for fifteen months, by which time, some estimates had it, the church had been visited by more than 100,000 outsiders, and as many as 20,000 petitions from all over the world had been heaped at the base of the statue. It was remarked that some church dignitaries deplored the publicity given the case, but it was stated that the pastor of the church regarded it as a spiritual phenomenon. And as such, it is wholly clear, rumor and a part of the press paraded it before multitudes of people.

Another widely circulated rumor spread the tale that when, under threat of Japanese invasion, the casket containing the body of St. Francis Xavier was opened at Goa, India, in 1943, the flesh was fresh-looking and blood seemed to flow from a pricked knee, although the burial had been done in Chinese fashion, with quicklime, and an arm of this saint had elsewhere been exhibited by the Church for nearly four hundred years. These and many similar cases that might be cited—usually widely circulated in the press and sometimes later repudiated by one or another Catholic authority—should provide thoughtful people with part of the reason for the slow and uncertain advance of sensible popular thinking. They illustrate a specific way in which the avenues of communication—whatever else those avenues may contribute of good—also and necessarily circulate fables of the miraculous, which help to rob millions of Catholic and non-Catholic citizens of a worthy outlook upon life.

A seeming climax to these reports of the miraculous comes even closer to the present. From its special correspondent in Rome, *The New York Times* published a story, on November 18, 1951, that included the following paragraphs:

The Vatican newspaper L'Osservatore Romano today published two photographs that it said were documentary evidence of a miracle that occurred at the Cave of Iria near the village of Fatima in Portugal on Oct. 13, 1917. The newspaper added that thirty-three years later "another surprising fact" occurred in the Vatican when the present Pope, while walking in the Vatican gardens, saw a phenomenon similar to one that had taken place at the Cave of Iria.

According to L'Osservatore's description of the miracle at the cave, the sun, shortly after noon, was seen to "revolve swiftly around itself" and then to dip down suddenly toward the horizon over which it hung for a few minutes. This miracle, states L'Osservatore, was observed by several thousand persons and was even photographed by some. Two photographs published by L'Osservatore show the sun in about the position it would normally occupy shortly before sunset.

On this same topic, an Associated Press dispatch from Rome, dated November 17, 1951, stated:

According to Roman Catholic records, crowds at the Cave of Iria witnessed the "revolving sun" miracle on Oct. 13, 1917.

Three Portuguese children earlier had reported seeing visions of the Virgin Mary, who implored Christians to strive for the salvation of Russia. On Oct. 13 the children said they again had seen the Madonna and church records say other persons present saw a white cloud over the children's heads and then witnessed the "revolving sun."

Federico Cardinal Tedeschini last Oct. 13—the thirty-fourth anniversary—at the services at Fatima closing the 1951 extension of the 1950 Holy Year, told a million faithful that Pope Pius had seen "this same miracle" on Oct. 30, Oct. 31, Nov. 1, and Nov. 9, 1950. It was on Nov. 1 that the Pope formally proclaimed the dogma of the Virgin Mary's bodily assumption into Heaven.

Nearly four months after these two dispatches, Rome quietly gave out another news release (March, 1952) stating that a photographer had mislabeled the photographs earlier used in connection with the sun's unusual performance at the miracle of the Cave of Iria. The release made no reference to the related, marvelous and law-breaking visions of the Pope.

This monstrous story of the Miracle of Our Lady of Fatima—minus the Pope's part in it—was immortalized in color film by Hollywood, in 1952. Every day and night for decades ahead—the end unknown—these films will carry their effective Catholic propaganda to free and captive audiences of the world.

The unreal and superstitious elements in all of the older revealed religions offer difficulties even to persons only lightly brushed by modern education. Yet many of these people find it simple and easy to assure themselves—and others—that the sacred phrases have only a "symbolic meaning" and are still quite true. Thus, according to the Associated Press (January 26, 1950), during a then current drought in India, a member of the Indian Assembly, in New Delhi, inquired of fellow members of Parliament "whether Hindu sacrifices were included among experiments in producing artificial rain." To this question the speaker of the Assembly replied that sacrifices are for "spiritual rain, not artificial rain." From this incident one gets an idea of the degree of clarity and service rendered by a bit of such holy writing—even where it was followed by a full three thousand years of continuous teaching to an extremely religious people.

On those not-too-rare occasions when a more or less prominent scientist or philosopher proclaims his rejection of the principle of organic evolution, his words are given much prominence in the world's press. The item is indeed "news" since it is quite unusual. But for nearly all readers of newspapers, the learned man's statement fixes an impression that the truth of the evolution principle is in doubt. To those readers the opinion does not illustrate the mere raw truth that an occasional human outlook is fearfully and wonderfully made. Sometimes the reporter or the editor is himself wholly deceived—a common result of the firm censorship of formal education along with wholesale suppression of critical news by religion. Two or three examples will be cited.

A column-length editorial in the Paris edition of the New York *Herald Tribune* of September 13, 1935, was captioned " 'Evolution' Not Dead." It stated:

There are scientists who say that evolution is dead, and others that it is not. It is a matter of opinion, for as Darwin himself said, "Evolution is a theory subject to future proof," and there is still no proof today, 75 years after the publication of "The Origin of Species." . . .

The controversy came up again at the meeting of the British Association which closed this week at Norwich. It was reopened a few months earlier by a physicist, Sir Ambrose Fleming, who in a presidential address to a meeting of scientists asserted that the Darwinian theory was "the product of imagination."

At Norwich it was a zoologist, Professor MacBride, who described natural selection as "a complete fraud." Both phrases let loose torrents of arguments from scientists and from others less qualified to speak. As usual, neither side came out victorious from this verbal battle, but each found compensation for wasted energy in the publicity given to the participants in the polemic and the meeting at which it took place.

Publicity is, in fact, about the only upshot of any controversy on a theory which interests everybody and which can neither be supported nor demolished by ascertained fact. . . .

But if "evolution" is not dead, it has hardly the force today which it had when Huxley wrote these [omitted] words. The issue has lost much of its passionate interest for the public. It is felt more and more that the theory of evolution is a question for scientists alone. The latter still argue about it as a useful hypothesis which is not fully supported by facts, but they have long given up the view that, even if established, evolution is going to demolish the fundamental beliefs of mankind. With physical theories crumbling about them, biologists will be the last to maintain that their own theories will stand the shock of new facts of observation in the future.

In this flagrant case, the reader is coached in misunderstanding by an editor who himself is childishly uninformed. The stray sheep is made equal to the flock; Professor MacBride is misquoted. An ocean of overwhelming evidence is unknown or ignored.

Another instance. Quite recently wide newspaper publicity was given to denials of the scope of the evolution principle by philosopher Mortimer J. Adler, of the University of Chicago. In a lecture given there, on May 21, 1951, sponsored by the university's (Catholic) Calvert Club, he declared that men and apes differ "essentially in kind," not in degree. They are as far apart, he argued, "as a square and a triangle. There can be no intermediates—no $3\frac{1}{2}$-sided figure." And, with no intermediate forms (no missing link), there can be no common ancestor. It was said that we can therefore continue to argue the possibility of man's special creation by God in His Own Image!

Professor Adler earlier wrote a book in which wholly similar ideas were expressed. Of that book, reviewer Harry Elmer Barnes remarked that "it is the most curious intellectual product, considering its auspices, which has ever fallen into my hands." A Harvard professor called this Adler lecture of 1951 "the kind of statement Bryan used to make in the Bible Belt." Few should doubt that the philosophy of Bryan is akin to the religion of Adler.

The further agreement of both Adler and Bryan with Father Neufeld also seems evident. When asked whether only humans have souls, Father Neufeld replied:

The animal's principle of life is its soul. It is essentially different from the human soul. The soul of man is a spiritual substance subsisting in itself. It is the principle of man's conscious life and its existence is independent of matter, i.e., the human soul can and does live apart from physical nature. The animal soul, on the contrary, is intrinsically and essentially dependent on matter. It is incapable of life apart from the body of the animal and dies with it.

This happy and uninhibited excursion of a priest into biology —though failing to extend its brilliant clarification of soul to the cases of plants and viruses—was likewise brought to the attention of many readers of newspapers, in 1949.

Among the writer's scientific colleagues, and among better-educated groups generally in America, it is easy to meet the view that religion is transforming and liberalizing itself in a fairly satisfactory manner—at least this is true for some sections of the nation, and most notably in New York and certain metropolitan centers of the East and North. Further, there is a familiar view that science now encounters no difficulties with religion that require aggressive action by scientists, and that the getting and reporting of significant results is the scientist's full and sufficient job. Into this frame of thought—while still conscious of the spread of grotesque religious views by newspapers—let us try to fit the following episode. At science meetings held in New York at the end of 1949, Dr. Einstein's recent calculations, which appear to unify, or bring into a single view, the fields of gravitation and electromagnetism, were announced. The Reverend Frank Curtis Williams, Protestant pastor of New York City's South Reformed Church, was reported by *The New York Times* to have used these words before his congregation on Sunday, January 1, 1950:

[The newly found] relationship between the atom and the star is a restatement of the ancient monotheism anticipated by Moses. . . . Einstein's mathematical approach to the unity of all things brings religion and science one step closer. The unified field theory ties our world together as one whole. The Einstein revelation at Christmas time is another manifestation of God's wonderworking.

It is possible that the pastor was misquoted. It is not possible that science and truth escaped frightful perversion in the words that are quoted. Nor do those words indicate that New York is an area of "satisfactory" progress in liberalized religion; for there is good reason to suppose that Mr. Williams' sermon was more forward-looking than most sermons and masses spoken in New York City on that same Sunday. The fact that he made any reference whatsoever to a conclusion presented at a scientific meeting during the preceding week suggests that his sermons are among the top half of such Sunday pronouncements. It is probable that on any Sunday morning equally bizarre assertions resound in a large majority of New York's churches and synagogues.

The long and censorious reach of religion into all the news media and entertainment agencies of this exceptionally communicative day and age must be registered as almost the most disquieting fact of modern life. When that fact is added to the tight prohibitions now placed by religion on formal education—the theme of Chapters 8 to 10—it is not possible to obtain a result that can confirm the present optimism or unconcern of many colleagues and fellow citizens. It is more reasonable to conclude that, against these formidable odds, only an immense effort can bring either our new evolutionary insight or the sinister powers of its religious foe into the consciousness of society.

Ours is a day of revelations of meanings, but as yet they enter but slightly into the stream of our intellectual lives. Even the man of today is mostly an eye and an ear—his brain takes more vacations than excursions. Our time still serves best its mountebanks, its mimics and its emotional giraffes.

This survey has indicated that where organized religion is fairly strong, it has all the resources of an Ariel and a Hermes to put wings to its blighting, backward-pointing words. There, both spoken and printed word—of a friend or of an agency of news—is religiously hushed or censored before it reaches eager youth or baffled grownups. There, a democracy must share with religions the power to reward and to punish its citizens. There, recent presidents and many other officers of the great Republic which long ago sought to separate civil and religious functions now unblushingly propagandize religion. There, society must shoulder and strive to carry the heavy burden of religion's accumulated error and prejudice. There, no one can be born free. There, an unborn society—one based on available knowledge—must await the burial of a mystical outlook which easily propagates itself and hovers always over politics and law.

These several restraints on the spread of vital knowledge and of liberating thought are a barbed threat to every democracy. Their persistence precludes all possibility that current society may graduate into a "genuinely modern" society.

<h1 style="text-align:center">12</h1>

Religion's Power to Divide

For I am come to set a man at variance against his father, and the daughter against her mother, and the daughter-in-law against her mother-in-law. And a man's foes shall be they of his own household. He that loveth father or mother more than me is not worthy of me: and he that loveth son or daughter more than me is not worthy of me.—*Matthew* 10:35-37

OTHER CHAPTERS OF THIS BOOK DISCUSS SOME PRODUCTS and by-products of religion that are considered untrue, harmful or dangerous. But the *divisive* nature of organized religion—its prevalence and its consequences to local and world-wide goals of society—is an unquestionable fact whose sinister meaning remains largely unexplored. Items from this area often tangle with the harmony of family or community, sometimes with national or international war or peace, and always with the happiness and progress of the race. Indeed, the past, present and future of the divisiveness of religion is subject matter for volumes rather than a chapter. Place here is found for an uncommon look at some aspects of religion's power to sunder the fabric of human society at this mid-century.

In the contest of naturalism with supernaturalism, organized religion is the stout opponent of evolutionary thought. The one is incompatible with the other, and society is making its choice—consciously or unconsciously—between them. These pages deal with a commonly overlooked group of consequences of religious thought and practice—including one wholly unrecognized outgrowth that now has tragic political importance. They review our present information on the extent to which naturalism and supernaturalism are now accepted by groups or by peoples. That review includes information concerning the very unequal extent to which the peoples of several Western nations now reject the idea of God. More specifically, it surveys the present trend toward unbelief in American college students, scientists and other professional groups. It scans the causes and the quite special consequences of a cleavage of thought among scientists.

307

PERSPECTIVE AND THE PAST

Race, religion, nationality and economics long have been mankind's main sources of irritating, and often tragic, divisiveness. To the reader himself may be left most of the task of giving relative importance to these four factors. Certainly the importance of each of these sources of disunion and enmity will vary according to place and time; and all of them have long histories. It may be recalled, however, that race was imposed by nature and that a growing biological understanding of it can and will help to erase its divisive power. Nationality, though it has firm roots, is largely defined by boundaries that may float away and vanish at some time in the future. Economics, though an area of universal and changing challenge to self-interest and to human adjustment, will probably make slow or rapid progress toward at least an equalization of opportunity among all peoples. On the other hand, if the divisiveness of religion has any terminal facilities, current religions do not point to a time or place.

Some of the strongest supporters of religion readily admit its divisive might. For example, the Reverend F. Howard Callahan, pastor of the Methodist Church of St. Paul and St. Andrew, New York, was quoted by *The New York Times* (June 21, 1943) as follows:

On every hand churchmen are demanding that the nations plan for greater unity and cooperation, while many of them are apparently blind to the fact that religion itself constitutes one of the great divisive forces in our very own civilization.

Lawyer Blanshard has observed that "segregation on the basis of creed can be just as damaging to American democracy as segregation on the basis of color." Perhaps nowhere more appropriately than just here may one record the shameful fact that, at this moment, immense crowds of American Whites quote or cite the Bible as their authority for assigning an inferior human status to the Negro. Too, in part, the Ku Klux Klan is rooted and nurtured in Fundamentalist religion.

From the point of view of history, the most deadly consequences of the divisive power of religion are those which have led to international strife and war. The same may hold for the future. Religious difference, or this *added* to difference in nationality or to some other irritating factor, has frequently tipped the balance in favor of war. Nevertheless—except for one immense and now world-threatening shadow—the sinister part played by religious difference in the international sphere must be dropped along with much else from these pages.

Even a short sketch may be expected to indicate the extent to which the more basic religious views are now rejected by the people of a single nation. If that task proves too large and impractical, it may still be possible to examine the extent of present rejection of basic religious thought within certain groups of a nation's population. And since this book deals largely with the struggle of a group to free society and science from religious obstruction, it is especially rewarding to examine closely the evidence that religion still divides the scientists—the very forces that should lead that crucial struggle.

The latter part of the foregoing sentence may well be reread. Perhaps the divisiveness of organized religion has never before been so meaningful and consequential to the race as in this fateful outcome, this lightly regarded or disregarded division among educators, professors and friends of science, which would seem to be unthinkable after three generations have passed since Darwin. If soul salvation and supernaturalism had not suffused the air of 1859 and later, scientists—biologists certainly—would then have agreed promptly that the personal God was liquidated by evolution, and that much or all of the supernatural element in all religious thought was thereafter indefensible. Moreover, if that militant and organized religious opposition had not everywhere exhibited its conditioning, punishing and rewarding power, those who heard and understood the message of Darwinism should have been able quickly to convince the entire range of thoughtful men —university colleagues, writers, and leaders of public education —that these implications of the evolution principle are valid.

The greater tragedy involved in these observations now becomes clearer. An *early* acceptance of the implications of Darwinism in most public education would then have taken from many Western nations—long before the Russian (Bolshevik) revolution of 1917—all traces of their trust in God to correct the ills of man. With that accomplished, the Communist ax would have been dulled before it was fashioned—for a different outlook upon religion was one additional way in which the ideals of that Socialist state must depart from those of other nations. And on this central point of intellectual outlook, the leaders of the Russian revolution could rightly feel that they were on firm ground, and far in advance of other nations. To what extent this solid and man-freeing conclusion—though drawn by many on undigested evidence—further influenced the power and fervor of Bolshevik sentiment for radical social transformation, for unrelenting opposition to the non-Soviet world, and for a contempt of ethics by Soviet leadership, no one can ever know. But to this writer it seems entirely probable that past and present attitudes of Soviet Russia toward Western nations are far more antagonistic and dangerous than would have been the case if other nations had dethroned the personal God before Russia did; and if thereafter those nations had taken the entirely natural step of reexamining their social policies to find what unaided, uncondemned and unanointed man can do to better his economic and social condition.

If this view is approximately correct, one must conclude that the now prevalent Communist threat to world order, and the related crushing costs to all peoples of this day, partly flow from the organized religions of the past half century.

BELIEVERS AND NONBELIEVERS IN THE UNITED STATES

A census of belief and unbelief in a sharp list of specific questions relating to religious ideas and to the supernatural would be of very great value if it included all adults of any large Western nation. That census would provide perhaps a better look into the collective civilized mind than has been obtained by any method heretofore. In particular, it would supply a closer measurement of the total impact of knowledge—of science—on the thought of a people and age than has yet been obtained. Incidentally, the word "knowledge" alone is meant here; not doubtful experience and knowledge together.

In view of the extent to which imperfectly registered beliefs and unbeliefs are motivating much of the activities of some persons we meet every day of our lives, and since these beliefs of others shape the rules and customs into which we must fit our own entire lives, it is truly surprising that we usually find abstract information about these beliefs uninteresting. Perhaps this is a natural result of an early-formed impression that we can do noth-

ing about the beliefs of others—not even about those of persons
we meet rather often. However that may be, when or if one becomes concerned with learning whether an idea or movement
within this field of religious thought is gaining or losing ground,
he will need to prize the scattered scraps that impoverished statistics can supply him.

Some data are now available for the more basic religious beliefs of the general population of the United States. Also for such
beliefs in several selected groups within that country. The information for the population as a whole is illuminating though
limited, and for some of the special groups it is deficient. Apparently, however, such information is in better supply for the
United States than for most other countries. In surveying it, the
reader may be assured that something better than guesswork is
being offered about the present state and the present trend of
religious belief in at least one Western nation. Moreover, a comparison of the beliefs of quite dissimilar groups of Americans
should provide some clues to present-day sources of belief and
unbelief.

According to a survey made by the National Council of
Churches, published in March, 1951, the fifty-four largest religious bodies in the United States showed a membership gain of
51.6 per cent between 1926 and 1949. In this same period the
population of the country increased slightly less than 30 per
cent.

That survey also established the percentage gains of each of
those fifty-four denominations, thus providing information on
the type of church—orthodox or liberal—that has grown least—
or most—during that period of twenty-three years. Only a few
of the many denominations can be considered here, and for present purposes the shorter list of seventeen "leading churches of
1949" which appeared in *Time* magazine (April 2, 1951) is reproduced. That table requires little comment. Clearly the very
large percentage gains—81 per cent and upward—are associated
with the denominations that have most retained Middle Ages
beliefs, including a hell. The Lutherans, with a 61 per cent gain,
are close kindred to the above groups in their degree of orthodoxy. The smallest gains found in this list, gains of 19 to 24 per
cent, include Congregationalists, Jewish Congregations, Northern Baptists, Episcopalians and Unitarians. Perhaps the Unitarians have discarded more of former Biblical and supernatural

thought than has any other group, with the probable exception
of the Reformed Jews, who are included in Jewish Congregations. The 48 per cent gain of Roman Catholics (as against a 30
per cent increase in total population) represents an intermediate
value that is near the average (51.6 per cent) for all of the fifty-
four groups. But several extremely large Protestant groups—
Methodists, Northern Presbyterians and Disciples of Christ—
made gains much smaller (26 to 30 per cent) than those of Catholics. The liberal, the educator or the scientist who complacently
assumes that the strength of a harmful supernaturalism is now
decreasing in the United States will find his view denied by these
statistics.

MEMBERSHIP OF LEADING CHURCHES IN THE UNITED STATES IN
1949 WITH THE PERCENTAGE GAIN OF EACH SINCE 1926

	(Number in thousands)	*(Percentage)*
Assemblies of God	275	474
Baptists, Northern	1,583	23
Baptists, Southern	6,761	92
Church of God in Christ	340	1,025
Congregationalists	1,184	19
Disciples of Christ	1,738	26
Episcopalians	2,298	24
Evangelical Lutherans	1,677	61
United Lutherans	1,952	61
Methodists	8,792	30
Mormons	980	81
Presbyterians, Northern	2,401	27
Presbyterians, Southern	653	50
Seventh Day Adventists	230	107
Unitarians	74	24
Roman Catholics	27,610	48
Jewish Congregations	5,000	23

In August, 1949, the *Christian Herald* published information
on the percentage of "church members" at some earlier periods
in the history of the United States: for 1880, it was listed as 19.9
per cent; for 1900, 34.7 per cent; for 1920, 39.8 per cent; and for
1948, 53.3 per cent. These figures certainly suggest rapid growth
in the power of organized religion during the precise period that
an entirely new concept of man's nature and origin has struggled

for a place in popular thought and public education. Further, it is notable that the rate of gain in church membership, relative to total population, during 1920 to 1948—the period of pressure for antievolution laws and for religious instruction in the schools—is much greater (about 0.408 per cent per year) than during the twenty-year period (0.255 per cent per year) which preceded that effort. For the year 1951, the National Council of Churches placed church membership at 58 per cent.

By personal interviews and Gallup-poll methods, Barnett[1] questioned in 1948 "a cross section of Americans from coast to coast." Since 76 per cent of his respondents acknowledged church membership, and the churches then claimed only about 50 per cent of the population, his sample apparently was somewhat "loaded" with that class. To the question, "Do you believe in God?" 95 per cent said yes, 2 per cent took an agnostic position, 2 per cent an atheistic position, and 1 per cent declined to reply. The report says that "one can infer that every one who believes in God recognizes Him as Creator" and that "three-fourths of those specifically asked the question also acknowledged God as their Judge." Belief in some kind of afterlife was professed by 73 per cent, while 15 per cent saw death definitely as "final extinction." To the question, "Do you think that life after death is divided into heaven and hell?" 52 per cent of believers in life after death said yes.

A rather similar survey of the opinions of youth of eighteen to twenty-five years in the United States was made by *Fortune* and published in the December, 1948, issue of that magazine. To the question, "Do you think there is a God who rewards and punishes after death?" 74 per cent said yes, 16 per cent said no, and 10 per cent gave no opinion. Nearly 84 per cent said they went to church occasionally or regularly, and 16 per cent said they never attended church. Perhaps this youth group expressed slightly less belief than did the "loaded" poll of the general population reported by Barnett.

Apparently, at the mid-point of the twentieth century, belief in supernaturalism and God is accepted by somewhat less than 95 per cent of the people of the United States. Only about 70 to 80 per cent, however, profess belief in a life after death. Church membership has shown a marked and steady increase in all dec-

1 Lincoln Barnett, "God and the American People," *Ladies Home Journal,* September, 1948.

ades since 1880. The Roman Catholic Church alone enrolled nearly 20 per cent of the nation's people in 1950. The clergy and associated church functions then made their direct appeal for supernaturalism and for an afterlife conditioned upon belief— regularly or occasionally and apart from radio and television— to approximately 85 per cent of the total population.

In 1950, Ross[2] published a more intensive and better analyzed survey of the religious opinion and practice of 1953 YMCA constituents aged eighteen to twenty-nine years. These individuals resided in nine states and in eleven cities of varying size located within 500 miles of New York City. Of this sample, 345 were women; 59 per cent were Protestant; 34 per cent, Catholic; 3.5 per cent, Jewish; 3.3 per cent, other religions; no religion, 1.7 per cent; not actual church members, 18.5 per cent. Half of the total number had attended or graduated from college. Fifty-six per cent had served in the Armed Forces, and 19 per cent were married. On the question concerning the nature and existence of God, a total of 72.4 per cent accepted "God as personal or as an intelligent and friendly Being successfully approached by prayer." Another 12.9 per cent believed "that one may discover the works of God's will in nature and natural law, and that we must act in accordance with these guides." Further, 8.3 per cent took the agnostic, and 3.8 per cent the atheistic, position, with 2.6 per cent dissenting from all of the six listed types of belief and unbelief. Of the college graduates, 18.4 per cent identified themselves with an atheistic or agnostic position as compared with 9.5 per cent of those with only part high-school education.

Further, 51.5 per cent considered Jesus divine, and 37.8 regarded Jesus as one of the great prophets or teachers of history. A belief in the divinity of Jesus was accepted by 55 per cent of the youngest group (eighteen to twenty-one years) but by only 44.7 per cent of the oldest group (twenty-six to twenty-nine years). The Bible was regarded as a sacred book with a divine message by 79.6 per cent of these respondents. The women were somewhat more religious than the men.

Immortality, in its religious and orthodox sense, was accepted by 52.3 per cent, considered uncertain by 29.1 per cent, and positively disbelieved by 11.7 per cent. Belief in immortality prevailed among 77.7 per cent of Catholics, but among only 43.7 per cent

[2] Murray G. Ross, *Religious Beliefs of Youth* (New York: The Association Press, 1950).

of Protestants. Belief that the soul will be rewarded or punished was accepted by 55.5 per cent of those with part high-school education, but by only 36.1 per cent of the college graduates; also by 59.6 per cent of the eighteen- to twenty-one-year-olds, but by only 43.5 per cent of those twenty-six to twenty-nine years old. The church was considered as "appointed by God" or as the "one sure foundation of civilized life" by 60.1 per cent of the respondents. Only 7 per cent doubted the usefulness of the church or were antagonistic to it. Apparently, all percentages for "belief" would be greater if the sample had included young women in numbers equal to the men.

Again, only 23.8 per cent of these respondents "rely on traditional religious practices as their primary source of help in solving a problem, and (even giving a broad interpretation to 'religious resources') only 39.2 per cent rely on such religious resources when deciding what is right or wrong." In the light of additional questions bearing on this point, Ross comments as follows:

The weight of evidence suggests that the real concerns of youth do not include very great interest in traditional religion, that the traditional resources of religion are not a primary source of help to young people, and that there is reason to doubt the effectiveness and usefulness of the vague religious concepts which do exist in the minds of the majority of the youth included in this study.

This YMCA study also included rather extensive interviews with 100 men aged eighteen to twenty-nine. That group included 69 Protestants, 22 Catholics, 4 Hebrews, 2 Greek Orthodox or Hindu, 3 no religion or a religion not indicated. Eighty were church members, 16 were not, and 4 did not clarify this point. Fourteen were married. From only 90 of this group was obtained adequate information on their belief or conception of God. Fourteen of them were deists with, for them, a clear concept of God unaccompanied by doubts; 70 had unclear ideas of God, with their "belief (in God) permeated with doubts and uncertainties"; 6 were agnostic or atheist. The published statements of twenty-six of these individuals, suitably described, illuminate public thinking on the concept of God in our day. Only half of this group discussed the question of immortality; of these, 50 per cent indicated acceptance of this as a possibility, 30 per cent were not sure, and 20 per cent rejected the idea.

Some studies on more highly restricted localities and groups will next be mentioned. Hollingshead[3] made observations on the high-school youth of "Elmtown" (somewhere in Midwestern United States) to whom he gave certain tests. His inquiries indicated "that 51 per cent of the high school students have no active connections with Elmtown's churches." Parts of his further comment are as follows:

> The impression gradually grew that religion to these adolescents is comparable in a way to the wearing of clothes or taking a bath. It is something one has to have or to do to be acceptable in society. . . . To a few, God, Satan, Heaven, Hell, Sin and Salvation are real entities that surround them at all times. . . . But to about 90 per cent of the boys and about 80 per cent of the girls, religion does not have this compulsive quality. . . .
>
> Church in Elmtown is an inclusive term that embraces dogma, doctrine, theology, edifice, members and beliefs about one's own faith and the faith of other churches. To be labeled a church member is very important, for it tells people where one belongs in the rather complex denominational structure. One may label himself a Methodist with approval and seldom, if ever, worship with the congregation; it is sufficient to be known as a member of an approved church. One can refer to himself as being of any Christian faith without inciting outright opposition, but, if he identifies himself as a Hebrew, a subtle web of distrust, suspicion, and possible hate will be spun around him. However, if he blandly says that he is an atheist, barriers will be erected around him by the devout, for the atheist and communist are two labels an Elmtowner must avoid if he desires to be accepted as a respectable member of society.

An unusual importance attaches to the present religious beliefs of college students. They are a segment of society whose present educational accomplishments, intellectual interests and prospects for leadership are somewhat above the average of society as a whole. Presumably, mass movements in the sphere of thought tend toward views first held and expressed by persons at least partially exposed to the tides of creative or of other thought. Now available is a report[4] upon the religious beliefs of 414 un-

---

[3] A. G. Hollingshead, *Elmtown's Youth* (New York: John Wiley and Sons, 1949).

dergraduates at Harvard College and 86 undergraduates at Radcliffe College. The views of these two groups—of men and of women—on the single question of Deity, or God, are here tabulated. Since 290 of the Harvard students had served with the Armed Forces and 123 had not, the beliefs of these two groups of men can be examined separately. For rough comparison the YMCA data previously described are placed in a final column of this table. The questions asked this latter group were not precisely the same as those used with the Harvard-Radcliffe groups, and the fitting of answers into this table involves some leniency in interpretation.

HARVARD-RADCLIFFE AND YMCA DATA ON THE CONCEPT OF GOD

Views of Deity	Harvard veterans (290) Percentage	Harvard nonvets (123) Percentage	Radcliffe (80) Percentage	YMCA (1,935) Percentage
1. There is an infinitely wise, omnipotent creator	17	25	40	37.7
2. There is an infinitely intelligent friendly being	25	27	19	34.7
3. There is a vast impersonal spiritual force	11	10	7	12.9
4. I neither believe nor disbelieve in God	23	17	12	8.3
5. The only power is natural law	8	7	9	} 3.8
6. The universe is merely a machine	5	2	2	
7. None of these alternatives	11	12	11	2.6

4 G. M. Allport, J. M. Gillespier, and J. Young, "The Religion of the Post-War College Student," *The Journal of Psychology,* 1948.

It will be seen that all three college groups are much more inclined toward doubt or denial of Deity than the YMCA group, which (mostly males) included only 50 per cent of college-trained youth. This is most pronounced in the men who had served in the Armed Forces; it is least pronounced in the women students of Radcliffe. Among the veterans, the God of the religions seems to be supported by less than half their number, 42 per cent; but if a belief in "a vast impersonal spiritual force" is to be added this proportion becomes 53 per cent. Clearly, at least 36 per cent of these college veterans reject or do not accept the idea of God; nonveterans follow with at least 26 per cent; and Radcliffe students with at least 23 per cent. These college students reject or question the God of the religions to a far greater extent than does any other of the groups thus far included in these pages. For the YMCA group the comparable value is 12.1 per cent. The indicated extent of doubt and denial of the existence of God among these college students is a matter of some significance. It seems clear that a drift from the most basic of all religious concepts—the existence of God—is present among significant groups of students in the better or best American colleges of today.

The views of the Harvard-Radcliffe groups on immortality and on the value of the church differ greatly from those of the YMCA group. Of the latter, 52.3 per cent believe in personal immortality (or in reincarnation) while this belief is held by only 21 per cent of veterans, 28 per cent of nonveterans, and by 34 per cent of the Radcliffe group. "Continued existence as part of a spiritual principle" is accepted by 10, 11 and 8 per cent of the latter groups, and by 5.4 per cent of the YMCA group. In this same order, 40, 34, 35, and 7 per cent believe that their immortality rests on their "influence on children and social institutions." Oddly enough, nearly one fifth of each of the three college groups reject both disbelief and belief in the six listed questions regarding immortality.

A smaller number of Harvard (170) and Radcliffe (63) students gave their estimates of the nature and value of the church. Only 6 per cent of each of these two groups think that "the church is the one sure and infallible foundation of civilized life," though ten times as many (60 per cent) of the YMCA group think so. "On the whole the church stands for the best in human life" is the view of 36 per cent of the Harvard and of 40 per cent of the

Radcliffe group. Thus fewer than half of these two groups—42 per cent of one and 46 per cent of the other—accord a clearly positive value to the church. "There is certain doubt. Possible that the church may do a great deal of harm" is the view of 18 per cent of the Harvard and of 13 per cent of the Radcliffe group. "The total influence may on the whole be harmful" is the view of 6 per cent and of 2 per cent of these groups. "Stronghold of much that is unwholesome and dangerous to human welfare" is the view of 10 and 6 per cent. "Insufficient familiarity [with the problem]" is expressed by 4 and 8 per cent. "A different attitude" is held by 20 and 25 per cent of these Harvard and Radcliffe groups.

The registration cards signed at the beginning of the academic year give Harvard students an opportunity to express their religious affiliation or preference. Of 7,500 cards thus signed in the autumn of 1948 "about 20 per cent" indicated no preference or affiliation, according to Dean Sperry,[5] of Harvard's Divinity School. Of other and greater significance is theologian Sperry's statement on present trends in Protestant thought in America and Europe. Parts of that statement are quoted here:

American Protestantism has been, culturally, "liberal"—whatever the formal creeds recited in church. It has believed in man's innate goodness and his ability to advance himself morally. . . .

But human nature in the last thirty or forty years has not been giving, the world over, as good an account of itself as the founding fathers of democracy and liberalism proposed. In Europe there has been a very general reaction against the whole liberal position. The case for religion is being referred back to God, who must save us from ourselves. This movement is generally known as "neo-orthodoxy." Karl Barth has been its European spokesman and spearhead. He stresses the moral impotence of man and rests the case for religion upon the grace of God. This is merely the revival of dogmas proposed by Calvin and Jonathan Edwards. It has had its repercussions in America, *and is much the fashion in certain theological circles* [italics mine]. . . .

We admit that we need a new doctrine of man which, without denying his innate worth and moral ability, shall be more realistic than has been our convention hitherto. We need a contemporary and credible account of human nature, which shall preserve the permanent values of liberalism, but rid that faith of its unwarranted romanticism.

Finally we have seen in the Western World in the last half

---

[5] Willard L. Sperry, "The Present Outlook for Religion," *Bulletin,* The American Academy of Arts and Sciences, December, 1948.

century a widespread revolt against the Jewish-Christian ethic, which had been for sixteen hundred years accepted as our ideal. The present problem in many quarters is not whether the Christian ethic is practicable, but whether it is even desirable. The totalitarian states have all repudiated it, and many of their tempers have subtly invaded our life.

A much broader, though less statistical, survey of trends in religious opinion in American youth over a period of three decades has been supplied by a social scientist whose earlier experience included posts as director of religious education in Denver and New York and as instructor in the Union Theological Seminary. These trends and a prediction based upon them are partly indicated by the following quotations from Professor Watson's[6] book:

The greater part of American youth after the first World War were only mildly aware of some changes taking place in their religious outlook. They condemned ecclesiastical hypocrisy, distrusted the authority of religious officials, and were impatient of denominational distinctions. They felt no need for some of the old doctrines and knew that others were being called in question. There was no sharp decrease in church membership or church attendance in the ensuing decades. . . . Religious attitudes moved slowly away from the orthodox position.

That condition or at least some aspects of that trend are said by Watson to have continued during the thirties and the depression—the period of "disillusioned youth." Of greater interest are his observations on the aftermath of World War II, and on two trends in religious thought of the church itself, which he thinks have become evident and will project themselves into the future. Of these two trends he says:

1. "Organized authoritarian religion will further lose its hold upon youth. . . .
2. A revival of fundamentalism, neo-orthodoxy, and religious reaction is in process. . . .
Even within the ranks of religious leaders there are those who question whether the attempt of liberals to accommodate ancient faith to modern social thinking has not lost its thrust. The church as an institution has not carried out the implication of its preaching. . . .

[6] Goodwin Watson, *Youth After Conflict* (New York: The Association Press, 1947).

The modernist emphasis of the 1920's is no longer thriving. A kind of bi-modalism of religious attitude is developing—on the one hand, those "secularists" who discard religion entirely; on the other, *the return to the ancient doctrines* [italics mine].

The evidence cited by Watson (and also quoted by Ross) for this last-named trend includes the growth of the "Youth for Christ" movement; the Evangelism Fellowship; the support by the wealthy of Fundamentalist sects; the conversion of well-known figures to Roman Catholicism; the preaching by former "liberal" theologians of the futility of man's efforts to bring the kingdom of heaven on earth; and "the movement at the University of Chicago to return to St. Thomas Aquinas and his doctrine of first principles and eternal truth." Further, from "a religious leader" he quotes the following:

The liberal tide has ebbed and something very closely approximating neo-classicism has appeared. In Protestant circles we call it neo-orthodoxy and in Catholic circles neo-Thomism. It is not the kind of movement that young people will "flame" over. It hasn't the thrill of emancipation. Yet some impressive demonstrations have been made of the way in which an orthodox revival can get the support of the young.

Others, too, might be quoted in support of the view that "the liberal tide has ebbed" in the field of theology and religion. If that proves to be true, its consequences are truly disconcerting. Science and education will be more steeply hedged; and all evolutionary thought, for the public, will be still further diluted. An outsider might suppose that it may bring additional clash and division among the supporters of religion. It seems quite incredible that the appeal of science is now declining in the United States. But the larger and more definite fact is that science has never had a strong appeal to any large percentage of Americans. Technology, yes. In this country, gadgets and the instruments of doing have so nearly filled hand and mind that they are usually miscalled science, and quite generally they have limited or precluded a grasp of the message and meaning of science.

Is religious influence now increasing? The answer of theologian Reinhold Niebuhr, writing in *The New York Times Magazine* (November 19, 1950), includes these two statements:

There is scarcely a college or university [in the United States] which has not recently either created a department of religious studies or substantially enlarged existing departments. . . .

The influence of the Catholic faith upon culture in Europe is wider and deeper than either Protestant or secular leaders of thought in this country are able to understand or are inclined to admit. That influence has waxed rather than waned in the past decade.

The kind of books now being read by Americans has some bearing on this question. Eugene Exman, who for twenty-five years has headed Harper and Brothers' religious-book department, thus described[7] the recent surge in sales of religious books:

When American booksellers added up sales figures for the year 1949 they discovered that four out of the five best-selling titles of non-fiction (excluding "Zoo" and Canasta books) were religious titles. Each year since then religious books have climbed to the shelf of the ten best-sellers of fiction and non-fiction alike. . . . On February 22 of this year, the New York *Herald Tribune*'s best-seller list had as its four non-fiction leaders: *A Man Called Peter, This I Believe, The Power of Positive Thinking*, and the Revised Standard Version of the Bible. Twenty-five years ago the book world would have looked with incredulity at such a record.

Recently available are at least the beginnings of information on the religious beliefs of teachers of nonreligious subjects in church-related colleges—mostly smaller ones—of the United States. A "Yale Study on Religion in Higher Education," conducted by Espy,[8] gives results obtained by questionnaire from certain faculties of 73 colleges. The colleges are widely distributed, though New England and California are not represented. Only undergraduate, four-year colleges of Protestant connection are included, and only teachers of four subjects—English, sociology, economics and physics—are involved. Sixty per cent of those who received the extraordinarily long questionnaire returned it, and of these returns 440 were considered valid for study. The 73 colleges represented 29 denominations.

Three fourths of the respondents were men, one fourth women. Fifty-four per cent held a master's degree, but no higher

---

[7] *Harper's Magazine*, May, 1953.

[8] R. H. Edwin Espy, *The Religion of College Teachers* (New York: The Association Press, 1951).

degree; Ph.D.'s or other doctoral degrees were held by 35 per
cent. Teachers from the grades of instructor to dean were included. English teachers numbered 210; sociology, 76; economics, 79; physics, 75. Among these 440 were one Roman Catholic,
one Jew, and 17 persons with no church affiliation. Type of
belief in God could be expressed in one of nine—usually elaborate—statements. Nevertheless, 8 persons, or 2 per cent, were
uncertain or failed to specify their belief. Six teachers thought,
"God is another name for natural law." For the other categories
of belief, it seems best to use the words of the author:

We may consider at least the first four categories, totaling 87
per cent, as within the orbit of Christian theism. It is interesting
that this is somewhat less than the 94 per cent who regard themselves as Christians. . . .
Also, it is apparent that lack of church affiliation does not mean
atheism. Among the 5 per cent who are not members of churches
only two teachers do not believe in God. On the other side, one
of the three professed atheists is a member of a church, while
another, who neither believes in God nor belongs to a church,
regards himself as a Christian.

Thus these teachers of *nonreligious* subjects in church-related
colleges of the United States seem much more orthodox, much
less "liberal," than any group thus far encountered in this review—except Barnett's data, which were obtained by Gallup-poll
methods on a clearly "loaded" sample of the general population.
Of course, it is quite possible that replies from the 40 per cent
who received but did not return questionnaires would materially
reduce the indicated degree of orthodoxy of these teachers in
denominational colleges.

Strangely enough, these teachers were asked nothing concerning their attitude toward immortality. But they were asked, "Are
there irreconcilable conflicts between the Christian religion and
some of the findings of science and history?" To this, 8 per cent
said "yes"; another 8 per cent were uncertain or qualified or
omitted an answer. "No" was the reply of 84 per cent.

Some years ago psychologist Leuba[9] obtained rather definite
and solid statistics on the beliefs of American scientists in personal immortality and in one type of God—though not in a God

9 James H. Leuba, "Religious Beliefs of American Scientists," *Harper's
Magazine*, August, 1934.

of any other type. In the standard work *American Men of Science,* a certain proportion of the 23,000 names entered there are "starred," indicating eminence in their fields. A very short questionnaire was mailed to a suitable fraction of these scientists of greater and lesser eminence within each of the sciences. And here individual names were chosen in every instance on a rule of chance. Replies from scientists concerned with inanimate matter (chemists, astronomers, geologists, engineers, for example) were finally classified as those of "physicists"; replies from scientists dealing with living matter (biologists proper, bacteriologists, horticulturists, and so on) as those of "biologists." The inquiry was sent to one tenth of the two large groups. Sociologists and psychologists formed groups of still other scientists, and about one half of them were included in the inquiry. Within the several groups, more than 75 and up to 90 per cent returned replies. Leuba claims the validity of his results for *all* American scientists of the year 1933.

Inquiry concerning the *several* ideas of God was purposely avoided. Leuba says: "I chose to define God as given above [given below here] because that is the God worshipped in every branch of the Christian religion." The complete form of the question was: "(A) I believe in a God to whom one may pray in the expectation of receiving an answer. By 'answer' I mean more than the natural, subjective, psychological effect of prayer. (B) I do not believe in God as defined above. (C) I have no definite belief regarding this question."

And further: "Many of the disbelievers in the God defined were annoyed that I had not provided a way for them to say in what other God they placed their faith. They feared that a negative answer to statement (A) would class them among the materialists. . . . Many of my correspondents said, 'God is not moved to action by my desires or my feelings; He acts according to His laws.' . . . Several returned the questionnaire with remarks intended to justify their refusal to answer: 'Most of those who believe in God will answer an inquiry like this. Most of those who do not believe in God will put it in the waste basket. How are you to draw any conclusion?' . . . Another wrote: 'I am refraining from complying with your request because I believe that real harm is done in announcing to the world the opinion of scientists relative to religious matters.' "

The accompanying table suitably indicates the several facts

that emerge from this inquiry. Clearly, belief in the Christian type of God, as defined above, is markedly greater among those concerned with inanimate matter ("physicists") than among those immediately concerned with living organisms, society and the mind. Clearly, too, within all four groups, scientists of lesser eminence were more believing than those who had won greater recognition. Clearest of all is predominant disbelief and doubt among worthy American scientists.

PERCENTAGE OF BELIEVERS IN A CHRISTIAN TYPE OF GOD
AMONG SCIENTISTS (LEUBA, 1934)

	Believers	Disbelievers	Disbelievers and Doubters
Lesser Physicists	43	43	58
Greater Physicists	17	60	83
Lesser Biologists	31	56	69
Greater Biologists	12	76	88
Lesser Sociologists	30	60	70
Greater Sociologists	20	70	80
Greatest Sociologists	5	95	95
Lesser Psychologists	13	74	87
Greater Psychologists	2	87	98
All Lesser Scientists	35	51	65
All Greater Scientists	13	71	87

There is good support for the view that the best-informed opinion about God should be held by those who, using trustworthy methods, have long and professionally studied phenomena and nature. Such are the scientists. Logically, too, among them the most eminent ones should more often arrive at sound opinions. This reasoning and the data of the present table have led to the conclusion, elsewhere stated in this book, that "a majority of best-informed minds" now reject a belief in God.

These data from scientists are certainly in very great contrast with those obtained for the several groups previously reviewed in these pages. For example, among teachers of nonreligious subjects in denominational colleges "at least 87 per cent are within the orbit of Christian theism." Here, the most eminent groups from sociology and psychology show only 5 and 2 per cent within that orbit.

Leuba obtained from the scientists who recorded their belief concerning God a further statement of their belief regarding immortality. On the results of this inquiry the author says:

In a general way the scientific men who believe in the God of the religions believe also in immortality; the two beliefs usually go together. The proportion of believers is nearly equal: 33 per cent for immortality and 30 per cent for God. . . . The several classes of scientists remain in the same order in the table on immortality as in the one referring to God: The "physicists" **head** the list with the largest proportion of believers (41 per cent) and the psychologists close it with the smallest (9 per cent).

It is notable that Leuba had made a similar study of opinion among scientists of the United States in 1914. In the article under review he calls attention to the fact that a comparison of the data obtained at the two periods shows a decrease of belief between 1914 and 1933. Of this he says:

In every group, without exception, the figures for [believers in] 1933 are considerably smaller than those for 1914. It should be noted also that, both with regard to God and to immortality, the order in which the four classes [of scientists] arrange themselves with regard to the proportion of believers is the same in the two investigations.

This declaration of a belief in immortality by many *adults* who are trained in one or another science is a truly remarkable matter. It contributes the small but topmost stone to the pyramid of belief. Here is acceptance of what some close colleagues may desire but cannot believe; here is belief that, presumably, has resisted the flames of serious questioning. Philosopher Santayana observed that "the fact of having been born is a bad augury for immortality."

A further and special word concerning these beliefs of scientists is reserved for the end of this chapter. Meanwhile, the basic religious beliefs of still other groups will be discussed. Two years after making his 1934 study, Leuba obtained similar and somewhat fuller information from leaders in four additional groups of Americans. Names included in *Who's Who in America* (one fifth of those named in each group) were the ones consulted in this inquiry. Bankers (and capitalists), other business people, lawyers (excluding judges and professors of law) and writers were the groups queried. Leuba thinks the statistics obtained are essentially reliable and has included them in a recent publication.[10]

BASIC BELIEFS OF FOUR GROUPS OF AMERICANS LISTED IN WHO'S WHO (LEUBA, 1935)

	Belief in God		
	Percentage of Believers	Percentage of Disbelievers	Percentage of Doubters
Bankers	64 (+20)*	29 (−20)*	7
Business People	53 (+30)	41 (−30)	6
Lawyers	53 (+25)	40 (−25)	7
Writers	32 (+36)	62 (−36)	6
	Belief in Immortality		
Bankers	71	11	18
Business People	62	15	23
Lawyers	59	16	26
Writers	40	28	32

* Reject a God-providence accessible through the current worship; accept a power not so accessible, but of spiritual or mental nature.

The tabulated results permit several worth-while comparisons. All these groups of *distinguished* persons believe *less* in God and immortality than does the general population of their coun-

10 James H. Leuba, *The Reformation of the Churches* (Boston: The Beacon Press, 1950).

try; also their score for belief is *greater* than that of any of the four groups of scientists ("physicists" possibly excepted) previously examined. All groups (comprising older or mature people) expressed more belief than did the Harvard students. Again, the bankers show most and the writers least belief in both God and immortality. The opportunity offered these groups to record belief in two types of God yielded a meaningful result. If one excepts the writers—the group with least belief—it will be seen that the idea of an impersonal God to replace the God of the religions had made (in 1935) only moderate or little headway among these *distinguished* Americans.

The general populations of different nations now differ widely in the extent of their unbelief in God and in an afterlife. Information on this point, for eleven nations, was obtained by the American Institute of Public Opinion (Gallup poll: release date, January 10, 1948) and its Overseas Affiliates. Replies relating to belief in God are indicated in the accompanying table. Brazil leads the nations in the percentage of believers, though Australia, Canada and the United States closely follow it. The home of doubt and denial among the listed nations is France, Denmark, Sweden and Holland. England also belongs in this group, but to English people the question was put in a somewhat different way: "There is a personal God" got 45 per cent of the votes; "There is some sort of spirit or vital force which controls life" got 39 per cent; and "I am not sure there is any sort of God or life force," 16 per cent. The clear evidence for much more of doubt and denial—that is, for more of independent and modern thought—in the cultured nations of Western Europe than in Brazil, Australia, Canada and the United States attests and parallels a basic intellectual and educational failure in these last-named countries.

Among the political parties in France, belief in God was found by the poll to vary greatly: Communists, 17 per cent; Socialists, 50 per cent; Union of the Left, 62 per cent. But even the most religious and clerical of all the six parties of France believes in God to no greater extent (93 per cent) than does the entire population of the United States (94 per cent). "In Denmark one in four [adults] under the age of thirty-five years either denies God's existence or expresses doubt. But among people over fifty years of age, one in twelve doubts or denies."

GALLUP POLL OF TEN NATIONS ON "DO YOU, PERSONALLY,
BELIEVE IN GOD?"

	Yes	No	Don't Know
	Percentage	Percentage	Percentage
Brazil	96	3	1
Australia	95	5	..
Canada	95	2	3
U.S.A.	94	3	3
Norway	84	7	9
Finland	83	5	12
Holland	80	14	6
Sweden	80	8	12
Denmark	80	9	11
France	66	20	14

A positive belief in life after death was declared much less often than was a belief in God. On the average, for the eleven countries, this was 21 per cent lower than for belief in God. Belief in an afterlife is held by 49 per cent in England and Sweden; by 55 per cent in Denmark; by 58 per cent in France; by 63 per cent in Australia; by 68 per cent in Holland and the United States; and by 78 per cent in Brazil and Canada. Surely these several huge, and probably growing, groups that reject all belief in an afterlife reject also the systems of salvation now central in Christianity and Judaism. One bargains for tickets to only accepted and credible destinations. If some of these disbelievers in an afterlife show no special hostility to those organized religions, it may be supposed that the other and ethical element of the religions persuades them to forbearance. But in most of these countries all of the disbelievers are taxed to support a Christian-creed. And who will doubt that a righteous rebellion against that supremely infamous act of government does not sometimes or often lessen loyalty to country and make even communism more acceptable than it would otherwise be?

Whatever else is indicated by these imperfect samplings of peoples in eleven Western nations, it is undeniable that large numbers of adults in every one of them—totaling very many millions of people—have considered and rejected both the God and the scheme of salvation currently proposed by Christianity and Judaism. Though their numbers scarcely attain a majority in any country, can any detached observer fail to conclude that their numbers indicate a revolt of *intelligence* against the supernatural element in religion—more especially since this revolt is most pronounced in the best-educated nations, and in the best-informed groups within a nation?

Further, in view of specific facts cited on these points in earlier chapters, can any unbiased person doubt that in all Western countries powerful agencies in education, in politics, and in government are now directed to the repression of precisely those unbelieving groups that look toward a saner culture? When, as now, we arrive at points where counts are made of current specific belief and unbelief, we meet the basis for our charge that the forces of organized religion curb and thwart the development of a genuinely modern society. No more serious or better-substantiated charge can be leveled against the democracies of the West. And religion could not strike a more telling blow than at the minds and ranks of those who must become leaders in a culture that accords better with newer knowledge.

Most of the preceding information relates to the population of the United States, or to smaller groups within that nation. Does the information support the assumption that socially harmful religious belief is waning in that country? Is that assumption by many writers, scholars and scientists warranted? Or is it dangerous? With probably more than 90 per cent of a nation accepting the supernatural (God), how good are the chances that American society will look and train toward human purposes instead of toward supernatural purposes? With church membership increasing as indicated, with religion itself giving some or fair evidence of a return to former orthodoxy, with increased infiltration of religion in the public schools, with subsidization of sectarian religious education by industry and business now in prospect, can a thoughtful person reasonably assume that no threat to a new and modern social outlook is involved?

If *political* leadership in the United States were proving itself untrapped by prevailing religious thought of the population, that, of course, would be a most significant fact. Where are we in that matter? This writer can supply no suitable table of percentages in answer to that question. Fortunately, most readers can get an approximately valid figure of their own. The two great political conventions of July, 1952, were heard over the radio by many millions of citizens. The bids and commendations of state and national political leadership that were expressed there, and were heard or read by this writer, provided—like our total population—appeals to the supernatural and prayer well in excess of 90 per cent. Speaking in St. Louis, April, 1954, ex-President Truman said:

It is only the people of religious faith throughout the world who have the power to overcome the force of tyranny. It is in their beliefs that the path can be found to justice, freedom and truth. Their religious concepts are the only sure foundation of the democratic ideal.

And quotations supplied elsewhere, like other well-publicized facts and acts, suggest that the recently elected President of the United States will officially support religion more extensively than did any one of his many predecessors.

CONSEQUENCES OF DIVISIVE BELIEFS AMONG SCIENTISTS

From the data already reviewed, and from familiarity with views of colleagues, every scientist knows that the study of science has changed and has reduced belief in the God of the religions among many or most colleagues of his acquaintance. Every scientist knows that study of one or another science has either left or has created among his colleagues a short staircase of belief and unbelief in a few types of God. But this spread and variety of opinion is not made clear in the table examined above. Mainly, that table indicated that a sizable minority of scientists —about 30 per cent of the sample studied—believe in a sort of God to whom one may pray in the expectation of receiving an answer. Having lived through childhood and often to adulthood with an intensely personal type of God, it would be surprising if some or even many scientists did not seem to struggle to retain something they may still call God—however unusual or unrecognizable the retained idea may be to others. And no one is surprised that close familiarity with extremely intricate processes of nature leads to quite new conceptions of Order. These orderly processes, indeed, may sometimes even be called Power—though a kind that heeds neither the sparrow's fall, the collapse of a civilization, nor the final incineration of an inhabited planet.

Such concepts, wherein Deity is deprived of participation in human affairs, all have essentially similar social consequences. Here, in any and every case, man is on his own. Within this broad area, who will venture to assign grades of value? The matter seems of little importance to anyone other than the holder of the view.

The consequences of a quite different and frankly conformist type of belief in a scientist are the things that require examination. And those consequences should be firmly registered and discussed. When that belief accepts a Being to whom prayer may be hopefully addressed, it accords, of course, with most traditional religions; and the scientist holding it is helping to sustain such religions. Whether that support helps or injures the society of today and tomorrow must depend upon whether organized religion is now helpful or injurious to society. Clearly, very many scientists have continued to support a religious view that, if proved either injurious or unscientific, gives them a poor place among present leaders of thought. From Leuba's study, it is some comfort to find that such scientists are largely restricted to the "less eminent" representatives of their science. But the *number* of the less eminent greatly *exceeds* the number of more eminent men of science, and the well-proved existence of that more-believing group means *a sharply divided front of science!* Thus the sweep and strength of evolutionary thought is being largely smothered by some of those presumed to support and embrace it. Society cannot now ask this respected source—science—for guidance on an acceptance or a rejection of the supernatural and receive a concordant or undivided answer. Society therefore fallaciously discounts the competency of science to provide a worthy answer. Thus the grave for mental release is dug and tended!

The more specific *source* of this fateful division of scientists is not difficult to find. An item taken from the abundant evidence will indicate that the faith of some scientists is no legacy of any science, but merely a residue of the faith of their tribe. The case of Alfred Russel Wallace, codiscoverer with Darwin of the principle of natural selection, is a notable and famous example. Though his ninety years extended into the year 1913, Wallace never accepted a natural origin for the intellectual and moral faculties of man—"for this origin we can only find an adequate cause in the unseen universe of Spirit." This famous biologist ardently championed spiritualism and opposed vaccination. He once presented an array of arguments to prove that the earth is the special object of divine favor—its size, its period of rotation, its distance from the sun, its tilt, its seasons, and its peculiar atmosphere were believed to be unmatched in any unseen world —and he urged that the universe was created for no other purpose than that of permitting man to evolve. Will a reasonable person err in ascribing this biologist's views on God and the supernatural more to accident of birth in a Christian England than to grasp of scientific law and fact?

Related reasons for a divided and faltering voice of science are well given in these words of philosopher Reichenbach:[11]

The elimination of meaningless questions from philosophy is difficult because there exists a certain type of mentality that aspires to find unanswerable questions. The desire to prove that science is of a limited power, that its ultimate foundations depend on faith rather than on knowledge, is explainable in terms of psychology and education, but finds no support in logic. There are scientists who are proud when their lectures on evolution conclude with a so-called proof that there remain questions unanswerable for the scientist. The testimony of such men is often invoked as evidence for the insufficiency of a scientific philosophy. Yet it proves merely that scientific training does not always equip the scientist with a backbone to withstand the appeal of a philosophy that calls for submission to faith.

[11] Hans Reichenbach, *The Rise of Scientific Philosophy* (Berkeley and Los Angeles: University of California Press, 1951).

13

Question Marks among Alleged Assets
of Religion

As I see it, the supernatural has no support in science, it is incompatible with science, it is frequently an active foe of science. It is unnecessary for the good life. And yet, the supernatural, in varying dilutions, is likely to persist in society for a very long time. The unconditioning of mankind in fundamentals has been a slow process in the past. It may go a little faster in the future. It is a matter of forgetting the hypothetical universe created out of ignorance and motivated by our undisciplined emotions; and of reconditioning to the actual universe, as gradually understood through controlled experience and experiment.—*Anton J. Carlson*

In their struggle for the ethical good, teachers of religion must have the stature to give up the doctrine of a personal God—that is, give up that source of fear and hope which in the past placed such vast power in the hands of priests.—*Albert Einstein*

ON NO OTHER TOPIC DISCUSSED IN THIS BOOK IS THE AVERage person so well informed as on the things commonly claimed as assets of religion. The exhibits and demonstrations of assets popularly ascribed to religion are found in every community, and usually they have been on exhibit there for ages. The showing and popularizing of these assets is unrestrained; every avenue of publicity and propaganda is usually open to a church; a flood of books in many languages appraises the assets of religion. Outside the Iron Curtain there is a flush of encouragement to the prevailing creed or creeds. Inside the Iron Curtain one or another church has enjoyed centuries of encouragement until the past three decades.

The title and main purposes of this book should lead none of its readers to expect to find in it a catalogue of the assets of religion. It is concerned throughout with the impasse resulting from religion's refusal to let society learn and use the now available, society-transforming outlooks of newer knowledge. These circumstances should warrant omission from it of any account

334

whatever of the assets of religion that are real and unquestionable—if they do not obstruct society in charting its course in
the light of our own times. Certain it is that ties to ethical acts,
however numerous, are not the grafted appendage of religion
through which it thwarts the cultural and intellectual maturity
of the race.

There are nevertheless valid reasons for the appraisal of some
elements often mistakenly regarded as assets of religion. Above
all else, in that connection, is the common and erroneous practice of apportioning to religion the values that belong to the
secular field of ethics. Too, there are questions of harmful alloys
intermixed with the gold of true assets. In addition, it is convenient to examine a suggestion that is shared with some others
on a way that society may more favorably utilize a trained clergy.
And, too, there is pleasure in pointedly acknowledging the existence of unquestionable assets.

It is quite impossible for thoughtful people to be unconcerned
with the assets, values, disabilities, drifts and dangers of organized religion. The institutions of religion—their power in education, in news and thought control, in social contacts, in politics
and law—now leave no one unaffected. And unconcern is especially hazardous at a time when, within large areas of the Western world, the actual trend—forward or backward—of theological
and religious thought is admittedly in doubt; when many able
persons quite carelessly assume that religion must and will "liberalize" itself in order to survive; and when religion has already
lost contact with impressive numbers of the best and the better
minds. This latter fact alone may permit Protestantism to join
the Vatican in a backward look—if it finds itself sufficiently subsidized, entrenched by legislation, and also relatively free internally from the type of mind that hitherto has asked for an
occasional nod to reason.

TOWARD A DISSECTION OF RELIGION'S ASSETS

"The Christian system is not primarily a system of ethics—it is
a system of supernatural salvation," says J. Hutton Hind, Leader
of the Ethical Society of St. Louis. The statement seems unchallengeable. Islamic faith is of quite similar nature. Earliest
Jewish faith neglected salvation while accenting the supernatural, race, law, history and ethics; to these components, however,

Orthodox Hebrew religion later added salvation. Brahmanism at first glorified the supernatural and disregarded ethics. Later, according to Hopkins—following the spread of the almost purely ethical teachings of Buddha—Brahmanism embraced both the supernatural and ethics.

Any examination of the debits or the assets of a religion—Christianity, for example—therefore requires a look into two distinct areas: its system of ethics, and its system of supernatural salvation. For the word "ethics," some American Protestant groups now like to substitute the term "social gospel of Christ." Just here the fact of first and central importance to this chapter can be stated to advantage. Society and evolutionary thought must now specifically wage a fight against the supernatural element (salvation included) of all religions. That fight is relatively little affected by or concerned with the purely ethical teachings of any religion. The defeat and disappearance of belief in the supernatural implies no loss to ethics. Loss or gain to ethics would depend chiefly upon the debatable extent to which education and other presently restricted or suppressed motivations can substitute, duplicate or excel the strong motivation that belief in the supernatural sometimes provides.

While continuing this consideration of the dual nature of Christianity, one recalls that numerous and significant evaluations of that faith—or of its relation to one or another social movement—have been made without regard to its two components. Illustrations of this are supplied from statements of two notable American women. Elizabeth Cady Stanton, social reformer and advocate of women's rights, said:

The Bible and the church have been the greatest stumbling blocks in the way of women's emancipation.

Mary Reed Henderson, member of a distinguished family and the mother of physiologist Lawrence J. Henderson, was reared in strict Calvinism. Late in life she wrote:

Theology is a thing which in the last 2,000 years has caused in the world more misery and suffering—woe of body and mind—than almost anything else, unless perhaps the inordinate pursuit of riches and power. I have hated creeds since I was a child.

Again, it is possible that Christianity's ultimate place in history may even involve something omitted—or perhaps not well reported—in its message. Certainly this was the view of Dr. M. M. Mangasarian, Leader of the Independent Religious Society of Chicago, who spoke in 1923, as follows:

Would you think me irreverent if I were to say that had I been on trial for my life, and Pontius Pilate had asked me, "What is Truth?" I would not have answered it by remaining silent. I would have replied: *Truth is that which admits of proof.* If Jesus had given an answer something like that instead of leaving the greatest of all questions unanswered, the history of the past two thousand years would have been gloriously different. Had Jesus defined the true as the provable there would have been no persecution in the name of Christianity, no Spanish Inquisition, no witchcraft barbarities and, greater than all, no opposition to or fear of science. What a benediction to our poor humanity that would have been! Europe would have been spared a thousand years of darkness; Galileo would not have been imprisoned, nor Bruno burned at the stake.

A more direct look at the ethical component of religions is warranted. The ethics of all prevailing religions are fairly strong pressures for order and for some acceptable behavior in large human societies—pressures that rather readily renew themselves under religious auspices in successive generations of men. Most world religions have helped continuously to make a few or many items of ethical behavior more acceptable to large populations. The ethics of Christianity are not all of one piece, very very far from it; but a better general level of excellence is difficult to find within the religions. Here are assets—recorded apart from adverse associations and costs.

But the ledger that lists these assets must report at once some firmly associated debits. The church's prideful pretension to an exclusive virtue and spirituality automatically condemns to an inferior position other agencies that supply human needs and demand acceptable behavior. Too, ecclesiastical emphasis upon personal salvation has resulted in shifting the end of moral action from benefits it should bring to the human community to the consequences for oneself. In general, even good behavior is obtained through deception and stifled thinking. Again, principles that cannot be applied are no principles at all, according to John Dewey. Those professing religious principles have demonstrated widely their incapacity and failure to render either critical or consistent judgments in the fields of morals, politics and art. These several liabilities and failures are, logically, in poor accord with the assertion of a privileged authority of the church. When philosopher Dewey reviewed these shortcomings, he concluded that if a religion of the supernatural is not adequate to deal with the problems of the contingent world, one must turn to nature and experience for sounder guidance.

In addition to assets already noted there are, of course, several services rendered by churches that fair-minded people will list as social gains—gains often only loosely tied to either the ethical or supernatural content of the teachings of those churches. Among the items on such a list are group urges to the establishment and maintenance of schools, colleges, hospitals and orphanages, where these have *not* been provided adequately by government; group pressures for one or another measure of social advance—some encyclicals of the Pope effectively urge an amelioration of the more common conditions of the workingman; aid given the poor and the socially rejected; and the country church as a place for social contacts and satisfactions. The opinions of fair-minded people may differ, however, concerning the church's proselyting, missionaries (in other than medical and agricultural fields), indiscriminate charity, money-raising methods, and costly ventures in church publications and buildings.

Most of these church activities require, moreover, further notice and extensive reservations. When church schools and colleges are built and maintained not to supplement the inadequate educational facilities of the state, but to propagandize divisive theological doctrines, this activity may seriously threaten both community needs and the integrity of the state. When competing organized religions establish their hospitals, schools and orphanages as parts of an aggrandizing program of the church, the ultimate outcome—good or harm to society—is at least partly determined by the long-term balance of good and harm wrought by the thus aggrandized church. It is notable too that this multiplied social activity—apart from the building of church, synagogue, mosque or temple—is not a true or actual part of religion; broadly, it is characteristic only of the proselyting religions. Indeed, even within such religions it is most utilized when and where a foothold or a dominance of the sect in society is at stake. And, in all cases, money used by churches, for whatever purpose, reduces the amount the community may take as tax and use socially for *all* citizens.

In Chapter 4 it was shown that ethics is wholly independent of religion. It is a matter of simple observation that equally moral persons choose the one and reject the other. The statistical study by Thorndyke, reported near the end of this chapter, confirms this view. On this subject historian David S. Muzzey[1] said:

The widespread assertion that moral excellence is a corollary of belief in religious creeds or performance of religious rites has been disproved throughout history by the hosts of noble, upright men who have made no profession of the faith of church or temple.

One may now look more closely into the supernatural element of religion. Do the systems of supernatural salvation that are present in some religions provide assets of their own? The answer must rest partly on the truth or falsity—on the riches or in the delusion—of the promise of salvation; and partly it must rest on results recorded in history and sociology. Surely the very existence of the supernatural is now highly questionable. Belief in it is now rejected by a large majority of the best-informed minds. On the quite different question of whether values may be derived from that illusory idea, there is, however, something further to be said.

True or false, that promise of contingent salvation—of happy personal survival beyond death, *if* a believer—has proved itself able to yield happiness to many people; they readily make great economic and family sacrifices to retain it. This happiness and this belief have some remarkable associations. From the sixteenth to the nineteenth century some but not all of the best of Western minds were able to discard Christian and other schemes of salvation, retaining sometimes a belief both in God and in an unconditional survival. But even today the common Western minds that remain practically untouched by science and learning rarely arrive at the conclusion that, if survival is a fact at all, it is a fact in nature, and therefore all may expect subjection to a like fate, whether they wish and prepare for it or not. Prevailing religious thought has interposed a widely effective block to that sensible deduction. Instead of developing this or a similar avenue of thought for their own purposes, most Christian churches continue to cherish their earlier inventions of un-

1 *The Standard*, December, 1948.

like destinations for the dead, and to repair their tollgates at thresholds to the beyond.

Again, with—and just possibly without—the persuasion of others, a sensitive person may construct for himself a vision of a Creator whose wholly kindly rule attends his own every breath. And, definitely enough, that person may obtain both a kind of peace in life and a comfort in death, from his notion. When we spread these solaces it becomes clear that from these notions of salvation and the supernatural at least part of a society derives a kind of happiness—a religious, mystical happiness that every person should always have full liberty to enjoy. These fragile fancies will not, however, bear handling by knowledge and logic. And least of all can they fail to evaporate in that long hard course the race inevitably runs from age to age. Indeed, in our own day the maintenance of this belief from childhood throughout a life is usually accomplished only through the omission, the careful selection, or the rejection, of an education. And for those who are early molded in that belief and must later drop it, a prolonged period of scarring self-struggle and anxiety is sometimes imposed. That such a group of strugglers is among and around us is easily observed by anyone. It is implied in the statistics provided in the preceding chapter.

It is true that many people have taken a dim view of an age without a God. Renan remarked that "the day after that on which the world should no longer believe in God, atheists would be the wretchedest of all men." That may have seemed true when and where a nearly universal belief in God fiercely coerced the inquiring but unbelieving mind; and when disbelief in God rested on grounds almost as insecure as those of belief. The informed atheist of our day gladly looks at the passing of God as part of the dawn of the day of Man. Otherwise he views it as dispassionately as others view the passing of the fairies and the devil. And no void is left by the relinquished God. That space is overfilled by regard for the surge and performance of nature, by a welcome to the stunning worth of an enlightened person, and by the wealth of private satisfactions unleashed daily in the new and ever-renewing adventure of living an earned life.

Only an infantile emotionalism pillows the worthiness of human existence upon the supernatural. It is failure to grapple firmly with the natural that prevents one's finding a saner satisfaction in naturalism. **Distrust of thinking is the lurking great** destroyer. The grasp and vision of the process of continuous change throughout that long path from the nonliving world to human society is a benediction to all thought—to every career. Here reason and research are repaid in the most worthy of all ways—in the sane comprehension of ourselves and of our place in nature. To learn early that one will never become an angel is swiftly to find reason for becoming a worthy person.

We may now lightly touch the answer that history and sociology give to the earlier question. Have systems of salvation and supernaturalism, as they are involved in Christianity, contributed advantageously to human action and behavior? The answer expands into areas of the favorable and into prolonged centuries of the terrible. Some of the indirect or associated assets were mentioned earlier. One now recalls that those systems provoked ages of bloody conflict that stretched from England to India. Those systems, and schisms within them, for long promoted in Europe the will to fight, kill and impoverish, to do judicial murder on the grand scale, and effectively to enslave the intellect. An almost modern page of that long, full story can indicate its reach into affairs of the moment.

In the year 1474, in Basle, Switzerland, a "rooster" that laid an egg was tried for witchcraft and publicly burnt. Not pagan but Bible teaching nourished that belief and act. Periodical flowerings of that belief snuffed out the lives of thousands of human suspects in Europe—and nineteen of them in Salem, Massachusetts. The Inquisition saw to it that not only the witches but many who denied a belief in witchcraft were speedily destroyed. All could quote: "Thou shalt not suffer a witch to live," and "They that believeth not shall be damned" took care of deniers. But that widespread tragedy is merely introductory to a continuous and continuing story. Regularly, since 1939, the legislature of the state of New York has considered a bill to permit a sane sufferer from an incurable disease—after these two points are attested by a board of physicians and agreed to by a court—to ask and receive an overdose of morphine to end his undesired life and suffering. Let it be granted that there are various arguments against euthanasia, and some of them are not of religious origin. But ultimately it is Bible-born arguments in the democracy of New York that now decide the matter, since their possessors have the balance of power—the big vote—and that group may long continue to deny this well-guarded form of merciful release to some persons who need and want it. When the argument against "abuse" of such an authorization has been suitably met by proposed legal safeguards, there always remain to a large band of voters a religious and irrelevant quotation of "Thou shalt not kill" (apparently *mis*translated from "Thou shalt do no murder") and also a contention—particularly from Catholics—that suffering greatly strengthens the soul and prepares it for heaven. Upon these arguments is still stymied the sane sufferer's right to end an intolerable existence and to cease to be a burden and expense to family and society.

The New York legislature of 1947 received the petition of one thousand physicians of that state, supporting the legalization of voluntary euthanasia. It also received resolutions adopted during the previous August by the National Catholic Women's Union —the plea of *many* thousands—that stated as follows:

The advocates of euthanasia disregard . . . the role which suffering can and does play in the achievement of sanctity. . . . Suffering is a blessing in disguise. . . . Many people would lose their souls were it not for the suffering they are called upon to endure.

That legislature, like others before and after it, found the religious argument—or the votes of the religious—the more persuasive. In nearly all civilizations, old and new, history and sociology find that religious elements of mysticism or of salvation have fastened themselves into society, where they tightly limit the use of reason, emancipating knowledge, and the growth of social policy. Those elements earlier made "witches" suffer; they still make many incurables suffer. Within a few decades, in civilized lands, the population has changed its age structure—older people have become predominant. In this elderly group, cancer and other protracted diseases claim their millions of victims. The long, lingering death, which formerly was rare, is now common. But erratic religious taboos still—now and today—condemn an enlarging procession of sufferers to unnecessary suffering.

The extent to which the notable philanthropies of this century originate in a "religious" motive is a moot and much mixed question. Certainly the early example and the great Foundations of skeptic Andrew Carnegie are not assignable to that motive. Nor in the field of charitable enterprise is that motive clearly predominant. The first Nobel Peace Prize went, in 1901, to Henri Dunant. Dunant was the founder of the Red Cross, but he could not become its first elective head—so it is widely believed —because of his agnostic views.

The role of religion and of religious sects in the building of modern democracy provides plus and minus items that belong in this discussion. The size and nature of that subject, however, largely preclude its pursuit here. An account of the part played —for and against—by the more important Christian creeds was recently written from the Protestant standpoint by church historian Nichols.[2] This study regards the Protestant Reformation as the "watershed" on which the political differences of present-day Christians had their origin. As the medieval system gave way to the new idea of political sovereignty, he says, two divergent streams of religious thought swept forward into the past century. In one group were Roman Catholics, Lutherans and Anglicans, who "taught generally the 'divine right of kings' with the correlative denial of the right of resistance by subjects." In the other group were the Calvinist or Presbyterian churches. A later and left wing of Calvinism helped build "Puritan Protestantism," which, says Nichols, contributed more to democratic ways and means than any other Christian faith.

The history of this same period, as recorded by Catholic authorities, is a quite different portfolio of facts, impulses and events. Indeed, the hierarchy—again like communism—directs the writing and teaching of all history, from the birth of Jesus onward, in terms convenient to itself. In their view, Catholicism has nourished the roots of liberty. However loose this association may be—and it is quite loose—it seems that two objective statements with some bearing on this matter are warranted. First, the notable dictatorships of our own day were established over peoples whose religious backgrounds were as follows: Catholic in Mussolini's Italy, Franco's Spain, Salazar's Portugal, Diaz's Mexico, Peron's Argentina; Lutheran-Catholic in Hitler's Germany; Russian Orthodox in Soviet Russia. Second, Catholicism is, or earlier was, strong among a large majority of those peoples whose cheerful and amiable outlook upon life is especially noteworthy, and whose customs and laws put less weighty *personal* restrictions of religious origin on themselves and their guests

[2] James Hastings Nichols, *Democracy and the Churches* (New York: Westminster Press, 1951).

from abroad. These cultural amenities apply to the French, the Italians, the Austrians, the Bavarians, the Irish, the Spanish, and to Latin American peoples. The world traveler of this century warms to these affable peoples. And if the church of their choice should claim only partial responsibility for these fine amenities of life, that church certainly distinguishes itself from most other Christian creeds by noninterference in *some* areas of individual freedom. Catholic peoples are spared many harsh abnegations of the Lutheran, Puritan, Hindu and Muslim.

The preceding account inadequately reminds the reader that only some Christian creeds promoted democracy, while others equally supported feudalism; that where secularized Western democracy has till now arisen, it was opposed by many of the adherents of *all* major creeds of that faith. Also that a worthy democracy existed not only in Greece and early Rome but in much of India four hundred years before the birth of Jesus.

Many people credit the churches with high success in informing man of his past history. In truth, however, the propagators of the supernatural commonly conceal an element of dishonesty through omissions from their swollen messages to men. For example, from their Bible the Hebrews and Christians all get and often repeat a story—of highly questionable truth and importance —of Samson pulling down a great house of the Philistines. But only rarely from sober history do they pick up the meaningful truth that, helped by a crack in the structure of the Roman Church, men like Copernicus and Galileo crumbled the great edifice of medieval scholasticism in Europe for a start toward the liberation of the West. Nor do they tell that much of that most costly release of intellect had to be won against the power— even the torture—of Bible-inspired religion.

Again, almost never do outsize sermons acquaint their listeners with the enormous and irreparable losses their own religions inflicted in still earlier times upon all mankind. One example of early thought-destroying vandalism of religion is best presented in the recent words of historian Dampier:[3]

About the middle of the third century (B.C.) the famous Museum, or place dedicated to the Muses, was founded at Alexandria. The four departments of literature, mathematics, astronomy and medicine were in the nature of research institutes as well

[3] Sir William Cecil Dampier, *A History of Science* (New York: The Macmillan Co., 1943).

as schools, and the needs of them all were served by the largest library of the ancient world, containing some 400,000 volumes. One section of the library was destroyed by the Christian Bishop Theophilus about A.D. 390, and, after the Muslim conquest in the year 640, the Muhammedans, whether accidentally or deliberately is uncertain, destroyed what the Christians left. But for some centuries the Library of Alexandria was one of the wonders of the world, and its destruction was one of the greatest intellectual catastrophies in history.

Similar losses to society through the badgering of individuals by the church are not merely unknowable but they are known to be vast, continuous and immeasurable. Perhaps the most brilliant and valuable personality produced by Italy in the more than twenty-five hundred years of its history was Leonardo da Vinci. Nevertheless, the first of his published books did not appear until one hundred and thirty-six years after his death. When the first edition of Vasari's biography of da Vinci was published in 1550, thirty-one years after Leonardo's death, it contained (Herbert Horne's translation) the following statement:

Leonardo was of so heretical a cast of mind that he conformed with no religion whatsoever, accounting it, perchance, much better to be a philosopher than a christian.

This passage was *omitted* in Vasari's second edition, of 1568.[4] Naturally no fingerprints of church manipulation in these events are available. It is clear, however, that the then all-powerful church was prepared to use this titan's painting, music and sculpture but not his liberating and irreligious thought, his many inventions, his approach to modern geology, or his near-discovery of the circulation of the blood.

The days of Christian-cultivated wrath toward liberalizing fact and thought are not yet finished in Central Europe. Ponder the following item from lawyer-reporter Blanshard:[5]

At the very moment when the Vatican was making a world hero of Cardinal Mindszenty in his struggle for Catholic rights in Hungary, the cardinal told an American correspondent that he regarded Darwin as "a dangerous heretic who should have

---

4 These items are from the Elmer Belt Library of Vinciana, Los Angeles.
5 Paul Blanshard, *Communism, Democracy and Catholic Power* (Boston: The Beacon Press, 1951).

been burned at the stake." This same cardinal in 1945 refused to alter the Catholic description of the French Revolution in the parochial-school textbooks as "that mob movement of the late eighteenth century in France which was designed primarily to rob the church of its lands."

Another rather wide approach to an assay of religions would lead to two different places—to their Scriptures and to their institutionalized churches. The Hebrew and Christian Scriptures —the Bible—have often been instigators of revolt against the worst forms of clerical and political despotisms and a bulwark of the poor and the oppressed. On the other hand, their churches have frequently been and sometimes now are the leading oppressors of the poor. On this point, observe persisting Catholic practice in Peru.[6] The Scriptures themselves, moreover, are the definite source of the socially paralyzing doctrines of the supernatural and the miraculous. Belief in these two doctrines by an *individual* is rightly called a privilege and not a sin; yet when that same belief is held by a governing majority it becomes, socially, the blight of blights. Through the centuries, the Scriptures have dragged unchanged, or but little changed, the influences that in our own day so fully cancel the assets of religion. By way of illustration one may look at only a minute point on the compass: in many communities, nowadays, if our footwork is fast and agile, we can tell school children the world was not made in six days; but still and everywhere, as when the elder Huxley noted it, "they are to hold for the certainest of truths, to be doubted only at the peril of their salvation, that their Galilean fellow-child Jesus, nineteen centuries ago, had no human father." Too, those Scriptures vividly spell out a most incredible and rationally outrageous story of the origin of us all. The Scriptures that tell of man's fall have him fall from a very great height —from an angel. Yet they carefully caught him while he was still in the clouds, so he remained very superior to any animal or earthly thing. If the constant tug of credulity involved any mental exercise at all, the religions could be easily recognized as the world's spiritual gymnasia.

Even this brief look into some of the "mixed" assets of religion must recognize that religions—particularly the proselyting ones—may claim a somewhat special association with fanaticism.

[6] Carleton Beals, *Fire in the Andes* (Philadelphia: J. B. Lippincott Co., 1934).

This, despite the fact that some congregations in advanced areas of the Western world are now largely free of fanatics, and despite the further fact that fanaticism is nowadays also rampant in nationalistic and partisan movements like Nazism, communism and McCarthyism. Biologist Haldane[7] counts fanaticism among the "only four" really important inventions of the period between 3000 B.C. and A.D. 1400. And this was a Judaic-Christian invention. Few facts seem more significant—more involved in great good and tragic evil—than that this "malady of the mind" early became an instrument for resurrecting ailing or dying societies, and also for a still unfinished series of conquests, crusades, inquisitions and world-wide wars.

The fanatic was thus portrayed in a recent book by writer Eric Hoffer:[8]

Only the individual who has come to terms with his self can have a dispassionate attitude toward the world. Once the harmony with the world is upset, and a man is impelled to reject, renounce, distrust or forget his self he turns into a highly reactive entity. Like an unstable chemical radical he hungers to combine with whatever comes within his reach. He cannot stand apart, poised and self-sufficient, but has to attach himself wholeheartedly to one side or another. . . . His only salvation is in rejecting his self and in finding a new life in the bosom of a holy corporate body—be it a church, a nation or a party. . . . It is doubtful whether the fanatic who deserts his holy cause or is suddenly left without one can ever adjust himself to an autonomous existence. He remains a homeless hitch-hiker on the highways of the world thumbing a ride on any eternal cause that rolls by."

A sincere belief that this life is essentially a place of preparation for the hereafter is a poor foundation for sane and effective efforts for social progress. For many men and women it is a ready-made foundation for fanaticism.

Numerous and cloudy extensions of the long shadow of supernaturalism fall upon contemporary individuals, communities and halls of legislation. Some of these clouds of confusion and of rooted error have rolled so close to this writer that he recalls and records them in this book: the church-subjugated, impover-

---

7 J. B. S. Haldane, *The Inequality of Man* (New York: Famous Books, Inc., 1938).

8 Eric Hoffer, *The True Believer* (New York: Harper and Bros., 1951).

ished, ever-toiling, and Mass-buying Indians of Peru; the women and men who, on knees, do their annual miles to the shrine of the "virgin of Guadalupe" near Mexico City; the tattered Hindu, scattering upon a dismal altar huge handfuls of rose petals brought on foot in hope and choking heat from afar; the insistent, global, and religion-hushed problem of birth control; the doubtful outcome of struggle everywhere for separation of church and state; the drugged and sleeping effort to free public education from religious dominance; the complaisance of peoples at the restraints of religion upon disseminating news; and the shadows covering the need to rub out the divisive influence of religions on men.

At this point, two social and political needs should be left entirely clear. For every individual, all thought, certainly including all "religious" thought and expression, whether in the depths or on the heights, should be always free from compulsion by government. More imperatively—not less—society and government must have free and full access to the tested experience and conclusions (science) of their day. Society *must* activate and use thought about itself—and components of itself—that is verifiable. Since uncertainty, too, is always present, every society needs the utmost freedom and incentive to find and choose the probable. How do these imperatives fare in nations sapped by an organized religion?

ON A PARTIALLY WASTED ASSET OF RELIGION

The trained and the untrained but dedicated men who serve religion are important assets; they can be regarded, however, as currently misused assets of religion. Certainly these men have capacities and urges that are of much value to any society. Whether the religious impulse now advantageously directs their capacities and urges is a most reasonable and socially serious question. The obvious fact that the clergy do many desirable things mainly reflects the neutral circumstance—largely unrelated to religion—that this group is part of the world's pool of expensively drilled persons.

It would seem that the human waste involved in current religious training, and in the later activities, of a trained clergy—and also in the activities of other little-trained or untrained but dedicated and often semi-isolated priests, monks and pastors—is nearly incalculable. A book written wholly on this topic—the change from priest and pastor to teacher—could note how greatly the needs of men differ at different places on the earth. And the many pages of such a book could perhaps indicate that a clergy trained very differently than now—free to serve the immense areas of earthy fact and truth rather than the alleys of emotional guidance and misguidance to a life beyond—might soon enrich the lives of many peoples.

Today, the more advanced groups of the Protestant clergy—except for frequent and marked success in church administration—seem to be traveling in low gear, in a partial vacuum, or perhaps in reverse. They have lost contact with many of the best minds, and they are maneuvering on an intellectual defensive. At this higher level, religion has come to fear truth about itself. During more than a generation, fair numbers of the graduates of some American theological schools have been graduates in doubt regarding much or all that relates to the creed that they must serve, and sometimes in doubt on much besides. The Sunday duties of some of this latter group, particularly of those placed in backward communities, become a saddening chore that, at the cost of much lost opportunity, leads them to desert the clergy for a vocation in which they are little trained.

The foregoing statement does not apply to the Catholic clergy. In high or low ranks of the hierarchy there is practically no defeatism, little discouragement; and an occasional desertion is a remarkable event. There, the march to success contrives to discount whatever happens in the intellectual world, while fully exploiting the universal and inexhaustible capital of human emotion and mental waxiness. That superb organization and firm faith may be called calculated fanaticism with a secure foothold. It is not beset with doubts. It is the world's bastion of supernaturalism. However numerous and showy the touchingly tender ministrations at the parish level, true success of the Roman Church is measured by the extension of Catholic power. And—apart from the Iron Curtain—the innumerable communities and states that include Catholics endlessly provide their own evident and individual examples of success in extending that power.

A further word is needed to distinguish sharply between the two rather distinct areas of Catholic effort, as these are observed through non-Catholic and this writer's eyes. At the one level—that of the local or parish church—members of the church perform the rites of worship and share in financial or other support of internal or community projects, notably a huge educational task. Some at least of these activities are actual social gains. In these projects and services the priest is usually or often to be regarded simply as leader and guide. This partially beneficent effort of the parish, like the similar efforts of other sects or religions, nevertheless propagates a man-submerging belief in the supernatural and enforces widespread subgovernment through the church. Above all else the parish is the recruiting post for the legions ceaselessly striving to shape society in a Catholic mold. Turning to the other area of Catholic effort one meets the alien and formidable hierarchy. That hierarchy was the thing partly described in the preceding paragraph. That hierarchy is the summit of enduring autocracy. Basically that hierarchy is a foreign one, now seeking to overthrow American ideals on such things as censorship and parochial schools. Its pressures are put continuously on some partly unwilling American prelates, and through all priests those pressures are put on every loyal Catholic. That autocratic and foreign hierarchy now shares only with communism a well-implemented urge to conquer the world; and the hierarchy is the more enduring threat. A more formidable menace to intellectual man and to self-propelling societies has not yet appeared on this planet. The Catholic hierarchy—all without malice—has built of supernaturalism a charmed vehicle in which the race may smoothly ride to ruin.

We next explore an item of quite important information. Who is it, qualitatively and quantitatively, that fills the ranks of the clergy at this mid-century? The general public looks upon the clergy as an educated group that knows what it should know. A little search shows that this is an unwarranted belief. Fortunately the illuminating evidence relating to the Protestant clergy of the United States and Canada was obtained and summarized in a friendly yet frank and impartial study by Professor William Adams Brown, of Union Theological Seminary, Professor Mark A. May, of Yale University, and several others. The results were published by the Institute of Social and Religious Research.[9]

Approved intelligence tests were given to the freshmen of sixty-two colleges, in the autumn of 1930. It was found that "those who had definitely chosen the pastorate for their life work were of distinctly lower intelligence grade"—their average

---

[9] *The Education of American Ministers* (4 vols.; New York, 1934).

score, 131; that of all others, 150. The future ministers examined there are now pastors in the prime of life. Everyone knows that many excellent minds are in the ministry. The result cited deals with averages, but on the question of average quality of mind it presents an unfavorable comparison with other professions.

Perhaps most informative of the results of this diversified study is the fact that *two* out of *five* ministers of seventeen of the largest Protestant denominations of the United States had graduated from neither a college nor a theological seminary. Some 224 seminaries were training ministers in the two countries. Of these, 176, including all the more important ones, were investigated. In 1930-31, they enrolled some ten thousand students. Forty of those seminaries admit only college graduates, ninety-eight admit also high-school graduates, and thirty-eight require neither college nor high-school graduation for admittance.

Another dark picture revealed by this report concerns the nature of the studies actually pursued within the seminaries. Rather more than half the ministers attend a seminary. To which subjects do they direct their time and effort? The detailed study and report upon fifty-seven seminaries indicated that 63 per cent of the total required time of the student is given to practical theology, English Bible, and theology and philosophy. And "in theology and philosophy the central course is systematic theology." Another 13 per cent of required time goes to Christian sociology, comparative religion and missions, religious education and psychology of religion. Twelve per cent is given to Greek and Hebrew. The Presbyterian Church usually, not always, requires "an acquaintance with Latin and the original languages of the Bible." One exceptional Lutheran institution in Missouri, with about four hundred students, "insists upon six years of Latin, four years of Greek, and two of Hebrew as a preparation for entrance upon the seminary."

No account of the training of Roman Catholic priests is attempted here. That group, however, is hardly in position to get satisfaction from the Protestant confession. Certainly they cannot match it in candor.

In many modern nations a very high proportion of the clergy has both an urge to serve man and some special training and ability to speak to men. Perhaps no people is rich enough to afford the waste—and certainly not the perversion—of these qualities. But the type of training and the type of service are the all-important matters. If instead of prolonged training in obsolete theology and in languages (Latin, Greek, Ancient Hebrew, Arabic, Sanskrit, and others) used almost solely as props to theological studies, these prospective community servants should go in for appropriate types of social, economic and technical training, and if such training led to literary, scientific and technological skill and outlook in those who are to serve backward rural areas, and if appropriate types of training were adopted by those who are to serve other cultural levels—then there could be only full days of productive effort for all such trained men. And perhaps there would be few communities too heavy for their outlook and labor to lift.

The trail of waste left by any proselyting religion whatever leads to all that the poor and backward community loses in having no suitably trained person or persons to serve it intimately and practically, though it does have one or more "dedicated guides" in "religious" feeling. Further waste arises from trust in religion to inspire charity to relieve the sufferings of the poor, while failing to use knowledge to banish poverty, disease and ignorance, and thus let dignity and opportunity be born. Who will say that these neglected needs—as demanding and persistent as the breath of men—are either less urgent or less worthy than the solaces for inevitable events now residing in the religion-built crutch?

Little related to the present theme but worth mentioning here is the enormous waste that attends the building of several churches where one or two would suffice. The world is well filled with examples—examples to which it gives little thought. Even where only Catholic churches are built, as in parts of Latin America, they are often built in profusion by extremely poor people. In that part of the world, this writer has visited many towns with populations of only two to five thousand but with five to ten Catholic churches. The impoverished little town of Cholula, Mexico, boasts more than a hundred. One of these, built on the most prominent and also the most inaccessible hill, long ago displaced and completely destroyed an ancient Indian temple.

One may note that it was left for a date around 1920, and for a schoolmaster and heretic, Plutarco Elias Calles—later president of Mexico—to start on his own plantation the first industrial school in his country. Again, one may recall the 40,000,000 very poor *rural* Brazilians who have—and for long have had—little schooling, few teachers, and many priests. Finally, citizens of the United States could, in time, become partly aware of the immense spread of leaderless communities within its borders—communities with priest, preacher, or preachers, but without a trained man of this day to speak to them and to live and labor daily with them.

"I believe that the priest must become a teacher if he wishes to do justice to his lofty educational mission," says physicist Einstein. Philosopher Montague concludes that "the most serious indictment of religion is its increasing irrelevance to the needs and interests of modern life." And psychologist Leuba says that "to prevent the religions from continuing to hamper, by false teaching and a false method, the intellectual and moral development of populations which have outgrown them, and to replace the religious method by other, more effective means of life, is one of the urgent problems before civilized humanity."

POSTSCRIPT

Only unreason or prejudice could fail to grant that some ingredients of current religious worship strongly support large areas of idealism and morality. Confession, contemplation, fellowship, example and esthetic enjoyment all help to develop those areas in many people. Requiring further notice here is the fact, and a corollary of the fact, that some elements or by-products of that worship also act as curative agents for a brood of psychically ill or maladjusted personalities. These curative agents are nevertheless wholly unrelated to the truth or falsity of the faith of the church—such as belief in a God or salvation. The cure is to be attributed to one or another quite human source: to a loyalty or a belief in something—something to replace a void of unawareness; to a body-mind relaxation; to release from a repression; to the needed stirrings in a lively song; to the re-education of a warped emotionalism; to an opened door of friendship with its sense of belonging. But it is especially important to observe that, as admitted by some clergymen, the churches have "played up the sense of guilt, stimulated it where it would not normally be, fostered it, taught it to children, cherished the concept of depravity, talked of unforgivable sin, until it has pro-

duced in lives innumerable a terrific burden of guilt, which is
morbid, artificial, and mentally disastrous." And psychologist
Leuba says, "Dependence upon God, Christ, the Virgin, fosters
an emotional infantilism and offers dangerous ways for ignoring failures, sickness and poverty." The churches cause much and
cure some mental upset and instability.

Of course, one may conceive of "renovated" churches—of
churches rid of their immensely damaging dogmas and their
frustrating trust in the supernatural—within which the present
support of idealism and morality would be continued and much
emphasized. That concept is familiar to many; some experiments
with it are already under way; and it seems certain that those
worthy experiments will be continued. In his last book Leuba[10]
presents an able argument for that new type of church. He
writes:

> The expression "spiritual hygiene and culture" may be used
> to designate the field of action of the renovated churches. They
> are to occupy that field in so far as it has not been adequately
> pre-empted by established professions and agencies—those of the
> educator, the social worker, the psychiatrist, the clinical psychol-
> ogist and others. . . . One should look forward to organizations
> that will give satisfaction to all kinds of people.

On the outcome and practicability of this suggestion these pages
offer no opinion. On a related matter—but one quite subordinate
to its guiding theme—this book records a personal belief that advanced societies of this day would soon be much more helped
than harmed by the abandonment and absence of the organized
religions of our time.

Historian Toynbee[11] attributes the failure and world crisis of
the last half century to Western failure to send Christianity along
with our secularized technology to peoples of the East. Apart
from the meager extent to which one system of supernaturalism
can be exported to supplant another, his argument is not logical.
He says:

> The truth is that, in offering [Asians] a secularized version of
> our Western civilization we have been offering them a stone in-

---

10 James H. Leuba, *The Reformation of the Churches* (Boston: The Beacon Press, 1950).
11 Arnold Toynbee, *The World and the West* (New York: Oxford University Press, 1953).

stead of bread, while the Russians, in offering them communism as well as technology, have been offering them bread of a sort—a gritty black bread . . . but still an edible substance that contains in it some grain of nutriment for the spiritual life without which Man cannot live.

Now if communism—with its obvious subordination of the individual to the state and a history of unceasing compulsion and cruelty—contains its "grain of nutriment for the spiritual life," why does this distinguished historian seem never to find any grain of the "spiritual" in naturalism? He always finds heaps of that quality in Christianity, which, in essence, is a system of salvation for a life after death. Naturalism upholds the dignity of the individual, derives morals and values from nature and warns against their neglect, brushes dusty minds, supports science as the fountain from which the desired technology flows, invites the formation of societies such as best serve their members, declares that human purposes are fully free to guide the human destiny—and *all* of this as a *new* word to men. Why can this historian find no "spiritual" quality in this—this, the endowment that Christianity has *denied* to the Western world? Thus, except for this vast injury from Christianity, the Western world—like the Russians—could have offered *its* brand of secular naturalism along with its secular technology to the Asians. Indeed, this could have been offered them long before the Russians had either communism or technology to offer anyone. Western failure does not lie in failure to export its Christianity to the East, but—in addition to offenses related to colonization—in its prolonged failure to accept a truer and worthier world view in the presence of that faith.

At the age of seventy-one, in 1927, psychologist Sigmund Freud wrote the booklet—*The Future of an Illusion*—for which he received so much abuse from Catholics and religious colleagues in Austria, Hungary and elsewhere. In that book he recited what the partisans of religion claim would follow the destruction of religion. He then wrote:

What a number of accusations all at once! However, I am prepared to deny them all; and what is more, I am prepared to defend the statement that culture incurs a greater danger by maintaining its present attitude to religion than by relinquishing it. But I hardly know where to begin to reply. . . . The person this publication will harm is myself. . . .

Religious ideas are illusions. They are ideas contrary to the practical world, cosmic ideas coming from the madness which science in the last centuries has done much to dissipate. I know that religion has rendered many services to human culture, but that is not enough and one sees throughout a thousand years of religious life that religion has not bettered mankind to any appreciable extent.

We have reached a stage when the intellectual development of humanity is endangered by the superstitions of religion. There are too many instances on record. But we are now at the dividing point when must come complete severance of the two, when one must disappear before the other, which is to say that science will remain.

The question of a "balance sheet" for good and harm of the various organized religions lies only partly within the scope of this chapter and of this book. Only release and use of evolutionary thought can promise or build a free and genuinely modern society—a society whose birth awaits the outcome of the world-wide contest between society and supernaturalism. Certainly most of the good of organized religions results from their ethics—a nonreligious element—and from their making ethics acceptable to men. Surely religion's incalculable injuries to present society could be avoided through an abandonment of the supernatural. The church, however, cultivates the view that men and women are better restrained from immoral acts by belief in the supernatural than by any other means. A large church membership is widely alleged to be an index of a state of well-being and of morality in a community or nation. To an astonishing extent this view is accepted by educated people who otherwise reject the teachings of the church. That claim, as applied to many cities and states of the United States, was recently and definitely refuted by the results of a study made by educational psychologist Thorndyke.[12]

That study is worthy of a summarizing paragraph. The cities studied were those having a population between 30,000 and 500,-000. Scores for "the general goodness of life for good people (G)," for "the per capita income of its residents (I)," and for "their personal qualities of intelligence, morality and care for their families (P)" were compiled for each city and each state. Data of the

12 Edward L. Thorndyke, "Amercian Cities and States: Variation and Correlation in Institutions, Activities, and the Personal Qualities of Residents," *Annals of the New York Academy of Sciences*, vol. 39, 1939.

1930 United States Census and the 1926 Census of Religious Bodies were used. The relationship of "church membership" to these items was calculated. For membership in all churches Thorndyke found, for the *cities*, a negative correlation of −.21 with G, and −.245 with P. For the *states*, the corresponding correlations are −.13 and .20. "The Unitarians, Universalists, and Christian Scientists are an exception, the correlations for them being .51 and .52 in the cities and .61 and .71 in the states. They, however, are notably unorthodox, and opposite in many respects to the general temper of the church, so that the positive correlations for them," he says, "may be taken to corroborate the conclusion from the negative correlations for the church as a whole."

Nor may this book be required further to map or sketch those means, services and motives that may supplant or supplement the several ethical and social services currently performed by organized religion. The past, present and future experience of our species must provide and further develop that guidance. But it is already clear that an ethics capable of stopping the present retreat from citizenship will have to be developed; clear, too, that henceforth our kind must have a hope for place and attainment within this earthly life. This book merely ventures the speculation that the strategy for both human advance and survival probably centers in mutual respect and mutual aid; that the tactics probably involve making humans more respectable **and more capable of providing mutual aid.**

PART III

OPINION AND OUTLOOK

14

Comment from Critics

It is impossible to conquer my own ignorance and frailty, to say nothing of overwhelming the vast stupidity and illiteracy of the world.—*Rupert Hughes*

FIFTEEN YEARS BEFORE THIS BOOK WAS WRITTEN MOST of its main thoughts and concerns were expressed all too briefly in the form of an address. It happened that parts of that address were carried to the public rather widely by newspapers dated January 2, 1936, and the whole of it was published by *Science* magazine under dates of January 17 and 24.[1] The address, delivered in St. Louis, was prepared as that of the retiring president of Section F (zoology) of the American Association for the Advancement of Science; it was thus a vice-presidential address of the latter organization. During the following weeks, this writer received some hundreds of letters—from a varied public, from the clergy, and from colleagues in various sciences—frankly giving their reaction to it. Some who have followed to this point the much fuller account given in this book will obtain profit or satisfaction—or perhaps a much-wanted revenge—from reading a few of those criticisms. For such readers the materials of this chapter have been assembled. Though that hour-length address included an attack on traditional religion and supernaturalism, it did not specifically discuss the God-idea.

Whatever the value or the lack of it that may attach to these personal letters, they indicate the attitude with which many will begin—and perhaps will end—the reading of the expanded discussion this book provides.

But they do more than that. These words and reactions of others supply present readers with outside testimony on the following points: the wholly unlike reactions of various clergymen;

---

1 Oscar Riddle, "The Confusion of Tongues," *Science,* vol. 83, 1936.

361

the way others apportion blame to educators and to the church for shortcomings in the teaching of modern evolutionary thought; the fiction or the reality of cleavage and antagonism between science and religion. Among the several opinions quoted are expressions from some of the top-ranking scientists of the United States and Canada, in 1936. Perhaps clearest of all, this comment shows the inadequacy—in some respects, the unfairness—of any *brief* discussion of the area of conflict between society (or science) and religion, and thus a need for doing this through one or more books.

In thus dealing with private correspondence, all names and identities are of course omitted, and no more than a general indication of the field of interest of the correspondent is given. The reader will further understand that only the possibly significant and relevant parts of a letter are here reproduced. A few excerpts from newspaper editorials, sent to the writer in considerable number, are also given a place in these pages.

Though it may look like vanity, included here are many letters of approval and indeed of much overpraise for something found in the address. These are included partly because of the understandable circumstance that fully nine tenths of the letters, aside from some anonymous blasts, were letters of approval. Clearly, the anonymous notes can only be ignored. Replies were sent to several correspondents, but these are now disregarded, since the various chapters of the present book are believed to contain answers to all of the serious questions then raised. Indeed, the size, contents and plan of the present volume were determined largely by this extensive correspondence.

———

From the rector emeritus of a Protestant church, adjacent to a university, Chicago:

This letter is respectfully and frankly written on the supposition that the enclosed clipping from today's Chicago *Tribune* quotes you correctly. If it does, and you made the statements about religion in this clipping, please let me characterize your language as inaccurate, unfair, unscientific and unworthy of a man in your position. You ought to know and you probably do know, that there are plenty of religious men who are respectful students of biology, and who are deeply read in many branches of science. You also should know that in the United States there are possibly 175 kinds of Protestants, and five or six kinds of Catholics. For you to lump them all in one slap-dash indictment as this clipping states, is most regrettable, and quite unworthy of the importance both of religion and of science.

These are days, as you ought to know, and probably do know, that many brilliant men in many branches of science are earnest Christians. I will supply you with names if you do not recall them and wish them.

My object in sending you this letter, which I hope you will read and digest, is not of course simply to annoy you. Neither you nor I have time for such trivialities as that, but it is to beseech you in the interest of fairness and of truth never again to make such a slap-dash and inaccurate indictment of religion in these terribly critical days.

From an economist in Chicago:

I want to congratulate you on your very clear and uncompromising demonstration of the true relation that exists between science and religion. It is almost a unique occurrence, as far as I have observed, for a scientist to be able to free himself entirely from traditional beliefs and then to have the courage to come out in the open and expose the methods religion uses for the suppression or warping of the facts that in any way undermine the foundations of religious faith.

Your paper as reported in *The New York Times* gave me the greatest pleasure. I hope this letter may offset some of the abusive letters you may receive. Please do not bother to acknowledge it.

From an undergraduate student in the University of California, Berkeley:

Without a doubt your contentions regarding the ignoble influence of theological tradition upon education, as expressed at the St. Louis meetings of the A.A.A.S., and appearing in the local periodicals this morning, are commendable. From intimate discussions with undergraduate fellow students of the University at Berkeley that is precisely the opinion of a large number of students who have pursued courses in paleontology, biology, and chemistry. It is too bad that so few enlightening statements of unstifled opinion are uttered by able minds.

From the rector of an Episcopal church, New York City:

I was very much interested in a reference printed in *The New York Times* on January 2nd of a paper you read in St. Louis. The newspaper gave excerpts, mentioning your reference to the influence of religion upon scientific subjects. I found myself in hearty sympathy with your point of view and I have been very anxious indeed to read the whole paper, if that is possible. Will you be good enough to tell me if this article is to be printed, and if so where I may somehow obtain a copy of it. I am wondering if you are susceptible to an invitation to lunch, say at the ——— or the ——— Club?

From a philosopher in Columbia University:

I have just now re-read the very interesting and valuable article which you were so good as to send me. I had already read it, but less attentively, in *Science;* now, with greater leisure I have had the chance more fully to enjoy and appreciate it. I have to thank you especially for two things—for your informing and suggestive sketch of recent biological progress, the grasp of which I thought admirable; and for the incisive criticism of present conditions in secondary education and the powerful appeal that followed it. The first instructed me; the second encouraged a spirit dangerously depressed by the unquestionable degradation of our public instruction. Efforts like this on the part of men like yourself may help much to turn the tide.

In a less effective way I have myself recently found the chance to do something similar—I'll send you the papers when they appear. And now I shall hope for the chance to attack this urgent problem more directly, in the hope that many others will do likewise. But today we need not only another Huxley but a dozen like him; for the people to be instructed and persuaded have lost the intellectual vigor of the nineteenth century. Still dangerous the influence of religious obstinacy may be; but I think that the situation is worse than they alone are likely to make it, for common people are becoming increasingly sceptical of traditional belief. Is not the danger more deep-seated? Has not our democratic sentimentality practically debased the secondary educational standards of the last generation; has it not already affected those of our colleges; is it not cleverly fostered by political chicanery everywhere?

Has not the educational theory of those who would prolong adolescence as far as possible and eliminate training as repressive, bred a horde of teachers to study formalized methods of presentation and to ignore the substance of education, to make jobs for themselves by inventing an endless succession of futile educational New Deals, and thus to turn our schools into asylums of self-expressive incompetence which democratic sentimentalizing makes lazily self-complacent as well, and where the blind lead the blind? Are not these Babbitts in secure control of a multitude of our educational systems? Are their brothers not shaping the curricula of our colleges, where required courses of study of the "survey" variety, which demand nothing of the student but a passive receptivity, are gradually replacing those that involve discipline and mental effort? In all schools are not a rapidly increasing number of students under the stupid guidance of such third-rate minds turning away from all studies that are in any sense rigorous, toward those which may be prepared for by the daily newspaper?

I do not want to assail your ears by mere Jeremiads, however. What is uppermost in my mind is that the first point of attack in the attempt to stay the progress of this definite cultural decline is a direct appeal to ordinary common sense against the sentimentalism that is deliberately inculcated by politicians and teachers alike. This might well be effective, for few ordinary men really believe this slush about individual equality and repressions. Those who have their living to make in a cold world, unprotected, and who have to live with their children, are quickly undeceived about it if they ever are temporarily influenced by it; they are merely too busy to study education, and in simple ignorance permit themselves to be misled by the sob-sisters and the racketeers in both politics and education. Begin, then, with the White House and Teachers College, not with the church. If common men are merely *informed* as to the trickery practised by those whom they naively trust, it may well be enough.

Within our educational institutions the vagaries of the local Brain Trusters are already being assailed vigorously by a few courageous men who are willing to risk their jobs in a good cause. But it is from outside, obviously, that the best work can be done; and most effectively, I am sure, by scientists working, like yourself, in independence of both public and educational politics. The industrial technicians next, and with them their managing directors; then the foremen, and so on. It is the *worker* who will most effectively silence the babbler.

Forgive this explosion!—but your paper was like a ray of sunlight in a fog.

From a professor (zoology) in Columbia University:

Please accept my heartfelt thanks for your honest and brave revolt against the devitalized brand of "science" teaching that leaves man's true place in nature entirely out of the picture.

The "reconcilers" prefer to stick their heads in the sand and deny that there is anything that needs adjustment. From a practical point of view, however, thorough-going evolutionists are a small minority, perhaps even among scientists.

From an item in the St. Louis Post-Dispatch, *January 4, 1936:*

Members of the Catholic Round Table of Scientists, meeting yesterday at St. Louis University School of Medicine, expressed the opinion that Dr. Oscar Riddle of the Carnegie Institution was "amazingly misinformed" when he told a group of zoologists Wednesday that "the present restrictive influence of organized religion on the teaching of the best in biology is intolerable."

Replying to Dr. Riddle, Dean Alphonse Schwitalla of the St. Louis University School of Medicine pointed out that the female sex hormone theelin and the related theelol had been isolated in St. Louis University under conditions of the fullest freedom in research and teaching.

"If it is true that organized religion is a restrictive influence on the teaching of the best in biology," Father Schwitalla said, "the statement cannot universally be applicable to all schools conducted under religious auspices.

"The worthy facts concerning man's origin and destiny as revealed by religion can be taught effectively without silencing even the most advanced scientific thoughts on the same subject. One need not repress biological fact concerning the origin, maintenance and reproduction of life in order to teach religion, but both must be taught simultaneously to give the student an adequate concept of the world in which he lives and the meaning of his own life."

The consensus of the Catholic scientists was that "men in authoritative positions who discuss the conflict between science and religion should carry over their research attitudes into the investigation of this real or alleged conflict and then limit their conclusions in accordance with the ascertained facts."

The Catholic Round Table is an organization of Catholic professors formed to promote advanced and research studies in Catholic institutions. It has a membership of 608. About 175 members representing 65 institutions attended yesterday's meeting.

From a professor (chemical engineering) in Yale University:

Allow me to congratulate you on your entirely sound and certainly courageous address before the A.A.A.S.! It distresses me to see the populace steeping itself in dogma to the complete exclusion of common sense and knowledge; but I rather expect that. When I hear eminent scientists upholding and encouraging the narrow tenets of orthodoxy I am positively alarmed. I am glad you took a poke at them. May you continue the New Year in the valiant manner in which you have started it.

From a professor (economics) in Yale University:

I was enormously interested in the report of your paper at St. Louis on bridging the gap between animate and inanimate matter.

Incidentally, I have always felt confident that something of this sort would be forthcoming.

I would be greatly obliged if you would send me a copy of your paper.

From a professor (biology) in Harvard University:

I've just been reading the 2nd installment of your address. And I want to tell you that after 30 years of reading presidential addresses yours impresses me as the finest effort of the lot. My heartiest congratulations.

From the superintendent of schools of a city in Minnesota:

I feel that I am reasonably familiar with the evidence supporting evolution, and I get perhaps unduly hot and bothered when the pastor of the only church I could conceivably attend, though he confesses almost complete ignorance of science, fulminates Sunday after Sunday against the findings of science, consigns most of our "higher institutions of learning" wholesale to the bottomless pit, and makes such (to me) outrageous statements as that most of our prominent scientists have repudiated evolution.

I believe that it is high time that scientists devoted a little of their time to answering some of these fanatics who are undermining our institutions. If they don't, the time will arrive when we shall reap the whirlwind for our neglect.

From the priest in charge of an Episcopal church, suburban New York:

You may possibly be interested in the enclosed copy of a letter which I have despatched to the Editor of *The New York Times* concerning your recent address at St. Louis.

May I say that I should welcome an opportunity to discuss with you personally at any time and at your convenience the issues represented in the latter portion of your address?

(The entire letter follows.)

The address of Dr. Oscar Riddle to the American Association for the Advancement of Science was significant for a number of reasons—for the occasion of its delivery, for the eminence of the speaker, for the universal importance of the issues discussed, and, not least, for the generous space assigned to it in *The New York Times* (January 2, 1936) and other metropolitan newspapers.

Of special interest to the general reader was the latter part of the address in which Dr. Riddle launched his vitriolic attack against those who are instrumental in keeping from our schools and colleges the knowledge gained by modern biological sciences on the origin and nature of life and consciousness. This attack was directed explicitly against "traditional religion." "In any consideration of this matter," Dr. Riddle is quoted as saying, "it is unquestionable that it was traditional religion that thus invoked the heavy hand of legislation. It is equally clear that elsewhere, without invoking the law but with its extended and varied influence, traditional religion is now effecting a widespread repression of the teaching of this central principle of biology in public schools throughout the United States and in practically all other civilized countries as well."

Dr. Riddle's formulation of his polemic in this indiscriminate fashion makes it necessary to point out that there are many religious teachers, representative of large organizations and influential schools of thought, who share sincerely his deprecation of the restrictions placed upon education wherever the established results of the general sciences are concerned. In fact, so considerable within the main body of organized religion is the attitude of cordial agreement with what science has to teach that Dr. Riddle's wholesale attack on "the Church" as a restrictive influence in scientific education does serious injury to the very cause which he is endeavoring to champion.

No enlightened Christian thinker will challenge for a moment a scientist's right to protest publicly against the erection of barriers to his teaching of the truth as he sees it. But Dr. Riddle's partial representation of this particular issue has the unfortunate result of helping to perpetuate the tragic and false dilemma that was crystallized in the popular mind by the Scopes trial, of which Professor Gilbert Murray said that it was "the most serious set-back to civilization in all history." The false dilemma is, of course, the notion that there is an inherent conflict between science and religion. It is a tragic dilemma because it is for many people an obstacle to their achievement of that integrated view of life which is the essential basis of a wholesome personality. It is false because the work of adjusting the new facts revealed by science to the universal hypothesis of traditional religious faith is continually progressing and, despite Dr. Riddle's statement to the contrary, has not failed. This false dilemma, moreover, has another aspect quite familiar to the parish priest though less familiar, no doubt, to its secular exponents. Because of the intimate relation between religion and ethics the notion of a conflict between science and religion furnishes welcome data for a legion of pseudo-intellectuals, particularly of the younger generation, whose undisciplined minds will permit them to find in such a careless statement as Dr. Riddle's a facile apologetic for their cynical rejection of established moral standards.

In this connection it is difficult to understand in a man of Dr. Riddle's eminence the marked antithesis which he draws between truth and tradition. One should expect the modern scientist, whose contributions are so manifestly the flowering of plants rooted in the accumulated wisdom of the past, to be the first to recognize the indispensable value of tradition. Indeed, the essence of the education for which Dr. Riddle pleads is tradition—the passing on to others of what has already been learned. The Church, it is true, often clings too tenaciously to traditional modes of thought in the face of new discoveries; but it is because the Church has so much to pass on of tried and proven value that tradition is so substantial a part of religion.

Dr. Riddle claims to be unconcerned as to whether religion is or is not important. But this pretended indifference on his part is contradicted by his assertion that "the worthy facts concerning man's origin and destiny come not from religious traditions but from biological investigations made within the time of men now living." Setting aside for a moment the question of the validity of this astounding claim, one ought to point out that by reckoning as of no consequence for our estimate of man's origin, nature, and destiny the beliefs which are the product of more than thirty-four hundred years of progressive religious experience Dr. Riddle adopts a viewpoint for the defense of which he will be required to concern himself very intently with the question as to whether religion is or is not important. Furthermore, while Dr. Riddle's illuminating discussion of the origin of life seems to leave the Christian doctrine of creation precisely where it has always been, many theologians will be exceedingly curious to know what light the biologists have been able to throw on the question of man's *destiny!*

It seems that Dr. Riddle has fallen prey to a fallacy common to the human mind even when it rises to great heights—that of assuming that authority in one field vouchsafes authority in other fields as well. Millikan, Dewey, and Freud are respectable thinkers in their respective subjects—physics, philosophy, and psychology. But all three have ventured to air their views on the subject of religion. And all three have therein committed themselves to statements which, from the standpoint of the religious expert, verge rashly on nonsense. Dr. Riddle should learn, when speaking *ex cathedra,* to respect the same discipline which, as a scientist, he rightly demands of other teachers—namely, that of confining one's published statements to the particular branch of learning within which and only within which one is avowedly expert.

From a religious leader in New York:

I have just read with keenest interest your paper on "The Confusion of Tongues." I should confess perhaps that not being a scientist some of your statements were too technical for me to follow closely but I am sure that I caught your meaning and felt the force of your argument.

Some of the statements I have underlined as especially germane to us. We must get people to realize that "vital processes are natural events, not capricious acts of some supernatural spirit." Our Religion fully recognizes the "ferment" that for all enlightened men has swept away fear and superstition and brought new conceptions of the origin and destiny of man. One of our chief objectives is to get "a great many more new ultimate consumers for the body of biological knowledge we now have." We would raise the slogan "Follow the leadership of science." There is no hope for religion or for civilization except as we are guided "by knowledge rather than by the dead hands of the past."

There is great confusion, without doubt. What can we now living do to help guide the world out of its confusion and away from its silly divisions into understanding and unity? I expect some day to gather a group of scientists for a public rally in the name of truth and progress—an appeal to intelligent men and women to aid in the spread of scientific knowledge, etc., etc. Can you offer any suggestions as to the selection of speakers, time and place of meeting, and how to get the cooperation of men and women who are in a position to furnish the backing the great cause merits?

I would indeed appreciate a few lines from you. Whether you have time for this or not I thank you for your splendid contribution to the cause of human progress.

From a retired admiral in New York:

Thank you for sending me the reprint of The Confusion of Tongues. I have read it with care, and find understanding of it through having heard you give some of its substance at the American Institute dinner.

Some of the quotations are amazing and would be ludicrous if the ignorance back of them were not so fundamental. Not in a long time have I been so impressed, as by what you and Dean Loomis said at the dinner—for the thoughts in themselves, and for the serious light in which they make modern educating and the present practice of democracy stand out.

It was a most pleasant beginning of an acquaintance which I hope may be renewed.

From a professor (zoology) in a university in Texas:

It is not my custom to write "fan mail," but the appearance of your "Confusion of Tongues" makes me break a good rule. I am pleased immensely with this address and am sending to *Science* for three extra copies of the January 24 number which I hope to use profitably here.

Would you mind if I have mimeographed (for my advanced students) perhaps a maximum of 45 copies (enough for three years) of the second half of your paper? I should be very grateful.

From a pamphleteer (with an autographed pamphlet) in To- ronto, Canada:

With regard to the opposition voiced in this City to the efforts of the International Christian Crusade against teaching "evolution," I am neither fundamentalist nor modernist, neither Protestant nor Catholic, but would willingly join any movement for placing in an asylum for the insane any teacher, professor, or clergyman who has leanings towards Darwinism and spoon-feeds them to unsuspecting youngsters.

The enthusiastic "evolutionist" is a menace to the community and the race. As a rule he is profoundly ignorant of what "evolution" is, means or implies. He is entirely unaware of the pernicious effect his views have upon the morals and general conduct of the rising generation. He is (not blissfully) unconscious that he is spreading abroad seeds of atheism, agnosticism, paganism and animalism, that he is destroying everything worth-while the race has gained through the efforts of Great Minds since the beginning.

The case for the "biologist and evolutionist" was argued before the American Association for the Advancement of Science on 1st January last by Dr. Oscar Riddle of the Station for Experimental

Evolution at Cold Spring Harbor, N.Y. His address was . . . titled "The Confusion of Tongues." . . . From the viewpoint of Exact Science "The Confusion of Tongues" is from beginning to end unmitigated rubbish.

From a professor (botany) in the University of Toronto, Canada:

I was delighted and yet saddened on reading your recent article in *Science* on "The Confusion of Tongues." You have directed attention to a matter that is deserving of the serious attention and effort of every one who is interested in the future of the biological sciences. I can assure you that your conclusions apply equally to the conditions in the schools of Canada as they do to those of the United States.

If you could spare me a separate or two of your article I think I could make good use of them here. A marked copy should go to every Provincial Department of Education in Canada, and I should like personally to draw the attention of the Minister of Education for Ontario to it.

Perhaps I should not have bothered you and have written directly to the editor of *Science,* but I thought that I should like to let you know that your work is appreciated.

From a professor (psychobiology) in Johns Hopkins University:

I should appreciate very much the possibility of getting two reprints of your recent address, "The Confusion of Tongues," one for my personal possession and one for the library. I am very anxious to see that my coworkers get easy access to the very excellent presentation of the issue of biology and its reaching into the psychobiological field.

From a professor emeritus (psychology) in New York:

I am sending you a line of appreciation on "The Confusion of Tongues." Not only have you presented very vital considerations in zoology, but you have done it in a manner that commands my admiration.

In a forthcoming article of my own I shall have occasion to cite a few high-light sentences from your address.

From a professor (psychology) in New York:

Thank you for your very thoughtful address. It is a statesman-like document.

The paper also shows that you are in the tradition of those few scientists who *can* write understandingly for men in other fields than your own. So, you must accept your share of the responsibilities which you have placed on biologists, to inform America regarding the profoundly important findings of the past twenty-five years.

From a professor (psychology) in Pennsylvania:

I was pleased to read your address on The Confusion of Tongues. We psychologists, many of us at least, feel as you do about the relation of biology to our science—the closer it is, the better.

Your forceful remarks upon the decline of the teaching of biology and, in general, of science in the schools, and of the antagonistic influence of religion are very timely. Too few of us are aware of the extent of the evil.

You may know that I have given much time to the study of the so-called "religious experiences" and done what I could to demonstrate their naturalness and the advantage there would be in replacing the God of the religions by a scientific use of natural forces—biological and psychological. The inclosed circular announcing my last book may interest you.

I should very much like to know the name of the text-book of biology from which you quote—also the publisher.

I shall quote some of your remarks on the evil influence of religion on science in a book I am now preparing.

From a professor (sociology) in New York:

Thank you for the reprint of your article on The Confusion of Tongues. I have read it carefully and taken great delight in the vigor and clarity with which you show the present critical need for a vastly increased body of scientific consumers as a defense against wilful obscurantism and as a means of making science an effective force in society. The social basis and import of science, as you so admirably indicate, need greater consideration from scientists than they have shown thus far.

From a professor (biology) in Alabama:

I have read with interest your vice-presidential address, on The Confusion of Tongues, and I would appreciate a reprint of it.

It is my opinion that the anti-evolutionists just use religion for a pretext, and their real reason is lack of intelligence. To understand evolution requires more mental exertion than the ignorant masses are capable of, and that is why they want to suppress it as that class comes more and more into power.

This state of affairs I believe is due largely to our educational "racketeers," whom you refer to incidentally. In their strenuous efforts to make more jobs for teachers (in which they encounter little or no opposition, but only inertia and friction), they have put a powerful tool or weapon into the hands of too many people who do not know how to use it properly, and these are electing men of their own stripe to public office. They also put too many students into and through college who have not the mental capacity to absorb an education, and when these get through they are still morons. Many are even pushed on further, through the Ph.D. mill, regardless of qualifications, just to satisfy present arbitrary requirements of autocratic committees.

From the state supervisor of education in a large Eastern state:

Your address has been quoted as having the following sentiment: "The present restrictive influence of organized religion on the teaching of the best in biology is intolerable. The tongues of the traditionalists are heard not merely from pulpits, but they echo also within our schools—curbing the tongues of biologic truth."

It seems unfortunate that you did not make more definite quotation as to the specific acts or attitudes that "restrict biology," for such general statements lead the uninformed to gather that all religions fear to face the new facts brought out by science. You erred in not stating some instances of such restrictions.

What is your reaction to Rev. Schwitalla who stated at the same meeting, concerning your paper, "If it is true that organized religion is a restrictive influence on the teaching of the best in biology, the statement cannot be universally applicable to all schools conducted under religious auspices.

"One need not repress biological fact concerning the origin and reproduction of life in order to teach religion, but both must be taught simultaneously to give the student an adequate concept of the world in which he is living and of the meaning of his life.' You speak as if there is a real and mutually exclusive teaching by biology and by religions. But you fail to make any item clear as to the basis for such conflict.

From a professor (biochemistry) in Missouri:

I was sorry I could not have seen you after your address to the zoologists to express my appreciation and approbation. Have you a reprint of the talk that I could beg of you? The form given it in *The New York Times* was good enough so that I read it aloud in the family. I wish the subject might have larger publicity in quarters where it would challenge those who are influential in secondary education. For example, it would delight me to bring it to the notice of the dean of our College of Education. Could an abbreviated form of it, with emphasis on the second half, find place in some educational journal?

From a professor (zoology) in the University of California, Berkeley:

I have just read with great interest the second part of your Section F address in *Science*. I think you have called attention to some vital points in the teaching of zoology and I hope you will not be too much bored by a few lines from me bearing on part of your comments.

I had begun to think that you, like so many eminent zoologists, overlooked a very vital factor in the lack of interest in the zoological sciences or appreciation of their value in our public schools, colleges and universities when I was relieved by your pungent reference to "an elementary text-book published in 1934." . . . I hope your paper will bear fruit in that direction.

From a columnist's column in the Christian Science Monitor, *January 4, 1936:*

Among the interesting published accounts of the proceedings of the annual meeting of the American Association for the Advancement of Science, held in St. Louis during the last few days, is that reporting the presentation and discussion of a paper by Dr. Oscar Riddle, of the Carnegie Institution. Dr. Riddle is a biologist, and a learned and inquisitive one, no doubt. It is to be regretted, therefore, that instead of utilizing his opportunity to inform his audience, and those who took the pains to read such portions of his address as were released for publication in newspapers, of the progress made by those engaged in biological studies and research, he sought to place upon others, and particularly the churches and schools, the blame for failing to appreciate the importance of the work he and his fellows are pursuing. He is quoted as having declared that "the restrictive influence of organized religion on the best of biology is intolerable."

Many thoughtful persons even among those who may have sat within sound of the speaker's voice probably are inclined to a more temperate and reasonable view of this lack of interest in even Dr. Riddle's theory of what he refers to and seeks to establish as the progress, through evolutionary processes, of a material creation. . . .

If the truth were known, as I suspect it is, it is quite probable that with the years there is less and less concern felt by either the youth of the world or their elders as to what natural scientists regard or conclude to be their place in nature. More and more clearly it is being realized that the origin of the real man is not in the molecule, or in a drop of sea water, or in the rays of the sun. It is of the origin and destiny of this man, and not material man as he is pictured, that the thought of even those who are not classed among stubborn religionists referred to by Dr. Riddle turns in serious contemplation.

But it is interesting, especially as one regards the fervor and determined eloquence of this scholarly defender of the materialistic faith that the observing layman is compelled to the surprising conclusion that through it all he has read or listened to nothing more than the expounding of a more or less interesting theory. The theory itself is as old as Darwin.

Thoughtful men and women, even those free from "ignorant prejudice," have gained that advanced position in thought where they do not any longer regard existence as an accident or themselves the children of chance or the product of some designed or accidental combination of chemicals or other inanimate matter.

From a physician in New York:

Your address before Section F of the A.A.A.S. at St. Louis was the kind the country has been urgently in need of for a number of years. It was thrilling indeed to know that there was at least one man with a scientific reputation who had the courage to condemn publicly those who would reconcile science with superstition and by their influence and prestige support those forces of darkness which would keep the world in ignorance. You certainly deserve the applause and gratitude of all those interested in the progress of science and freedom of teaching. I hope your magnificent example will serve as a stimulus to others to enlist in a great cause.

From the superintendent of schools of a city in Minnesota:

In my saner moments I am quite philosophical about the general uselessness of martyrdom, but I am nevertheless quite unable to "pull my punches" on a number of subjects which are anathema to a considerable portion of the people I serve. If I must sound off, I want to be sure of my ground, for my own sake. Your address answered a number of my questions.

I can agree with everything you say about the present situation. There is a long, tough struggle ahead of us, and it is heartening to know that there is at least one voice crying in the wilderness. The solution, after all, is largely in the hands of men like you. Your statement that Darwin and Huxley met men's minds more effectively than the present crop of scientists is precisely to the point. I believe that your call for an active, aggressive campaign is sound. There has been too much pussyfooting on this issue. It should be as dead as the Ptolemaic system by this time.

I hope that you will have occasion to reiterate your position on evolution and the teaching of science frequently, and that we may get some real action as a result. We are more fortunate in Minnesota than in many states, but the situation still leaves much to be desired.

From an editor (biologist) in Washington, D.C.:

I want to congratulate you on the "Confusion of Tongues." You have put your finger on the basic problem of our times, it seems to me. There is no use talking about eugenics, a "planned society," or any of the other hopeful ideas until we succeed in educating the educable part of our population in the fundamentals of biology.

I find in checking up among my friends that not enough of them read the important things in *Science*. I wonder if you would be so generous as to let me have two or three extra copies of this reprint. If a few more are available I believe that I could place them from time to time where they might do some good. Personally I feel pretty pessimistic as to what can be done, but if enough intelligent people realize that the beginnings of effectively applied human wisdom involve capturing that stronghold of institutionalized reaction, the National Education Association, located at 16th and M Street in this fair city, possibly a beginning will have been made.

From an editorial in the Danbury (Connecticut) News:

Whereas Darwin brought his idea of life's beginning back to inferior sorts of animal life, or at most to forms of vegetable life precedent, Dr. Riddle insists that life and consciousness originated in the action of sunlight on water, carbon dioxide, sugar, nitrogen and similar inorganic substances.

As a hypothesis there is no objection to Dr. Riddle's theory. As a fact, it has nothing to sustain it. Darwin demonstrated that lower forms of life evolve into higher. Dr. Riddle does not demonstrate the evolution of inorganic matter into life and mind. He can produce no instance of it. . . .

We cannot assume that the sun originated in a fortuitous supply of water, sugar and such like. We do not know how the sun happens to exist. It is easier to assume that the forces which ordered the sun are at least nearer to the sources of life and mind, and that life and mind are rather the cause of the inorganic materials, than they are of life and mind.

This latter idea, borrowed from revelation, seems to be supported by the circumstances that all of Dr. Riddle's materials resolve into those electrons, protons, neutrons and such like items which can scarcely be distinguished from electrical charges.

Dr. Riddle's inorganic substances are certainly built up of those electrical elements, hence he might as well say that this is where life begins.

The electronic substances may be as far away from life and

mind as a stone wall is from them, but in these identical bricks of stuff which is neither matter nor not matter, like Dr. Riddle's virus which is "neither living nor not living," we seem to be coming close to life and consciousness, which probably are the antecedents of everything in our world.

It is true that religious leaders have sometimes discouraged biological investigation. It is also true that from religion, through revelation, is derived that hypothesis regarding the origin of things which has most evidence to sustain it.

The universe is mentative. The inorganic proceeds from the spiritual and creation was "by the word."

From a professor (chemistry) in Columbia University:

Permit me to thank you for "The Confusion of Tongues" which I have read with interest and pleasure, and with which I am sure most scientists will sympathize heartily.

From an industrialist (engineer) in Pennsylvania:

Now that I have had some time to think over your article in *Science* for January 17 and 24 I still have the same thought: The next logical step is for you to write a suitable book, have it edited by a journalist who knows good English but *nothing* about evolution (except perhaps the evolution of poker chips) and can write for a mental age of 13 to 15 years.

From a professor (geophysics) in a university in North Carolina:

I have just read your paper in *Science*. Permit me to congratulate you. I have long suspected that in and before World War days high school biology training was superior to that of today. I am rather far from your problem and subject in training, but understand your difficulties and sympathize with your objects. Your science is the most important one.

From a woman official (department of education) of a New York museum:

I have read your address with considerable interest and agree with you that the greatest cause of lack of biological training in our schools today is the stand of traditional religion. Just as long as legislation can be controlled by religion and superstition, we shall not see real biology teaching in high schools and colleges in many parts of our country.

From a professor (zoology) in a college in Ohio:

After a year's discussion, a committee consisting largely of deans and of non-scientific men—with the viewpoint of the Teacher's College fairly prominent—pushed hard and adopted a new "General College" curriculum, which changes greatly our first two years of student instruction. Though it has its good points in integrating some of the instruction it has some bad features in which the science departments suffer most. One such loss relates to dropping the requirement of a laboratory science (8 credits, a one year course) for graduation. This is "replaced" by a purely lecture course—of "survey" type—which presumes to cover all the sciences.

Your reference to weak or insufficient biology teaching touches mostly the secondary school, but your points are decidedly pertinent to us.

From a professor (geology) in Connecticut:

Some time ago I listened with interest to a tirade against science by one Father Gillis. I secured copies of the radio talk and am sending one of them to you. I have read your paper in the current issue of *Science* and know you will be interested in this as further evidence of what is being done.

I consider this attack made in "The Catholic Hour" as definitely dishonest and am wondering what, if anything, can be done about it.

From an A.A.A.S. member to the Minneapolis Tribune:

Your reporter makes Dr. Riddle say on the "origin of life": "The influence of religion in education has definitely and materially prevented the attainment of adequate knowledge of this fundamental problem of science." And again, "The present restrictive influence of organized religion on the teaching of the best in biology is intolerable."

It is rather unscientific that such a diatribe should be injected into a discussion on the origin of life, and especially at a meeting under the patronage of A.A.A.S. For many years we have been a member and supporter of the A.A.A.S., have met some of its notable members, almost all men of strong religious convictions and some active Christian workers. In the days of the discussion over the Tennessee antievolution law, it was said by some of these truly scientific biologists: "We have gone too far in forcing scientific theories against the religious sentiments and faith of a sovereign state whose morals arise out of religion."

Mr. Editor, think of the liberty that religion has given to science. Science, like any other institution, has to be regulated by law, or it becomes lawless, and endangers life, property, government and all free institutions. Thus it will be found by historical study that many of the great religious rulers were the very best patrons of the scientific advance of their age. Remember Germany under the Hohenzollerns from Frederick down, all remarkably religious, all scientific.

Karl Marx was a scientist simon-pure—he was nothing else. The political and social science, as he taught it, was restricted in Germany, and chiefly for religious reasons; and what a blessing for Germany. There is hope for Germany today in spite of the penalty of Nazi rule. The Marxian political and social science had fell swoop in Russia. All religion is banished from education. There is no tyranny in all the ages like the bloody tyranny of a lawless science. It out-Frankensteins Frankenstein.

Dr. Riddle must be reminded that religion has its data as to the origin of life on this planet, and when we remember the antiquity and authority of the documentary evidence of this, no truly scientific mind can ignore it.

Religion does not combat science. It will and ought to combat theories and assumptions that are employed by pseudo-scientists to destroy the foundations of religion. We hesitate to say it, but we believe it is true that present-day science is far more bigoted in some of our universities and foundations than any present-day religion.

From a chemist in the Bureau of Standards, Washington, D.C.:

Ignorance and bigotry are still widespread, even in our "enlightened" land, and these elements do not hesitate to apply economic and social pressure to subdue those who would undermine their superstitions. They would use the methods of the jungle if they dared to do so, despite their boasted Christianity.

Though a chemist I have had some training in biology, including a course on "organic evolution." And while, of course, I cannot speak on the subject as an expert, I agree with every word that you have said in this paper. Truly, as you say, intelligent men no longer debate the reality of evolution, but merely the method and the course of evolution. I was particularly glad to note that you called attention to the essential dishonesty of the argument that a belief in evolution is quite compatible with "a" religious faith, and the fact that renowned physicists and astronomers may be quite unqualified to speak authoritatively on biological subjects.

I believe that it is high time that scientists assume a more militant attitude against the forces of ignorance, superstition and priestcraft which, in order to maintain their racket, presume to tell us what constitutes "real science" and arrogantly obstruct progress in the only field which gives promise of any kind of "salvation" for the human race.

From a Universalist pastor in Maine:

We need to remember that the first speck of life was a creature that had to be supplied with organs necessary for the continuance of its existence and enable it to leave descendents. A chunk of matter, regardless of its composition, could not be a creature of this kind. The first tiny life that was capable of survival and propagation was far too complicated to have originated in a dead world by chance.

From a consulting chemist of Chicago and New York:

The trouble with these "anti" laws is that cowardice permitted their enactment, and cowardice will tend to keep them on the books along with the other Blue Laws.

Education is the only answer and you say that is slipping in our schools. Maybe the daily press, even if it does not really do the work of the schools, will whip them into line.

The Church was a quarter of a millennium getting around to the removal of its ban against Galileo's teachings. But with "The Confusion of Tongues" making the front page and getting columns of space, we ought to be able to clip a little off that time.

From an Englishman living in New York:

From the latter section of your article I deduce that the educational system in the U.S.A. is neglecting this vitally necessary study [biology] in its schools. Having been educated otherwise in England I, myself, cannot write with authority on this matter. My recollection of the English system would prompt me to say that mouldy, insidious tradition has England more in its smothering power than it has America. This is not surprising when one contemplates the English scene and realizes that the whole foundation of the English social and political system is *tradition.*

The difficulties encountered by curious souls such as myself in a search for information portrayed in a non-technical manner tend eventually to discourage a search for *truth.* This reason quite probably applies to the scholastic portrayal of biology in this country. From chance remarks made by various high school students of my acquaintance, biology is generally taught by staid and strictly technical professors, who teach as they think, unemotionally, and in so doing fail to use the most effective weapon they possess; that is, they shun the human interest and romance of the story they have to tell. If unobstructed, and told with warmth, the stories of the "life sciences" would seem to have few rivals.

———

What the reader could expect to obtain from the preceding assembly of criticisms was forecast at this chapter's beginning. It remains for the writer to lay claim to something that this unasked and partly quoted correspondence has meant to him. That he treasures it may not be doubted. But it also well supports some statements which were made in earlier chapters without full documentation. There is importance in the fact that several recognized authorities thus voiced their great concern—or their dismay—at a weakening, a smothering or a prostitution of our educational procedures and possibilities; at the tragic failure of citizens to see and defend the sources of their liberties; at the limited extent to which scientists themselves are either "thorough-going" evolutionists or sensitive to the suppression by religion of the full and free thought on which democracy and an advanced society must rest.

Said at this point, it could be put as the joint assertion of a distinguished group that, in the America of 1936, no other class of professional men could match the clergy in the amount of misrepresentation and untruth that it continually spreads to all who will listen. And no one joining in that assertion would think of applying it to all clergymen. Well known, important and highly hopeful is the circumstance that a few pulpits venture to lead in the earthy fields of actual life and common sense. Somewhat larger numbers of the clergy can and do limit their traffic in untruth.

There is also something else. This long ride down the dusky valley of criticism has been richly rewarding because it has led this pilgrim to a theologically trained editor who can speak his meaning clearly and unhedged: "The universe is mentative. The inorganic proceeds from the spiritual and creation was 'by the word.'" Our fellow scientists, right and left, who hug the view that there can be no conflict between science and religion may—for fit—try that one out on their own science.

Finally, a situation that was already apparent was put by this large correpondence into the foreground of this writer's consciousness: never again can any or all of the organized religions hope to attract more than a few of the best-informed minds of any educated people. Among the growing group of persons best trained in the key sciences, the organized religions can only rarely salvage even a belief in a nearly unrecognizable God. Further, a widely shared opinion regards the creeping invasions of government by religion as a foremost threat to American ideals. Or, said more comprehensively, when governments do anything more than protect believers against political disabilities—as asked on political grounds by Thomas Paine—they elect to fortify an obsolete structure, and to pursue a course directly opposite to that traveled by a significant part of the intellectual leadership of the United States.

15

The Broader Battle

In spite of the lull in the storm, the final battle between the mediaeval and the modern systems of thought is yet to be fought out. The affair of Galileo was a matter of outposts; that of Darwin a heavy skirmish; but the main issues, in spite of all smooth words, remain unreconciled.—*Julian Huxley*

The genuinely modern is still to be brought into existence.
—*John Dewey*

POET AND SCIENTIST, EACH RELEASES HIS CRYSTAL OF thought to his fellows. That of the poet may wear a shine and dress that put it—true or false—past forgetting. The thoughts of the two are sometimes in severest conflict; he who collects the distillates of mind must further sort and appraise them. Thus Goethe insisted that "gray is all theory and green is the rich tree of life." But no scientist could accept that vivid and beautiful expression. For, as biologist Dahlberg has noted, the *facts* of everyday life are of relatively little interest in themselves. The leverage, value and creativeness of the scientist center precisely in finding *theories* to explain the seemingly disjointed facts; only the later task of testing those theories may become a gray and monotonous chore. So, in all conscience, a confirmed theory is a well-rooted and a blossoming operation of man's mind; and the gods would have to do their best to outdo it.

Men—or surely those who lead men—will have to make up their minds on how much they care for truth. Then that leadership must offer it to an age or generation. The Western world has reached the point where that decision must be made, and where truth and reality must become accessible and acceptable to peoples. This is the central, insistent and prodigious need of our age; incidentally, it is also the ample theme and entire reason for this book. Failure or faltering here means the extension of instabilities much too dangerous for peoples who must meet the impromptu stresses of a crowded world and an industrial age— for peoples acquiring almost cosmic power while still wearing the mental habits of the child. When beauty erases truth, may men yet flirt with beauty? When the presumed good outshines it, shall men still embrace their good? Whoever searches for a unity of goal and master can doubtless find it—be it beauty, good or truth. But those who thus elect truth must also cherish its unbending quality. Truth can wear no identifying label; it may require a price; it usually must be sought before it is found; and—though now compulsory for society—the *individual* may often in complete safety either accept, neglect or reject it. There is little wonder that no people seems ever to have worshiped truth, despite the fact that the cultural levels of peoples to date roughly correspond to their grasp of reality. Truth is the unpaved detour usually bypassed in all religion. Truth is a master too unmanageable for him who himself wants to look like a master in the eyes of obedient followers. Truth seems most at home in a hardy race of men. Of which fibers is our tribe built?

We need not think that facts and insights must fill and guide the minds of *all* persons in order that such knowledge may render all men a magnificent service—and even build a new society. That view is disproved by the familiar detail that absolutely all of us are born and pass some years quite without a grip on reality. And yet, throughout that erratic and vagabond period, we all profit immensely by the precise medical and hygienic knowledge of a small group of persons and by much else that is merely custom and outlook in our community. These surpassing benefits also attend the entire life spans of those few among us who must remain wholly and forever mentally incapable or little capable of grasping reality. Rather must we acknowledge that facts and reality serve humanity best and most through a twofold medium —by specialized training of groups of individuals, and by a widespread and common social heritage which lifts one whole culture above another that has a feebler grasp on reality. For example, this lifting power of community culture alone will doubtless result from the mere loss of a common view that this life is but one of a series of reincarnations—a present view of hundreds of millions; and, likewise, a loss of that other view that the sin of a "first" man is passed on to all, and that a Savior was required to redeem a species—a view of several other hundreds of millions.

Clearly, then, although a few individuals must first explore the maze of fact and implication, and though they and others must fight valiantly to establish their findings in social policy and community life, such is the combined power of leadership and of established custom that, when this is done, the great masses of men are thereafter spared their share of this lifting chore. From a platform thus raised by leaders, plus the hard fight that, in democracies, may now require a majority vote, the minds of men of following generations become heirs to a new level of existence. Then, at this new level, human diversity and individuality may find opportunity for rich and harmless expression. Yes, for the masses thereafter—as with all young children and the mental misfits cited above—beauty may then without peril to others become more satisfying than fact, contemplation and poetry become superior to test, activity become more gratifying than knowing, and love often become personally more consequential than truth. On platforms thus raised—but on such platforms only— both the best and the diversity of the truly human in us can breathe and grow.

The materials already examined in this book lead to four conclusions: (1) Firmly entrenched attitudes of organized religion toward evolutionary thought constitute a dangerous cultural impasse in this generation. (2) Always incomplete, science has now advanced far enough to make any imaginable view of the supernatural unacceptable to a high proportion of best-informed minds. (3) The organized religions of the world all render many notable services to their followers while consuming and hoarding much wealth and while enlisting and misdirecting much human effort. They also repress thought, freedom and progress while pointing individual men and society to unreal goals and socially ruinous hopes. Their relation to modern society is such that government should protect fully their right to believe but grant them no single additional favor. It is of still greater importance that government ensure all other citizens a total aloofness and uninvolvement in religion, and ensure society full and free access to the tested knowledge of its time. (4) The acceptance of now available solid knowledge of ourselves by all or most leadership, together with a substitution of human purposes for supernatural purposes by a majority of the race, is mankind's big unfinished task of thought about itself. This unmasking task is sizable and tough, not because of the limitations of contemporary knowledge or of science, but because of the limitations of man—his assorted endowments, grooves, entrenched institutions, habits, and earthly origin. Further comment on two aspects of this unfinished task is warranted.

THE PRESSING CONFLICT

What course may one expect the conflict between naturalism and supernaturalism to take in the Western world during the decades or centuries immediately ahead? If the word "supernaturalism" merely implies belief in a personal God and in a life after death, it will be recalled that Chapter 12 provided evidence for a prevailing but moderate retreat from that position, at least in Europe; this, despite a relative increase of church membership in some countries. In America, that retreat was observed to be led by scientists and by college students, and to be more pronounced among Protestant than among Catholic peoples. But in most Western nations the number and power of Catholics are showing a steady increase relative to Protestants. It was also noted that both Protestant and Catholic theology and pulpit preaching already show new movements toward what are called neo-orthodoxy and neo-Thomism—opposites of the "liberal" trend under which the observed slight or moderate retreat from supernaturalism has occurred. This newer movement may or may not affect the rate of recruitment to the smaller forces of naturalism. Again, within the United States, such things as the growing movement for religious instruction in the public schools, the very recent and still continuing establishment or expansion of departments of religious studies in colleges and universities, and the momentary beginnings of endowment of sectarian colleges by industry, are fresh and serious threats to the maintenance of even the present modest level of naturalistic views. In sum, this survey of the influences now in evidence and hitherto discussed leaves in some doubt any further increase in the percentage of those who, in either Europe or the Americas, accept naturalism.

Hitherto unconsidered, however, are the probable effects of the still spreading political doctrines of socialism and of communism on belief in the supernatural. Whatever the other threat or the other value that may attach to these movements, there can be no doubt that an unusually large percentage of those who accept these political and economic views also reject supernaturalism. That this is true is indicated by the statistical data supplied in Chapter 12 for the political parties of France. And those same statistics further show, contrary to the daily pronouncements of many politicians and advocates of religion, that belief or disbelief in God is not actually an essential of either communism or socialism. The Gallup poll in France indicated that 50 per cent of the members of the French Socialist party and 17 per cent of the members of the Communist party are believers in God. Of the total population of France, only 66 per cent were classed as believers. Communist and Socialist may or may not believe in God. Another statement of objective fact regarding these political movements would note that they give evidence of further expansion in Europe, Africa and Asia, and also in the Americas south of the Rio Grande. From these sources therefore, as from the colleges, it is probable that supernaturalism will receive further and heavy blows.

At this point one may ask: What part does wealth play in the struggle against supernaturalism? Many thoughtful persons of today and of the past hundred years have maintained that wealth is a powerful foe of change in both politics and religion. They have not failed to cite good evidence for this view. Within the United States at the present moment the teaching of Fundamentalism is being rather heavily endowed by wealthy individuals. And even the power of quite modest wealth is otherwise and effectively expressed. In widely scattered communities, securing or holding many a minor job, for example, is still contingent upon willingness to join a particular church. If such practices continue or if they develop further, it may be necessary to conclude that wealth is aligned with the forces of supernaturalism.

It is, of course, a commonplace of history that in Catholic countries like Spain and Mexico the large landowners and others of wealth have been for long and now are leading supporters of the church. In England similar groups are the prime props of the established church. But it is also true that with almost each succeeding year of the past half century a few men of very great wealth have rendered a magnificent, enduring and growing service to all the sciences, to all learning, and indeed to much of education and to general social progress. It is the view of this writer that in advanced countries these benefits perhaps outweigh the harm to intellect accomplished by all contrary uses of personal wealth. The various foundations associated with the names of Rockefeller, Carnegie, Harkness, Guggenheim,

Ford, Nuffield, Kaiser Wilhelm, Carlsberg and others sufficiently recall this incalculable service. Nevertheless, wealth is power, and if the balance of that power should be used to support supernaturalism, the consequences to the struggle ahead are not negligible.

Again, quite recent events (1953-54) suggest that one of the great foundations named above (or its fund-granting subsidiaries) may now be curtailing its usual immense aid to basic science under religious pressure. Precisely at this period direct religious criticism and pressure was exerted upon the Rockefeller agencies to terminate a ten-year support of the Kinsey Institute for Sex Research. Whether related to this pressure or not, that support ($40,000 yearly) was withdrawn as of June 30, 1954, and almost concurrently other grants of some two million dollars were made to religious education.

It was earlier indicated that only the unwary may assume that books and the printing press will win this particular battle in thought. First of all, already in the United States the article or book manuscript that effectively attacks religion may meet with insuperable difficulties in finding a publisher. Next, readers of earlier chapters will be reminded of some or much distrust of the printed page—distrust arising from the circumstance that the more potent of those pages are censored and scissored before they are set in type. The vital facts of human origin and existence, and the full implications of those facts, should find their place on pages where they could exercise their greatest power—in textbooks for the scores of millions who use them in schools and colleges. But in these most important of all books the pages that treat these topics usually are rewritten and edited by the politician to conform with what he thinks his constituents want, or perhaps by a sales department to conform with what it thinks will sell in a belief-ridden nation. With too much cooperation from authors, castrated script is now spread as widely as the world of crippling belief. On these most vital topics the total of illuminating information that is obtained from schooling—and from all other guarded avenues of news and learning—is too slight and too shallow to permit the public to know that unfamiliar depths exist. In the present battle for freedom to educate, there is nothing more immediate, and nothing on which the outcome depends upon the active participation of more kinds of citizens, than the fight to secure and to use textbooks that explore the meanings of new knowledge, instead of ravishing them to protect an ancient faith.

Many who well understand and discard supernaturalism make the big mistake of taking it for granted that the logical weakness of that belief will eventually destroy its popular support. That view, too, may easily prove disastrous. It can be considered probable that vulnerable and spurious arguments for supernaturalism are not a serious handicap to the perpetuation of that belief. This conclusion rests on the fact that the already well-entrenched churches maintain a large army of paid and skilled advocates, and on the more decisive circumstance that the majority of human beings cannot easily distinguish between the winner and the clever evader of the crucial but not-too-simple arguments. This writer has read articles that so skillfully present the case of supernaturalism—presenting it in reply to the printed word of those who most competently attacked it—that perhaps not more than one among five or ten adults would detect either error or crucial omission in the argument. One may conclude therefore that neither the censored printed page, nor the logical indefensibility of supernaturalism, nor again the coexistence of these two things, is likely to defeat supernaturalism. Something more is clearly required. Only the active opposition of informed people—including scientists, educators, writers and enlightened college students—can eventually defeat supernaturalism. And it is only through definitely aggressive action that even our state schools and colleges can be made or kept free of meddling and dominance by supernaturalists.

Matthew Arnold said that "knowledge and truth, in the full sense of the words, are not attainable by the great mass of the human race at all." The statement seems unchallengeable. But culture already has advanced the rational thinking and much of the deportment of the masses. The attainment on a national scale of a still more advanced culture, in which a sufficient proportion of the masses may weaken their urge for the supernatural, is not denied by human experience to this moment. The persistent advance of that culture should permit insight and truth to rule, although not a possession of the majority. An alternative to attainment is a respect for attainment.

The areas of truth and of existing conditions reviewed in these chapters have led to the view that the enduring hope of mankind lies largely in becoming free to educate, in learning how to educate, and in educating endlessly. Thus the whole problem is greater than the part concerned only with the hard task of becoming free to educate. And it may be well to point out that this is no lonely view. That service, and still other merit, attach to these words of writer and editor Ferril:[1]

Our only course, whether it ever works or not, must depend on fullest exploitation of our best minds to offset if possible the damage our worst mob thinking does to us. School must be a place where the student will be obliged to learn much that he doesn't like about the long, long faring of mankind on the million-year road that led to the modern mob 150 years ago. He must be compelled to reconstruct what some of the old-timers like Jesus Christ or Leonardo would be saying today if they'd had the experience of Niels Bohr. If the Jukes children can't take it, let their only penalty be our refusal to lower our educational standards to their level. Let them hew wood and draw water while we go ahead with the children who can take it. The Jukes themselves will profit.

Whatever else may be required of future primary- and secondary-school training, in our civilized societies and in each generation it would seem clear that the emphasis must go to civics and manners and to science and discipline. What rich blessings would flow to a nation whose lightheaded youth had been firmly taught the not-too-clear *duties*—civic, social and economic—of their particular day and locality! Ultimately, what a lift to education itself once it can bypass the gods and unreservedly teach the unqualified primacy of man and society! Nevertheless, here and there nothing today is easier than for the four above-named central themes to be displaced by several of the more than two hundred subjects now taught in high schools of the United States. The Jukes children ask for and receive the larger half of these. Play and sport, and a dozen dodges of mental effort, are frequently accorded central instead of peripheral posts in this competition for emphasis of subject matter. Which peoples will be able to attain a sensible perspective of these competing items —some centered upon preparing youth to help an insecure society to survive, and some trying to make the schoolday a bit more agreeable to Buster Brown or perhaps to his indifferent parents? When will it become acceptable, and be sustained by

[1] Thomas Hornsby Ferril, "Western Half-Acre," *Harper's Magazine,* December, 1945.

actual school practice, to tell teen-agers that the reason for the seemingly harsh and undemocratic laws that compel them to attend school is nothing less than to prepare them to preserve the national or collective life and to explore possible paths to personal happiness and the survival of the race? When can they be told firmly that mankind must train to live together? Here indeed is spot and place for character-building in education. Whoever thinks seriously about social problems—about citizens capable of both honest thought and useful effort—must look hopefully to the schools. But without at least one entire generation's free and enlightened attention to central areas of education, and without the permanent enlistment of a fair proportion of a nation's best minds in its schools, that hope is not likely to be fulfilled.

For Americans, some parts of the needs and obligations of citizenship—of training in civics—were discussed several decades ago, by historian Francis Parkman:

And now it remains for her [the United States] to prove, if she can, that the rule of the masses is consistent with the highest growth of the individual; that democracy can give the world a civilization as mature and pregnant, ideas as energetic and vitalizing, and types of manhood as lofty and strong as any systems which it boasts to supplant. . . . She must shun the excess and perversion of the principles that made her great, prate less about the enemies of the past and strive more against the enemies of the present, resist the mob and the demagogue as she resisted Parliament and King . . . and turn some fair proportion of her vast mental forces to other objects than material progress and the game of party politics.

If Parkman were writing today he would surely note that "the rule of the masses" is being displaced by the rule of pressure groups; that democracy itself is under heavy attack; that very few enemies of the present may be even mentioned in the nation's schools; and, though confronted by nearly insoluble problems, her vast mental forces are leashed and undeveloped. Serious training for citizenship would seem to be a seriously neglected responsibility.

On the place of science in education this writer will add nothing here. And whatever may be the rightful place of evolutionary science in particular, this book need not remind readers that the long-delayed contest that would attain it is immediately around and ahead of us. That matter merges with the practical struggle to become free to educate—to brush away the restraints of organized religion. Without this freedom to educate, our ducking and shadow-boxing in the classrooms of the world may continue to the end of time. In many countries there is now no slight hope that those parts of the social and life sciences that are vital to thought and social insight will be taught in significant amounts in the foreseeable future. In those countries there is no hope of soon breaking the strangle hold of religion. And perhaps in no nation of the present world is an easy or an early victory predictable. The prime certainty is that the tough fight is on.

Of the existing social and political need for science, former university president and publicist Glenn Frank wrote, in 1926, in his syndicated newspaper column, as follows:

We must somehow find ways and means for thrusting the results of scientific research into the stream of common thought and for making them the basis of our public policies and social procedures. . . .

The victories of intelligence will remain insecure and liable to periodic defeat by strange revivals of superstition and intolerance unless we can manage to match the evangelism of superstition with an equally effective evangelism of science.

Because of all this, the most useful man in America today is the occasional scholar-genius who combines the burrowing qualities of the mole with the singing qualities of the lark, the man who is master alike of the science of research and of the art of expression.

This clear call of publicist Frank is a call to everyone. Where public policies and social procedures are involved, no citizen escapes a share in arriving at solutions. The informed and the thoughtful should find avenues to helpful action if those qualities are still effective in governments of today. The entrapped educator and teacher can *add* to his restive effort; and, with support from others, his leverage on thought is multiplied. The scholar is in the midst of a conflict that either buries or crowns his ideals; he is placed for effective action and his responsibilities were never greater. Scientists, some of them especially, are at the center of the continuing battle whose boundaries may seem indistinct but whose defeats and devastations are beyond doubt.

Will these several groups find further ways to cut crooked thought of its persisting shreds of respectability?

Let it become known that the scientist's most worthy vision is his when farthest from his laboratory. For it is then that he may see, through Galileo, the human mind tangling with orderliness in the skies; the change that has come over thinkers and writers because of Darwin's work; the effect of Freud on the novel, the drama, and on medicine; and, clearest of all, he may then see the kindling certainty that science is not mainly for him but for everyone. Let this scientist have will as well as power to contest the rape of the common mind—to meet with aggressive strength the rough challenge to the spread of truth.

At the moment, the reactions of the individual citizen to suppressed and tethered thought seem limited to restiveness, rebellion and complacence. Daring and adequate leadership for a generation graced with sane ideals beyond self, and with little love for appeasement, is the supreme need of the hour. While it continues to be true that complacence is promptly rewarded by more and surer promotions, by prized positions, and by other financial advantage, the ranks of the restive do not rapidly overfill. When or as those conditions of personal profit from complacence disappear, the religious outlook will weaken. Where or if those conditions remain or worsen, the scientific outlook can only do battle or wither on the vine.

It was probably in 1930 that English philosopher-historian Dampier[2] revised the last section in the first edition of his book to say that, apparently, "the greatest danger to science is the growth of such movements as popular anti-evolutionary 'fundamentalism' in the United States." In the still later revision of 1942 the status of greatest menace was accorded to "the rise of Nazi power in Germany." If Dampier were writing in 1953 he would probably reassign, for quite similar reasons, that unenviable post to now current communism. To the present writer it seems, contrariwise, that a completed "blueprint" for the Fundamentalism and supernaturalism of our times would indicate that these have remained more formidable and more durable enemies of science and society than Nazism and communism.

This writer well knows that a great many scholars of our day—

[2] Sir William Cecil Dampier, *A History of Science, and Its Relations with Philosophy and Religion* (New York: The Macmillan Co., 1943).

including scientists in both academic and industrial pursuits—are restively conscious of the conflict between the scientific and religious outlooks. This certainly holds also for very much larger numbers of informed and aroused people who lay no claim to scholarship or to science. It must nevertheless be granted that all this kindled consciousness of a minority is still held closely on leash. Only on many college campuses and in certain socialist communities are high percentages of people who think in this way thrown together. And these two dissimilar guilds are now of highest strategic importance in the hard fight against supernaturalism. The campuses are especially significant in this mounting struggle, despite the fact that the student fraction of that population has not yet shown the confidence and urge that spark the turning of thought into action. But a long and momentous step toward a genuinely modern society will have been taken when large numbers of these student groups do become aggressively active against the tethering of thought and learning. That date and procedure may mark the rift between two main epochs in total human history—presumed supernatural purposes having dominated the one, and human purposes the other.

THE ENDURING EFFORT

An upheaval in Western thought concerning both God and the Christian scheme of salvation began about two hundred years ago, and already that upheaval has proved itself of vast consequence to the society of today. All too frequently, however, those who refer to that eruption of reason neglect to note that only or mainly the top of any culture then shed some part of its skin of faith. Within the past two hundred years, the masses of no nation have turned against supernaturalism—though various peoples have revolted against an unbearable and oppressor church. Nevertheless, several peoples—both great and minor— have founded nations of a new and tentatively secular type. Some writers declare, despite these demonstrations of immense though partial social release, that "scientific materialism has failed," "naturalistic philosophy has failed," "the rejection of religion has failed."

Two hundred years ago the leaders among Western men might surmise something, though they could know nothing, of the natural laws that led to life—to living things including man himself. But many of them became aware of the real character of our solar system, and of the liberating fact that natural law rules both there and in the nonliving world on the earth around us. At that period, too, literary investigations of the New Testament deprived that document of much of its time-honored halo. The total of this insight and lively knowledge began at once to split all Western thought on religious faith. And that firm knowledge soon powerfully helped to bring truly basic *political* and *social* gains to Western peoples. In America, it produced leaders who—though Deists—had renounced the Christian scheme of salvation; leaders who could write a Constitution without a God; a nation which, in an early treaty, could note that "the government of the United States of America is not in any sense founded on the Christian Religion." In France it supplied leaders for a revolution against deeply entrenched feudalism and church power, and for liberty, equality and fraternity. Clearly, therefore, the recognition of natural law in the inorganic world was a source—one of two intellectual parents—of very much that is best in now current Western society.

Only much later, through Darwin and associated developments, did investigation effectively remove the foundations of theological structure. Only then could intellectuals cleanly sweep the supernatural from thought. That decisive extension of solid knowledge began with Darwin nearly a century ago and continues to this moment. It carried the rule of natural law squarely through all that lives, through man and society. In doing this it uprooted the essentials of the ideas of Providence and of God —however much or little this is or will be recognized by the masses of human beings. "It was geology, Darwin and the doctrine of evolution, that first upset the faith of British men of science," says Bertrand Russell. The same statement would apply to scientists of the United States and most other countries. In few nations other than France had intellectuals already widely adopted Diderot's earlier and more sweeping verdict—now just two hundred years old—that "we must put theology to the sword."

The curious and unpredictable course of popular thinking since Darwin's day well deserves much more than a passing word. A more shattering blow to theology than that delivered by the extended and now established evolution principle is scarcely imaginable. But what, in terms of history, has been its effect on the ranks of the believers? It is now generally conceded that evolutionary science weakened religious faith in advanced nations during the period 1859 to 1900. It is widely conceded also that a reversal of that trend has occurred during the present half century. This book provides many instances of the suppression of evolutionary thought and of religious victories at this mid-century and during the last decade or decades. What light does this reversal throw upon the long contest ahead? Why has the full and growing strength of scientific insight been a weaker liberating force during the last half of a ninety-four-year period? There are two parts to the answer to this question—two in addition to whatever may have been the influence of the adoption by some theologians of a much revised and less crude conception of God. This latter influence, in the writer's opinion, is now becoming important in the college populations of a very few countries.

First of all, the informed reader can little doubt that it was only or mainly after the year 1900 that the total of *community* religious thought—rather than the thought of the trained teacher —became really decisive in suppressing or emasculating the teaching of evolutionary science in the public schools. Public schools were yet to be formed in some advanced countries in 1859, and only a rather small and select group of persons continued through several decades thereafter to attend what public schools there were. However, by 1900, public schools had been established in all progressive countries, and thenceforward larger proportions of the uneducated but religious masses attended them. And by that time the antagonism of evolutionary thought to religious thought had become so widely evident that community sentiment, under increasingly well-organized religious guidance, was put to work to stifle it. New means and abilities both to suppress evolutionary thought and to infuse religion into public education have appeared since then, with each succeeding decade. When the entire core and shell of supernaturalism became challenged by evolutionary science, the temporarily staggered hosts of the faithful slowly but effectively learned to use the entrenched beliefs of the uneducated masses, together with new and ancient law, to block the offending science from public education and communications. These are alarming but unassailable facts.

A second item fitting into this frame is the circumstance that, somewhat prior to 1900, it had become clear that not merely religion but social theory was involved in the evolution principle. Fuller realization of this fact gradually put many industrialists and other supporters of the *status quo* on the side of religionists. They tend to remain there. These two wholly unlike groups, for unlike reasons, prefer and prescribe the application of the wet blanket to the evolution principle at all levels of public education. Not even at the university level will either group ordinarily concede truly free examination of evolutionary and social insights. Finally, it is in rich America, and to an extent unmatched elsewhere on earth, that these two groups, independently of the state, now unite in sustaining—and at least have opportunity to determine the policies of—the large majority of a nation's colleges and universities.

What have been the results of these restraints and compulsions on the educational practice of peoples? Consider the extent to which your community permits your teachers in primary and secondary schools to discuss the reach of evolutionary thought into religious and social performance. Is your state university—or the college of your choice—alive with discussions of these challenging problems by the young men and women who should be wrestling with and contributing to their solution? Are present legitimate criticisms of college education at all related to college restrictions upon the critical examination of the most basic questions of religion, social theory and politics? Is lack of real moral training in *all* schools—parochial and other—at all related to the fact that politics, and the basic scientific facts on which morality depends, can only rarely or never be broached and discussed there? The American who asks these questions owes a most humiliating apology to one of his country's founders. When, with perhaps less thrilling avenues then open before him, Chancellor Thomas Jefferson first invited scholars and students to the University of Virginia—where church and chapel were foreign to the campus—he said: "This institution will be based on the illimitible freedom of the human mind. For here, we are not afraid to follow truth wherever it may lead, nor to tolerate error so long as reason is left to combat it." What may be the result when, because of the flowering of thwarting, jealous, divisive religions among us, the educational program of a nation may not aggressively discuss those precise matters which most affect national and racial survival?

A tree may put its feet suitably in the mud. A man—the being hoisted highest in nature—may not walk upon the winds but his feet fit well on the very surface of the soil. Only a reasonably lifted head, however, can keep them there. The species that has power to plan and to think constructively of its future may now have, too, an amazing and hopeful view of its own remote past. Parts of that prophetic past were thus scanned by psychiatrist Cameron:[3]

It is true that our adaptability is immeasurably greater than that of any other living creature. We are the animals which have taken a chance; we left the shelter of the trees and then the protection of the caves; we played with fire; we crossed the perilous seas; we have dared to leave the earth. Other animals have chosen to adapt themselves to circumstances and have produced protective coloring, speed for escape, armor for defense, and prolific fertility for survival. We have never made that choice. We have come to our problems without ready-made solutions, relying on our ability to work out answers on the spot.

While this has given us great gains, it has been at the cost of great anxiety and insecurity. It has been necessary to build up worldwide reassurance illusions, which in turn have slowed the progress of science. We can see this at a glance in our fairy tales, in the folk stories of powerful figures, in our father-image myths, beliefs, and creeds.

Well fitted to meet major problems of existence without ready-made solutions, man—or one or another of his subdivisions—may prove himself capable of solving more severe problems than any he has yet met or imagined. Certainly he is forever enlarging his collection of instruments—the all-serving mechanical, the social, the political, and even the tools of thought itself—for use on still unmet problems. It is not unreasonable to assume that much of his past and present anxiety and insecurity—the parents of his self-encouraging myths and creeds—can be left behind as his arsenal of devices and a freer intellect shape a society that is more resourceful and productive, more just and more humane. Quite clearly the heavy load of mystical beliefs is already being cast off by large numbers of the better-informed members of his species. Never again can a majority of the best-informed minds of any advanced culture give support or countenance to a belief in the supernatural. Whatever their load of anxiety and inse-

3 D. Ewen Cameron, *Science,* vol. 107, 1948.

curity, that growing group already looks to society or to self for solutions to all social problems and for encouragements to living. "Faith in the intellect . . . is the only faith sanctioned by its fruits," asserted Santayana.

The circumstance that will next be mentioned can be partly introduced with the reflection that few should find it embarrassing that man has surpassed the ape—since so many have so long regretted the evident and ugly residues of the toad and the louse.

Although man typically approaches his most basic problems without ready-made solutions, his natural responses to his more commonplace *social* problems are perhaps not such as to give him an unqualified guarantee of a happy or a moral future. These responses are partly guided by animal inheritances so unlike as to lead men to prefer—and to take—at least two different roads. One of these roads involves effort, is usefully productive, and tests the individual's powers; the other permits inaction or dalliance, or is perhaps negotiable on the back of a fellow traveler. In human societies men of the two inclinations must travel together, and the training and curbs imposed by the tribe seem to have proved that such unlike groups can travel—or, at least, tarry—more or less together. Sloth and obliquity may here be used as names for those gene-controlled traits, ingrained in many persons, which are unfavorable to sociality and moral living. All society is parasitized by such traits. Their presence and transmission by heredity is alone sufficient to disprove the "inevitable moral progress of man"—a fetish of the last century that is a mere relic in our time. To speak or to write nowadays of "the moral law" instead of "a growing moral need" is a sure means of telling that the speaker or writer has not learned the basic realities of evolutionary science. In brief, humanity is compelled to cope with old and fresh problems of sociality and morality though it has a biological equipment that is imperfect for those ends. A human species definitely and reliably adapted to social and moral purposes has not yet appeared on this planet. But the importance of that fact could easily be exaggerated. The outstanding capacity of the human species now inhabiting the earth is its educability. Fortunately that trait has, or at least it once had, a survival value that implies the possibility of increasing it. Partly for this reason man's foremost hope lies in that direction.

Man's political achievements are impressive and they seem

linked to his moral growth. Democracy seems to be founded partly on a belief—or on a hope—that man is a rational animal. The hope exists, though common experience joins with history, psychology, sociology and biology in declaring great lumps of him partly irrational; and though always present are lurking dangers of emotion-led mobs and mistaken views of majorities. But experience too has taught the defects and risks of other forms of government; so democracy accepts the obligatory risks. In theory, risk and responsibility may lessen irrationality, and the exercise of power may train in the uses of power. Again, a more widespread understanding and assertion of human rights has required the adventure of self-government. Though democracy is as yet scarcely mature, who will contend that the long-range welfare and outlook of our species have not been advanced through its tests, trials and risks? And if risk and responsibility have themselves had a positive share in this lifting power of democracy, may not the race hope to deal successfully—from its present higher levels—with still greater risks and responsibilities?

It may be true that more men must become more just. Nevertheless, a further word relates to that matter and to the conclusion stated above. When men of later generations shall have increased their politico-social training and understanding, and when some social frames of behavior shall have become both more sane and more acceptable, problems of morality may not prove too heavy for men in general. Though those men may develop little interest in either a tidy civilization or an unusual or obvious morality, they at least will have vestiges of an urge to personal effectiveness, worth and sanity. The attainable qualities of honesty, courage, pride and understanding are mighty sources of social strength.

Jungle law was concerned with survival, not justice. At some point in his history, man began to give casual thought and concession to justice—a task that he may never hope either to complete or to abandon. As early as a few thousand years ago he had found it well to put some limits on killing, stealing, sex indulgence, lying and other practices found or thought in those days to be contrary to the interests of the tribe. When he learned writing, these thoughts and taboos were thus set down, as were also his mystical views about nature and himself. Sooner or later, mystical-minded persons among several large populations incorporated those taboos into their current mystical thought. In this way were founded the several religions—things which, like customs and literatures, are usually more durable than nations.

Most surviving religions were crystallized before democracy was rooted; before an industrial age had been hinted; before either health or disease was comprehended; before experiment and controlled observation had been accepted as the path to truth. Neither the aged taboos nor the best and gentlest of mystical teachings that were merged into any religion center upon the prevailing injustices of the kinds of lives we already live. Indeed, each religion magnifies a fragment of the past, and into its mountain of taboo and tale, of miracle and legend, it scatters the moral ideas of a tribe, a teacher, or a time. Though all religions present their urges to "right living" and to "loving thy neighbor," they often base their urge upon a wrong, an unacceptable, or a mystical reason. Missing is the now understood reason and basis for morality; missing are the fingers that should point to now prevailing injustices at the doorways of trade and of the professions, at the gates of the impersonal corporation and the labor union, at the doors of officers and servants of governments, at the rife injustices of the home, and at the fierce injustice and indignity of ignorance itself. Only education and social policy that mean business—and bow neither to dalliance nor to myth, neither to mystic nor to selfish interest—can serve adequately an advanced people in its present struggle toward justice and morality. We are learning that "without equity there is no justice, and without justice there is no morality." And while displacing the religions of the past it is a permanent task of peoples to learn and to like the worth and ways of justice. That long and endless road from the jungle leads toward free societies—societies worthy of enduring.

In the field of political and social sciences it has been learned that justice is not a by-product of liberty, that the home of a worthy democracy is constructed by the opposed but finely balanced hands of liberty and of order. The fact that this supremely delicate task—clearly calling for the utmost of reality and of reason—has had to proceed under the confusing shadows of the supernatural, and of clashing schemes of soul salvation, is the costly curse self-imposed till now on civilized societies.

After its long apprenticeship to the past, the race should graduate into the present. Great gains may come of wooing the past but of wedding the present. It seems incredible that anyone should profit over the course of a lifetime by being blind to the obvious and allergic to the inevitable. Again, there is so much of the present—the immediate—that it is expedient to make of it a prized possession. Partly of this novelist Cozzens[4] wrote:

There'll be deaths and disappointments and failures. When they come you meet them. Nobody promises you a good time or an easy time. I don't know who it was who said when we think of the past we regret and when we think of the future we fear. And with reason. But no bets are off. There is the present to think of, and as long as you live there always will be.

Like it or not, some serious social jobs—jobs now partly arrested by religion—must be done soon by some peoples, and later and forever afterwards by all peoples. One such problem, with such imperative demands, is contained in a small biological item called human fertility. Too many people are added daily—about thirty millions a year, or rather more than thirty-four hundred an hour.[5] The earth's population has doubled in the past 135 years, despite a moderate use of birth control within some nations. Sir John Orr, head of the Food Division of the United Nations, recently stated that in order to feed adequately the present population of the world the present food supply would have to be practically doubled. In March, 1952, statisticians of the United Nations reported in the *Monthly Bulletin of Statistics* that in view of declining death rates and high birth rates nineteen countries could expect to double their population in twenty-eight years. Should this period prove to be longer than thus predicted, the threat is not less real.

Now loosed upon the earth is a species that, more and more, knows how to defy its expected death rate, and whose uncontrolled fertility—82,000 *additional* mouths per day—must bring starvation to later generations. A species, too, whose past has shown little regard for the food supply of its future. The Mexico seen by Cortez in 1520 was probably 85 per cent forested. In 1947, its forested area was 16 per cent, its climate is becoming

4 James Gould Cozzens, *The Just and the Unjust* (New York: Harcourt, Brace and Co., 1951).

5 Robert C. Cook, *Human Fertility* (New York: Sloan Associates, 1951). Charles G. Darwin, *The Next Million Years* (New York: Doubleday and Co., 1952). Also, data for 1954.

progressively drier, and its land areas are seriously eroding. Couched around and upon its wasted Apennines, distracted and industrious Italy now has forty-seven million people, and a yearly increase of 400,000. Not long ago an Italian economist said, "Take 400,000 people a year off our hands and Italy would need no other economic help whatever." Peoples already pay heavily for overpopulation. If the race approaches the future without making wide and effective use of controlled or spaced parenthood, it embarks upon a ghastly adventure.

Spokesmen for religion are accustomed to assert with an air of finality that religions have always existed and therefore may be expected to exist always. All of this, it is alleged, chiefly because men carry an inner urge to religion. This is a spurious argument. It is better to outline the facts in truer form. The entire chronicled past of mankind presumably represents only a minute in man's full day. The religion or irreligion of most human societies during that moment is partly unknown, though a magic or a religion—often more a tribal than a personal affair— was fairly prevalent in those societies which are known. But more and more the supernatural is shrinking in size and power in recent societies. Greater and greater numbers of persons discard it altogether. The age of firm belief in the supernatural was the prescientific age. Practically all of the scientific age is still ahead of us, and there is insufficient reason to assume that belief in the supernatural can long survive that age. Religious beliefs are cloaks built from social heredity—from the blowing winds of fear, of untouched reality, of unexplorable threats to callow cultures. Natural selection, which helped to build man's body, cannot help to preserve those early trappings of a jungle-freed species, which built its first city less than fifteen thousand years ago. The ethical element associated with the various religions is not a truly religious element but is drawn from a common and growing pool of human experience, mutuality and understanding (ethics). Many humans draw upon that pool who firmly reject every religion. Religion disappears along with the supernatural. Or—if one overstrains the word—the "religion" that dispenses with the supernatural and is found only in the deeper personal feelings of awe and admiration does not organize itself. And those feelings, one hopes and presumes, will persist in sane persons.

PURPOSES AND OUTLOOKS

In the biological sphere we arrive at a fragment of space and of time in which there are persons as well as events. And in this area, dotted with persons, it is possible for aims, ideals, philosophies and purposes to arise. Here the word "arise" means that the ideals and purposes *begin* their existence.

Though the conclusion just stated is well grounded in sound deductions from modern knowledge, much of this knowledge has been too recently acquired to have become well known to the public under even the most favorable circumstances. The circumstances, however, have been quite unfavorable. Hurdles and censorships at community and higher levels have continually checked the spreading of the specially relevant facts of science. Too, the public has been continually coached against drawing the conclusion that purposes arise and begin their existence within this biological area. In its stead, the religions still propagate a revised form of the crustiest of all myths—"a meaningful universe." Ignorant primitive peoples pulled the original from the thin air in which all myths are born; and though trimmed, changed, curbed and reshaped by civilized thought—from early Greeks and Hebrews to the present—a kernel of that ancient myth is almost everywhere retained to shield the religious mind and institution against the advancing thrust of science.

Born into a flood of problems, the enlightened peoples of the world may now make and scatter their purposes. Scatter them because the problems are actually several at any moment, and because purposes have necessarily varied during the million years required by earliest man to become man of today. Modern learning presents humanity with nothing more virile and heartening than an assurance that purpose or purposes are not to be found in nature but are made by men. It would have been a calamity of the first order for our species if the purposes made and followed by man during his million years of untutored growth turned out to be mostly contrary to some purpose that only a metaphysician or theologian of recent years could sense and discover. Through this assurance of a sound and practical basis for man-made purposes—there being no other—the sphere of human activity is increased, the human being is dignified, and a realm of freedom is entered. This gift of modern learning breaks a chain to an intellectual past that was more fear and dread than purpose.

We can now recognize, with philosopher Russell, that "if it is the purpose of the Cosmos to evolve mind, we must regard it as rather incompetent in having produced so little in such a long time." Others have discussed the nature of purpose more directly. Philosopher Edman[6] said:

I do not believe that life in general or the world in general has any meaning. I do not think there is any meaning in saying that they could have. But many things, all things in nature may have meaning; and any life may generate its own purposes or ends. Life itself is what St. Augustine would have called grace, what—in language that, however different, means the same thing —I should call "so much velvet." Not what life means, but what meanings it may have, is what counts. It is possible that, short and doomed though it be, it may be brilliant and varied as well as smooth. It is not the prelude to glory, but the occasional vehicle and revealer of it.

And biologist Huxley[7] wrote:

The purpose manifested in evolution, whether in adaptation, specialization, or biological progress, is only an apparent purpose. It is just as much a product of blind forces as is the falling of a stone to earth or the ebb and flow of the tides. It is we who have read purpose into evolution, as earlier men projected will and emotion into inorganic phenomena like storm and earthquake. If we wish to work towards a purpose we must formulate that purpose ourselves. Purposes in life are made not found.

And these words of writer Ferril[8] belong here:

The fugitives from the beauties of reality say that our times are mechanistic. They would be happier in a spiritual climate. So they go back to the cloister, and what cloister is it? . . .

Our only hope, if we want a truly spiritual life, is to go further, without compromise or evasion, on the nature-searching course we are taking. Call it materialistic, mechanistic, or what you will, beauty and prophecy will come in obliquely.

---

6 Irwin Edman, in *Living Philosophies* (Cleveland and New York: World Publishing Co., 1951).

7 Julian Huxley, *Evolution, the Modern Synthesis* (New York: Harper and Bros., 1942).

8 Thomas Hornsby Ferril, in *Harper's Magazine*, October, 1946.

The purposes, hopes and philosophies of men, like the solution of some of their problems, must all be shaped by the kind of universe in which we live and by the nature and characteristics of the principal thing that men have unconsciously and consciously created—namely, society. It is well therefore again to glance momentarily at both man and society in their relation to a partly known universe.

In the fullest meaning of those words, both the human being and society are at once structures and organizations. The many separate parts that are integrated to establish those two wholly unlike organizations can be fairly well identified and comprehended. But the new and unpredictable properties that arise through the putting together of their many components are things much less easy to grasp. After all, the task of visualizing and understanding the *properties* of water that result from putting together two atoms of hydrogen and one of oxygen greatly differs from the simple task of picturing and understanding the three entering atoms of hydrogen and oxygen. This remains true even though one remembers that the simplest atom is itself already a community and built of moving parts. Both encouraging and meaningful in evolutionary thought is this reminder: the total complexity of the human organism is perhaps scarcely greater than that of the social organism—a thing that, under integrative law and the long arm of time, men themselves have created. Both those structures—man and society—stretch with history, both trace their remote parentage to movements within the atom, and both—though offspring of enduring law—are inconceivable in their entirety apart from cradles of chance and of unfolding time.

Society, though measured in time by a scant million years, spreads on restless human feet and by quickened thought; it is spurred or muted by genes born in the lowly ancestors of man; it is an infinite web of seen and unseen mutuality or dependence, and even the dead may affect and change it.

Man is an organization that could be built only through the course of a few billion years. First steps toward him were taken through the formation of the 98 or more elements. The later trails of atoms, earth-cloud, molecules and simpler organisms that lead to man are either left in rocks, in other living things, or they are uncovered by the newer tools of thought. Man is that most advanced living unit as yet built by the blind and ever-integrating processes of the earth-cloud—the being lately fashioned upon the crest of countless living things that pregnant cloud slowly formed from its own restless atoms.

Science has illuminated the distant past in a way that no other branch of learning could hopefully attempt. But it is clear enough that science cannot be or become equally dominant in the practical affairs of societies of the present and future. The rules under which men work, live, own property, train and relax are largely man-made rules; they are based on human experience, urges and concepts, and are framed and fortified by the humanities and the arts. Even the mere enforcement of them is a heavy task of law and government. Science teamed with technology will increase and determine the productivity of many kinds of labor. It will tap many sources of energy to spare human effort, and will produce new foods and fibers to sustain additional human lives. More and more it will guard and augment the health and the efficiency of individuals. And, increasingly, those several lifts will populate the world with older people. Endlessly it will bring new problems to society. Progressively it will add an element of grandeur to humankind. With the help of the humanities and arts, of educational and related agencies, it *could* give man an objective acquaintance with himself. If, when, or as that is done, the supernatural will cease to frighten or direct the minds of worthy peoples. Success in that much-shared task would resolve the cultural impasse of this generation —the subject of this book.

To leadership of the Soviet Union is now widely ascribed a deadly threat to the lives and freedoms of Western peoples. And that indeed may prove the most serious threat of the past thousand years. Yet it must be said that only during the past five or six years have Western peoples, particularly the United States, permitted that problem to attain its deadly proportions. Communist Russia did not have predominant land and air power at the end of World War II. But despite its knowable nature and aims it was permitted quickly to attain such power. Western peoples thus give immediate proof of failure—from whatever combination of reasons—to guard adequately their own survival. With failure at this most elemental level now demonstrated, there can be no firm assurance that these peoples will not fail also to deal with other and still more durable threats to their survival and their democratic ideals.

The strengths and weaknesses of the partly tamed species that rather thoughtlessly takes the untrod road ahead are likely to be tested fully and severely. No particular people is insured against disaster or decline. But where or to what species may we look for a record of equal past achievement? Or of like equipment for tomorrow? The species is greater than its parts, and he who sells mankind short should traffic in individuals or nations only.

Man's growing danger from other men, the protection of the public interest against private interests, the attainment of full freedom to educate, and still other social and political problems are examples of issues that areas other than science must do most to resolve. But in these tasks science co-operates vitally through increased production of goods and services, through the promotion of health and sound social outlook, and by spreading a respect for truth and reason. To restrain science from performing these substantial services is to hazard the future of peoples. One recognizes that, through atom-hydrogen bomb or otherwise, society could perhaps come near to obliterating itself and the human brood. That, however, is an improbable outcome. And while employing measures to insure its own survival, society should make itself assuredly worthy of survival. Indeed, the two assignments may prove to be essentially a single task.

Some able writers and critics of the adequacy of a scientific philosophy erroneously assert that the recent history of Germany and Russia is, in effect, a warning to a people who would put its trust in reason and intellect. Quite otherwise: that history provides two examples of fanatic hate unleashed in politically immature societies trained by the Western world's most repressive religions—Lutheran and Catholic for the one, and Russian Orthodox for the other. Neither those peoples nor their leaders knew the rational story of man and society or sensed the dignity of a free and rising human being. Without that background of rational social and political maturity, those nations could make no test of the worth of reason and intellect; they could merely adventure in racial and class prejudice, in mass ignorance, and in whetted hate. An affirmative base for a worthy society, and for a worth-while individual existence, should be sought in the meaningful whole—not in a confused part—of the very long and tested trail that leads to the advanced but still groping man of today. That base, whatever its spread, seems likely to include a trust in things that free and venturing men may rationally do together—and that is or was a part of what has sometimes been called the American tradition.

Just now nearly all crowded societies are subjected to a weighty threat whose poorly recognized origin is in fact partly traceable to religion. The stifling social climate that gave birth to communism included very broad areas of economic servitude, of deprivation, and of hopelessness. Within that climate, prevailing governments stalled in ineffective action, and prevailing religion still talked of a helping God who provided no deliverance. All of this was carried into late nineteenth and early twentieth centuries, when science and philosophy had already disposed of the personal God, and when technology, industrialization, colonies, and a brief breath of freedom had already given help and hope to *some* favored and foolishly complacent peoples. It is far from strange that thinkers within the stifled climate should then resent the myth of Providence and militantly declare that—far above all else—"*men* must change the world." That parts or much of that thought was shared by many thoughtful people everywhere is also neither droll nor remarkable. Any truly liberal and thoughtful person was already prepared for theoretical communism (aggressive socialism, at least) *if* he could regard the economic, political and social concepts of Karl Marx as a practical avenue to the new and better society. And, after the Russian experiment has been observed in action, he could become and remain a Communist *if* either duped or further prepared to dedicate himself fanatically to the faith that the new society must be built even at the cost of breaking any and every moral sentiment of man. Marxian doctrine is without comprehension of the deep biological origins and essentiality of morality. Equally flagrant, the religions—from Darwin of 1859 to this moment—have prevented any people anywhere from learning those origins and admitting their complete separateness from religion. Thus the communism that has spread knows hate and some of the hoaxes of religion but not sociality or mercy. In its dedication and fanaticism it mimics a religion. Societies of today are paying in tragedy for their own earlier failure to accept and *quickly spread* the full story of society and of unaided, aspiring man. He who fully absolves his own immediate ancestors and the world religions and then places on Communists the entire blame for this powerful menace is far too little acquainted with evolutionary science, sociology and history.

Ominously, through a welter of alluring or of vengeful words, the new myth and mirage of salvation through communism already begins to enmesh many crowded peoples. The cold lead —the heavy burden of each individual of the swarming crowd—is being melted by communism into a hot lava, which spreads and sinks and sears and burns. Cooking the other man's goose is nowhere a lost art; everyone's freedom is a price that hunger and hopelessness can always pay. Denying worth to all that does not serve a foisted idea is easy for aggrieved overpopulations that are unaware of the amazing history and the varied triumphs of man. Where full, unhesitant truth is not supplied, the mirage may be accepted. Most cruelly and quite unconsciously, the religions persisting since the Darwin of 1859 have contributed to the spread and virulence of communism.

To individual and community the full bloom of science signals a soul-shaking deliverance. Already, for many millions of freedom-loving persons, disciplined thought has neatly displaced the jungle-born fears of man with courage. In losing the fears long associated with the supernatural we lose the nightmare they caused, while we acquire the courage that only knowledge gives. Others, Bertrand Russell and T. V. Smith, for example, have observed that what stands in the way of knowledge stands in the way of courage, and what stands in the way of courage stands athwart the path of effective endeavor. No peace of soul can ensue to those who uncourageously contaminate the soul with fear.

Again, confidence and hope are better tactics than humility. Though humility could be born of the immensity of man's knowledge, that quite subordinate trait flows plenteously from favored glimpses of the wider immensity of common ignorance. For marching men, worry and humility are excess baggage. For mere ambulators, ulcers and self-pity are costly equipment. If there is much rough road ahead, may the race not hope to chance upon the occasional titans who are equal to their tasks? A titan or two for each thousand years would not violate history and could perhaps repeatedly salvage a civilization. In any case, peoples do not travel alone; the race's towering personalities get recorded in its goals and quickened steps.

It is well, too, to prize the ways of chance and circumstance in human progress. A rugged Martin Luther, ungracious and half demon to the end of his stormy life, could perhaps save a civilization by throwing heavy doubt into his cup of faith. He, his doubt, and a new means to spread a daring doubt, all came at a moment when a spark of reason in religion meant a continent afire. He was nine years old when Columbus took the corners off the earth. The frozen lethargy of Christian centuries choked his youth, instilling a fear for his soul and a hatred of God. This cruel conflict within a restive priest soon resolved itself through acceptance of an unseen and slightly amended God of grace and mercy, along with utter rejection of most religious practices his eyes could see—papal infallibility, celibacy for the clergy, the sale of indulgences, and still other fat dogmas of the Roman Church. His torrents of speech on these boiling subjects were just then able to be alarmingly multiplied by the newborn printing press. But it was doubt and denial that kindled the fires of action and release. Esoteric talks directed only to celestial grace and mercy would have been mere ashes—ashes for the tomb that for so long had held the intellect of Europe. Fortunately for us all, the slow march of our hardy tribe is speeded both by man's good purposes and by his earnest doubts.